Given World and Time

Given World and Time

TEMPORALITIES IN CONTEXT

Edited by
Tyrus Miller

C E U PRESS

Central European University Press
Budapest New York

Published in 2008 by

Central European University Press

An imprint of the
Central European University Share Company
Nádor utca 11, H-1051 Budapest, Hungary
Tel: +36-1-327-3138 or 327-3000
Fax: +36-1-327-3183
E-mail: ceupress@ceu.hu
Website: www.ceupress.com

400 West 59th Street, New York NY 10019, USA
Tel: +1-212-547-6932
Fax: +1-212-548-4607
E-mail: mgreenwald@sorosny.org

ISBN 978-963-9776-27-2

Library of Congress Cataloging-in-Publication Data
Given world and time : temporalities in context / edited by Tyrus Miller.
p. cm.

ISBN 978-9639776272 (cloth : alk. paper)
1. Time. I. Miller, Tyrus, 1963-
BD638.G25 2008
115--dc22
2008030064

Printed in Hungary
by
Akadémia Nyomda, Martonvásár

Contents

Acknowledgments

The various encounters, projects, and exchanges that ultimately generated this volume have extended over a number of years and have been the result of intensive individual, group, and institutional cooperation. This process has included the following main events: *Given World and Time: Temporalities in Context*, a conference co-sponsored by the University of California at Santa Cruz (UCSC) and Central European University's (CEU) Pasts, Inc. Center for Historical Studies, held in Budapest at CEU on November 28–29, 2003; *Cosmologies of History: The Symbolic Organization of Time*, a summer course for advanced graduates and postdocs, part of CEU's Summer University, held in Budapest, July 3–17, 2004; *Materiality/History: The Materialization of Historical Time*, a conference co-sponsored by UCSC and CEU, held at the Wissenschaftskolleg zu Berlin on November 26–27, 2004; *Saving Time: Interdisciplinary Perspectives on Memory and Memorialization*, a conference sponsored by the University of California Presidential Chair's Group on the Philosophy of Time, with Cowell College, held at UCSC on November 18–19, 2005.

I would like to acknowledge the exceptionally generous support, over several years, of David Hoy, whose University of California Presidential Chair provided funding for the original core of faculty in the research group of the Philosophy and History of Time at the University of California at Santa Cruz, as well as substantive funding for the conferences in Budapest, Berlin, and Santa Cruz and for the publication of this volume. Wlad Godzich, Dean of Humanities during the period of these activities, helped to provide funding for graduate participants in the CEU Summer University course, on topics related to this volume. In addition I would like to thank Jocelyn Hoy and Deanna Shemek for their active participation in and support of the activities of the research group.

I would also like to acknowledge the generous support, institutional and financial, of CEU's Pasts, Inc. Center for Historical Studies and its director, Sorin Antohi, as well as CEU's Summer University. My sincere thanks go to the leadership and staff of the Wissenschaftskolleg zu Berlin, who contributed to the success of our second conference.

I wish to give heartfelt thanks to Darren Waterston, whose beautiful painting adorns our cover.

Editorial assistance was provided by Harold Strecker and Linda Kunos, and by the CEU Press copy editors. Their patience and efforts are greatly appreciated.

The Editor

Introduction

Tyrus Miller

In his 1991 novel *Time's Arrow*, the British writer Martin Amis, playing a postmodern "narrative game with time" (as Paul Ricoeur would say) explored the implications of a very simple narrative twist for our historical and moral perception. How would some key twentieth-century historical event, an event as ineluctable as the Nazi seizure of power and the unfolding of Nazi genocide, appear from a radically different temporal perspective? And what might this perspective shift tell us about the way our historical and moral judgments carry along with them, enfolded into their conclusions, assumptions about the relation of time and its nature to the meaning of history? To what extent are we obligated, ontologically and morally, to yield to the necessities of time and to what extent is it legitimate to give time over to a play of fictions and interpretations? It is on this dangerous, but consequential borderline between freedom and necessity, between guilt and evasion, between responsibility and play that Amis's novel plays out its thought-experiment.

The novel raises a seemingly absurd question, but one fraught with psychological and cultural symbolism, conjuring the various evasions and denials, collective and individual, that this critical moment in contemporary history has provoked. It asks: how would the life of the concentration camp doctor Odilo Unverdorben, who has fled atonement and punishment, changing his identity and emigrating under this cover to the United States, look if it were narrated backwards, with the arrow of time reversed? Regressing backwards in time, the "normal" surgical labors of the post-war American physician appears to Amis's "third-person" narrator like horrifying violence and torture: wounds are opened in patients, sutures are removed and blood flows out of bandages and sponges, recovery becomes disease. Whereas in contrast, those "exceptional" perversions of medicine in the concentrations camps now appear as miraculous acts of mercy towards the favored recipients of the Nazi "charity," the Jews: the dead are reanimated, the beaten and starved are returned to health, prisoners are released with the happy prospect of repatriation and restoration of property. Amis ends his book with the pathos of Odilo's approaching infancy, with birth paradoxically marking the moment of his death and disappearance from time's regressing line.

At the same time, however, Amis's narrator anxiously hints that this fiction is unsustainable, that the very premise of his character's life, this whole effort to reverse time and cancel history, contains at its innermost core the traumatic force of collective denial that once allowed the obscene, violent, scatological, necrophilic crimes of the concentrations camps to take place under the cover of night and fog and perhaps will allow them to recur in new forms. With this nameless force of denial, the narrator is himself complicitous: the third person narrator and the unspoken first person of his character appear to have the same aim, to regress, to erase, to undo. When the arrow of time threatens to turn around the right way, the narrator "corrects" it, insisting on his favorite character's childlike innocence, his "Unverdorben-heit" (unspoiledness) by history, to the very end. Everything that took place—earlier in the book—was, we are given to understand, all just a perverse insistence of time running forward, leaving the storyteller stranded in untimeliness and forcing him to tell his story imperfectly, at least *this time around*:

> Only a moment. There are no larger units of his time. He has to act while childhood is still here, while everything is his playmate—including his *ca-ca*. He has to act while childhood is still here before somebody comes and takes it away. And they will come. I hope the doctor will be wearing something nice, something appropriate, and not the white coat and the black boots, which surely. ... Myself. Mistake. Mistake. ... When Odilo closes his eyes I see an arrow fly—but wrongly. Point-first. Oh no, but then. ... We're away once more, over the field. Odilo Unverdorben and his eager heart. And I within, who came at the wrong time—either too soon, or after it was all too late.[1]

Amis, clearly, is playing a Samuel Beckett-like narrative game with time to suggest the *limits* of our capacities to reshape time—or at least, the moral desirability that these limits be observed. Yet this question of limits is not merely a question of individual responsibility and choice, nor just a matter for literary fiction. It is also a problem for our inherited and contemporary social ontologies: a matter of the degree to which our social being is fungible by acts of political will, solidarity, organization, rationalization, technological mediatization, utopian imagination, ideological compulsion, and organized terror. It is hence also a profoundly political and ideological problem, linked indissolubly to the myriad political and ideological fictions supporting so much of contemporary history.

The Russian emigré art critic Boris Groys, for example, has suggested that post-communist culture, both in its spontaneous everyday forms and its reflective manifestations in art and literature, can be understood as a set of responses to the collapse of a particular form of communist ultramodernity, which was above all a way of politically organizing time. Formulated summarily, in its original utopian aims, arguably carried to hyberbolic actuality under Stalin, communist ideology and state organization sought to liquidate the past and utilize the energies and resources released in this process to fuel accelerated progress into the future. The present was viewed as a moment of break with a compromised past and as a point of transition to a radically transformed future. Communist movements saw themselves as the anticipatory embodiment of the future within the present—at once a disruptive force that had to burst open the closure of the present and also the necessary goal towards which subsequent history was heading. This orientation could be seen from the very moment of stabilization of the Bolshevik regime following the Russian Civil War, in the debates of the 1920s around Soviet industrialization. The goal was full industrial modernity, and time was the enemy: the calculation was made that without accelerated accumulation it would take decades to achieve even the industrial capacity of pre-World War I Russia. Not just the violent and victorious Stalinist imperative to "complete the Five-Year Plan in four years" and the devastating war on the countryside and its inhabitants, but the whole spectrum of positions in the debate, from the left-wing "socialist primitive accumulation" to the more moderate range of accommodations of the socialist sector with markets and peasant economies, was fundamentally about the economy of *time* and the extent to which the socialist state could actively shape that economy or would be shaped by its necessity. In turn, post-communism, Groys suggests, has had to experience a retreat from communism's ideological fiction of the future present. Post-communism, like Amis's fictive post-fascism, regresses back along the overextended arrow of communist time, or at least along the symbolic, cultural, and psychological traces left in its wake. "Once they have dispersed," Groys concludes, "what such modern, yet closed, communities leave behind them is not the past, but the future. ... [T]he post-communist subject... moves... against the flow of time. ... [T]he post-communist enters the present from the future."[2]

Of course, Groys's Augustinian vision of post-communism is linked to the anti- and retro-utopian orientation of contemporary eastern European arts, and thus emphasizes the most spectacularly utopian facet of the time-politics of the Communist system. Groys takes communist time-ideology hyperbolically, grasping its most fantastic premises literally and turning its implications ironically against itself, so as to chastise the classical artistic avant-garde for its complicity with the destructive results.[3] Yet, however satisfying this malicious troping of official doctrine might be—and however energizing it has been for a wide range of post-socialist artists and writers[4]—the empirical picture for historians and ethnographers of post-socialism is necessarily a good deal more nuanced. Several commentators have noted (see, for example, among our contributors to this volume, the essay of Lisa Rofel and Bill Nichols's discussion of the Romanian film *Reconstruction*) that communist time was itself complex and heterogeneous, often contradicting the official emphasis on a unitary vector of modernization. One facet of this complexity was communism's monumentalizing of the past, rather than its liquidation—an ironic result of the constant revision of the past as required by the changing vicissitudes of legitimacy for the latest personnel and policies of the centralized authorities. Though the past could be and was changed at will, as demanded by current needs, at the same time the past could not be subject to interpretative debate. Some parts of the past thus became petrified as fixed myths monotonously enshrined in everything from textbooks to architectural ornaments and monuments; others simply fell out of representation and were conveniently consigned to oblivion (such as the complicity of the "anti-fascist" nations of the socialist East Bloc with anti-semitism and the Holocaust). There was only one indivisible past available at any given moment: a web of authorized "facts" that was always seamlessly knit, however many stitches might be dropped or added to it. At the same time as the past was frozen in a parody of history, so too, as the eschatological horizon of world revolution retreated and terror gave way to stabilized bureaucratic power, did the future become far less charged with existential and social potentials. The past, repristined to legitimate the present, at best served to lend a bureaucratically organized present a bit of ceremonial pomp; while little would lead the average "socialist man" to believe that his future would be much different than the life he was enduring at present. Whether on the macrological scale of national defense or the micrological struggle to obtain daily goods, state socialism took on the dimensions of a gray present rendered eternal.

The examples discussed above, we hope, indicate the topicality of time and its multiple contexts for contemporary Europe, which has been so crucially affected by the events of World War II, the Cold War order, and their afterlife throughout the continent. This theme, however, is of much wider import, insofar as it touches on key questions of cultural, historical, and disciplinary difference for humanistic thought on a global scale. The interconnections of time with historical thought and knowledge have also come powerfully to the fore since the 1970s through a number of phenomena often viewed not merely as historical events that can be mapped by an otherwise stable historicizing framework, but instead as themselves frame-establishing or frame-breaking events for social and historical understanding. The rapid convergence of these phenomena, it is argued, has precipitated a new condition of post-modernism or global post-historism, in which the meaning-conferring frameworks of modern, Eurocentric history no longer suffice to comprehend the extended present. These phenomena include the abandonment of the gold standard and the development of a new order of "floating" global monetary exchange; the energy crisis and the growing fiscal and political power of oil-exporting nations, especially in the Middle East; the strong emergence of global ecological consciousness and the rise of environmental politics, ranging from global non-governmental organizations such as Greenpeace to the convergence of Baltic nationalism and ecological protests in the Glasnost years to recent animal rights campaigns and activism around the ecological effects of globalization (see also, Wai Chee Dimock's contribution to this volume about the nascent politics of global warming); the Iranian revolution and the emergence of fundamentalist Islam as a state ideology; the final unravelling of European colonialism in Africa, Asia, and Latin

America and the end of apartheid in Rhodesia and South Africa; the political discourse of "exterminism" and the "nuclear winter" in the 1980s; the Chernobyl nuclear disaster, which awakened popular consciousness to new borderless risks and led to a new theorization of "risk" as a socio-political factor; the weakening and eventual collapse of the Cold War order in the 1980s and 90s, bringing everything from liberal "velvet revolutions" to the return of atavistic nationalisms and bloody civil war; the increasing importance of "biopolitics," which charges issues of health, aging, disease, sexuality, and reproduction with unprecedented importance in national and international politics; the definitive manifestation of global warming and the urgency of international cooperation to address the problem; and the mutation of terrorism from guerilla movements, often linked with or covertly supported by states, to the diffuse "acephalic" networks that came into view with the spectacular events of September 11, 2001.

Just as this catalogue of phenomena and trends is extremely diverse in nature and scale, so too are the theoretical discourses that reflect on these developments extremely varied. Just taking, for example, four of the most widely-discussed theses: Jean-François Lyotard's resonant pronouncement that the grand narratives of modern history had given way to many overlapping *petits récits*[5]; Francis Fukayama's triumphalist declaration of the "end of history" and the beginning of an eon of liberal amelioration world-wide[6]; Samuel Huntington's prediction of increasingly violent clashes of large-scale civilizational blocs in the post-Cold War period[7]; and Ulrich Beck's bold argument that the temporally, geographically, and socially differentiated distribution of "risk," which operates at scales both more local and more global than the nation-state, has replaced national rights, political structures, and social strata as the fundamental socio-political pivot of contemporary life.[8] Otherwise utterly divergent in their theoretical presuppositions and ideological commitments, all four of these thinkers agree in their sense of a tectonic plate-shift in the very foundations of contemporary society and history, whose first rumblings could be sensed in the events of the 1970s and which became impossible to ignore by the 1980s. For our purposes, the symptomatic importance of this agreement is crucial. Whether or not one accepts the argumentation and conclusions of any of these theorists—and all of them are deeply problematic—each nonetheless demands not simply an evaluation of a set of theses about recent historical developments, but also reflection on the very conditions and possibilities of historical understanding in the present day. As these examples and others suggest, time—psychological time, narrative time, social time, anthropological time, theological time, aesthetic time, ecological time, techno-scientific time—has increasingly emerged as a dimension of historical meaning that can not simply be taken for granted. Time must be reflexively considered in the construction of social and historical interpretations. So too, the essays in this volume are strongly marked by this contemporary horizon of thought in the social and historical sciences, and they seek to further our appreciation of the ways in which time and interpretation of historical phenomena are integrally linked.

***

The apparently paradoxical title of this volume, *Given World and Time: Temporalities in Contexts,* points readers to the enigmatic and potentially risky core of our subject matter: time, historical and cultural context, and ontological world-making, along with their intricate and variable interrelationships. Time and its contexts matter to us, to our knowledge and to our moral judgments about what happens in our world. Moreover, time and its contexts are only partly "given" to us: to the primordial donations of time and world correspond our epistemic, moral, and practical modes of receiving what has been granted. Thus, in the title, "given" is to be understood not as the self-evident data of a piece of worldliness at an instant of time, but rather as a conditional marker, indicating: *if* the world and time are given and given such, then other things may follow from these data. Several essays in this volume even call into question the seeming transparency of the word "and," the grammatical link smoothly connecting time to world. The contributors raise the question of whether time is to be understood as a contextual element of historical and cultural worlds, a *result* of a constructive world-making through thought and social practice; or alternatively, whether any possible "world" that we might construct depends on a particular way of appropriating "given" time, of distributing the features of that world into temporal divisions of past, present, and future. Some essayists imply still a third possibility, that world and time are complementary aspects of an on-going process of historical making and remaking, intertwined voices in the interminable dialogue of the world's timeliness and time's worldliness.

Similarly, the volume's essays critically interrogate the key problem of context and contextuality, connecting it to the problem of time. Contexts, the contributors suggest, are themselves not timeless, and hence, our apprehensions of time are entangled in productive ways with our apprehensions—or even active constitutions—of context. Not accidentally, even when it is not specifically thematized as a methodological issue, "context" is a major concept—or "superconcept"—shared by a whole spectrum of historical and human sciences, including history, classical and modern philology, cultural studies, linguistics, gender studies, ethnography, music history, ethnomusicology, comparative religious studies, and historical psychology. "Context" may stand, first and foremost, as the necessary background against which the figures of humanistic *disciplinary* knowledge can be rendered visible. Yet as the essays suggest, the notion of context may also have radically different parameters in different historical, cultural, and disciplinary situations. Contexts come into being and disappear, they emerge or are constructed in ways that subject them to the effects of time. The relative clarity and stability of contexts—or alternatively, their discontinuity and fuzziness—affect our sense of duration, in both its qualitative nature and its quantitative extent. Knowledge and experience are temporal in contexts, and their contextual temporality sets limits to their communicability and translatability into new contexts in new times. Accordingly, our contributors suggest that reflection on the interactions of context and time may be fruitful in making more conscious the conditions of historical knowledge and in facilitating interdisciplinary and intercultural dialogue.

In questioning the timelessness of context, some essays even subject to scrutiny the fundamentally spatial metaphors that govern our ideas of context. Context is ordinarily seen as a periphery of events and structures that serve to explain the historical character of some fact, event, person, or other focal point of historical investigation. We speak of "putting" facts "in" context. But what is a context, that it can have an inside and an outside, that it can function as an environment for something happening, that it can surround an event and suffuse it with meaning from the environs? Are there other ways in which we might think of contexts and their effectiveness than that of a cross-section at an instant of time? We think, for example, of "traditions" as temporally extended patterns of form and meaning, which in turn affect the form and meaning of present cultural artifacts, texts, and performances. Could we formulate a notion of context closer to that of traditions, contexts that would have not only a virtual spatiality, but also duration and temporal dimensionality? Could context include not just temporal extension, like tradition, but also the effective power of temporal discontinuities, catastrophes, erasures, repetitions, and leaps? Several of the essays in this volume ask us to imagine this methodological possibility and offer practical instances of the consequences of thinking of context more complexly and with more internal dimensions than we are typically accustomed to do.

Finally, as humanistic scholars committed to historical understanding, the contributors and editor hope that the essays in this volume will offer other scholars directional markers towards a project that can only be an on-going and unfinished one, so long as time continues to pass: a *history* of time, as one of the fundamental dimensions of the human spirit, of material nature, and of the spiritual and material history of humanity. We are not so immodest as to present this volume as even a "short history of time," to quote the title of physicist Steven Hawking's well-known popular lectures on the new cosmology. Our contributions are too heterogeneous in their approach and concerns for this to be the case. We do, however, believe that we have constituted a distinguished set of interventions into the problem of time's own historicity by an international, interdisciplinary group of scholars who, over the course of the past five years, several conferences in three countries, and innumerable conversations and arguments, have defined a partially shared site of investigation, to which no one can lay claim to exclusive possession or mastery.

* * *

I have divided the essays in our volume into four main topical groups. The first group is entitled "Temporality in the Long Run" and considers the effect of *longue durée* and other forms of large-scale dimensionality as contexts for historical thinking. The essays in this section explore such topics as the deep anteriority of antiquity, the timeless time of eternity, and the vast spans of planetary time as peculiar frameworks in which historical thought may be estranged and refashioned. The second group is entitled "Historical Figures: Mediations,

Quotations, Narrations," which examines the ways in which historical thought is affected by rhetorical expression and material transmission. The contributors explore such media as song, image, film, oral narration, and legal discourse, which articulate time and utilize history in medium-specific and pragmatically significant ways. The third group, entitled "Shapes of Modernity," considers both formal philosophies of history and the forms of histories implicit in individual and community experience, ranging from eighteenth-century thinkers of modernity such as William Robertson to the lives of individuals in a putatively secularized, post-modern, and globalized society. The volume concludes with a section entitled "'To the Planetarium': From Cosmos to History and Back," which includes two studies (including Karl Clausberg's detailed archeology of proto-relativity in an obscure nineteenth-century thinker) of ways in which cosmic speculation and historical thought have interacted, leaving both productively transformed in the encounter.

The volume's selection is intended to allow disciplinary readers to draw inspiration from the individual contributions, while seeing them in a new context provided by studies from other fields and with other approaches and methodological assumptions. We hope that the interplay between the essays, precisely in their disciplinary breadth and topical diversity, will prove fruitful for further explorations of the contextuality of time in humanistic discourse.

Notes

[1] Martin Amis, *Time's Arrow, or the Nature of the Offense* (New York: Vintage, 1991), 172–173.

[2] Boris Groys, "Back From the Future" in *2000+ Arteast Collection: The Art of Eastern Europe*, ed. Zdenka Badovinac and Peter Weibel (Vienna: Folio Verlag, 2001), 11.

[3] This is especially true in Groys's influential book on the links between the avant-garde and Stalinist totalitarianism: *The Total Art of Stalinism: Avant-Garde, Aesthetic Dictatorship, and Beyond*, trans. Charles Rougle (Princeton, New Jersey: Princeton University Press, 1992). Cf. Philippe Sers's more recent study *Totalitarisme et Avant-Gardes: Au Seuil de la Transcendance* (Paris: Les Belles Lettres, 2001).

[4] For further discussion of this artistic dimension, see Tyrus Miller, "Rethinking Central Europe: The Symbolic Geography of the Avant-Garde," *Modernism / Modernity* 10/3 (2003): 559–567; and Tyrus Miller, "Retroavantgarde: Aesthetic Revival and the Con/Figurations of 20th-Century Time," *Filozofski Vestnik* (Ljubljana, Slovenia), Summer 2007.

[5] Jean-François Lyotard, *The Postmodern Condition: A Report on Knowledge*, trans. Geoff Bennington and Brian Massumi (Minneapolis: University of Minnesota Press, 1984).

[6] Francis Fukuyama, *The End of History and the Last Man* (New York: Free Press, 1992).

[7] Samuel P. Huntington, *The Clash of Civilizations and the Remaking of the World Order* (New York: Simon and Schuster, 1996).

[8] Ulrich Beck, *Risk Society: Towards a New Modernity*, trans. Mark Ritter (London: Sage, 1992).

Temporality in the Long Run

Walking Backwards into the Future

The Conception of Time in the Ancient Near East

Stefan M. Maul

If we regard the Akkadian (i.e., Assyrian–Babylonian)[1] terms that designate "past" and "future" as more than simple equivalents to the corresponding English terms, we make an astounding discovery. An examination of temporal terms such as "earlier" (Akkadian: *pāna*, *pān*; *pānānu*(*m*); *pāni*; *pānû*(*m*)) or "former times, past" (Akkadian: *pānātu*; *pānītu*(*m*), *pānū*) shows that these are all related to the Akkadian *pānum*, or "front," plural *pānū*, or "face." The Sumerian[2] equivalents to the Akkadian terms for the past are formed with the word *igi*, which means "eye," "face" and also "front." In the Akkadian and Sumerian terms for the past, the underlying word "front" is used in the sense of "something that lies before/faces the observer." It is a similar case with terms that denote the future. The Akkadian (*w*)*arka*, (*w*)*arkānu*(*m*), (*w*)*arki* in the sense of "later, afterward," (*w*)*arkû*(*m*) in the sense of "future (adj.)," and (*w*)*arkītu*(*m*) in the sense of "something later, later days, future" are all related to the word (*w*)*arkatu*(*m*), meaning "reverse (side), behind." The equivalent Sumerian terms (*eger*; *murgu*; *bar*) also originally mean "behind" and "reverse (side)." Although here we cannot enter upon a closer examination of Mesopotamian terminology which is so important for the understanding of the culture of the Ancient Near East, it is nevertheless clear that for a Babylonian the past lay before him—it was something he "faced"; whereas that which was coming, the future (*warkītum*), was something he regarded as behind him, as at his "back." In the mental world of our own modern society the exact opposite is, of course, the case. When we look "into the future," we firmly believe that our gaze is fixed straight ahead. Nothing can shake our conviction that the past is at our back, that it lies behind us. While we advance along a time-line that has us "facing the future," the Mesopotamians advanced along the same time-line but with their eyes fixed on the past. They moved, as it were, back-to-front—backing into the future. Without belaboring the image, it would indeed suggest that Mesopotamian culture was focused on the past, and, ultimately, the starting point of all existence.

The concern of Mesopotamian culture with the past was, shall we say, omnipresent. In the remains of Assyrian and Babylonian culture in the first millennium A.D., one can easily recognize the extreme normative power of tradition, which permeated every aspect of life.

Languages of the Ancient Near East

The numerous inscriptions of the Mesopotamian rulers of the first millennium B.C.—which were left behind for posterity in the foundations of temples and palaces or made visible on reliefs and stelae—were composed in an artificial language that stood aloof from the demotic and took its cues from an ancient form of Akkadian. This language was regarded as a classic language and was spoken at the beginning of the second millennium B.C. and even then was full of archaisms. Other texts as well (religious and scholarly, epic and mythological) used this elevated form of language, which we may call "Standard Babylonian." With its archaic sound it conjured up not only the venerable reign of Hammurabi, the ruler who united all of Mesopotamia and parts of Syria into a powerful empire in the eighteenth century B.C., but this Standard Babylonian also evoked that linguistic form of Akkadian which in the early second millennium B.C. was the first Semitic language to be widely written down (which was then passed down to the end of cuneiform culture). Two thousand years after it had ceased to exist as a spoken language, Sumerian, the oldest known language of Mesopotamia, was still regarded as a sacred tongue used to address the gods. Sumerian songs, hymns and prayers that had their origin in the third millennium B.C. were always being copied down and accompanied by Akkadian translations. Together with later re-creations from the first and second millennia B.C., these songs, hymns and prayers still played an important role in the Babylonian cult of gods in the final centuries of the first millennium B.C.

Moreover, and very similar to Latin in our own culture, Sumerian survived as a scholarly language. Just as the Renaissance humanists latinized their names, Babylonian and Assyrian scholars translated their Semitic names into Sumerian. Along with countless lexical lists and grammatical paradigms, these scholars transmitted myths, proverbs, wise sayings, fables, omens, incantations and texts of exorcism in this ancient language. It was above all in those large cities which had been the centers of the early Sumerian culture that Assyrian and Babylonian kings of the first millennium B.C., following the old traditions, had their building and dedicatory inscriptions composed in Sumerian.

Writing

The royal inscriptions of the first millennium B.C. were not infrequently written down using very antiquated cuneiform signforms, which had gone out of quotidian use fifteen hundred years before and which could certainly not be deciphered by the less educated. The use of ancient characters, however, was not a phenomenon limited to the first millennium B.C. Hammurabi of Babylon (eighteenth century B.C.), whose era would later come to be characterized as "classic," had the text of his famous Susa law-stele inscribed in a writing gesture representative of a cuneiform whose developmental stage in a paleographic sense was then six hundred years old. In the sixth century B.C., on the magnificent edifices he constructed in Babylon, Nebuchadnezzar II followed Hammurabi's example by employing those cuneiform signforms which were typical for inscriptions composed in the Old-Akkadian period, twenty-fourth century B.C. The inscriptions of the late Babylonian kings not only were often copies of the old language of the Hammurabi era, but also regularly used the obsolete orthographic conventions.

The learned scribes of the first and second millennia B.C. compiled paleographic lists of characters—like modern Assyriologists. They studied old texts and fashioned clay-tablet facsimiles based on them, that were such faithful replicas that even Assyriologists are sometimes fooled as to their actual age. Thus the learned Neo-Assyrian King Ashurbanipal (669–627 B.C.) could boast that he was able to decipher inscriptions "from before the Flood."

Material Remains

But the interest of Mesopotamia in a past regarded as "classic" was manifest not only in the implementation of an ancient language and script. It can also be shown that in the material culture of Mesopotamia there were constant borrowings from periods that reached far back in time. One impressive example should suffice. As a modern scholar, one is astounded at how frequently one encounters in the royal inscriptions of Neo-Babylonian kings (sixth century B.C.) accounts of massive archeological excavations, undertaken at the behest of the ruler, in the millennia-old temple grounds so as to uncover the remnants of ancient and sometimes long-forgotten cultural venues.[3] In their search for old building foundations, the Babylonians—similar to modern-day archeologists—happened upon inscriptions, clay tablets and other artifacts. Specially commissioned scholars studied the old and hard-to-decipher texts, which were carefully preserved, copied and exhibited. One of the labels for just such an exhibit has survived to this day:

> Here are copies of the writing found on bricks discovered in the ruins of Ur, the work of Amar-Su'en, King of Ur, uncovered by the governor of Ur in searching for the ground plan [of the temple]. I copied them so as to amaze onlookers.

The head of a statue of the Old-Akkadian King Sargon of Akkade (ca. 2350 B.C.)—whose legendary conquests were still renowned in the Neo-Babylonian period—, damaged in an excavation, was restored and placed on display by King Nabonidus (556–539 B.C.).[4] He thereby situated himself as a direct recipient of his predecessor's glory, who, as Nabonidus believed, had reigned 2300 years before him. In an inscription, Nabonidus declares:

> The foundation stones of the Temple E'ulmash in the city of Akkade from the period of Sargon, the King of Babylon, my predecessor, who reigned 2300 years before me, went unseen by anyone until the reign of Nabonidus. Kurigalzu, a Babylonian king[5] who preceded me, had sought them, but did not find the foundation stones of E'ulmash. Nebuchadnezzar,[6] my royal predecessor, dispatched work brigades in large numbers to search out those foundation stones of E'ulmash, he took great pains, dug deep, repeated his efforts, but the foundation stones of E'ulmash he did not find. I, on the other hand, Nabonidus, the King of Babylon, during my lawful reign, fearing Ishtar of Akkade, beheld a face in a dream. Shamash and Adad assured me that I would find the foundation stones of E'ulmash, a favorable sign for the stability of my kingdom. My men I sent in large numbers to search for those foundation stones. For three

> years I dug through shafts sunk by Nebuchadnezzar, the King of Babylon. Right and left, forward and back I sought but found nothing. Thus did they then speak to me: "For the foundation stones we have searched, but we have not found them."

It was only in a later attempt, after the temple foundations of a comparatively recent building phase from the fourteenth century B.C. had been uncovered, that remnants of Sargon's edifice (constructed about 2350 B.C.) were found and Nabonidus could lay the new foundation stones atop the older ones with "not a finger's breadth of deviation."

The goal of such excavations was to identify the oldest foundation of a temple. The remnants of more recent overlays were carted away until it was believed that one had found the earliest form of the god's house, based on its foundations. But the interest of Babylonian (and Assyrian) kings in the temple architecture of their "antiquity" was not primarily of an antiquarian nature. Rather, the building plans were required to nullify the changes that had accreted over time and restore the temple to its unadulterated form and to deviate "not a finger's breadth"[7] from the original plans. Characteristically the Akkadian (and also the Sumerian) expression[8] found in the dictionaries for "restore" is literally "to lead something back to its planned/predetermined place." Clearly at work here is the Mesopotamian notion of each thing in the world being allocated its own fixed, unshakeable and eternal place. This divinely willed but historically altered place was to be restored with the reconstruction of the old temple. Myths that have grown up around Babylonian temples recount how these were not built by human hands but were erected by the gods themselves as part of the work of creation at the beginning of time.[9] Restoration of the temple according to the undistorted divine plan was intended by the Babylonian kings to transport both the state and its subjects back to their original, pristine, hallowed beginnings.

Hence, the search of Babylonians and Assyrians for "antiquity" emerges as a striving after the unsullied original order of a "distant yore," to which the gods themselves had imparted form through the act of its creation. Mesopotamian culture was ever focused on the origin of all things.

A look at the mythical texts of Mesopotamia shows very clearly that all the cultural achievements—in architecture, writing, goldsmithery, carpentry and so forth—were viewed as revelations of Ea, the god of wisdom, who had bestowed them upon humanity at the beginning of time. Even Berossos, a Marduk priest of the third century B.C., whose Greek-language work *Babyloniaka*[10] acquainted the Hellenistic world with the history and culture of ancient Mesopotamia, thought the following myth essential to any understanding of Babylonian culture. According to Berossus, in the first year of the world *immediately* subsequent to the creation of heaven, earth and humans, a fish-shaped creature called Oannes[11] rose from the Persian Gulf and taught

> humans writing and the manifold techniques of the arts, the building of cities and the construction of temples...whatever availed the domesticity of life in the world, it [i.e., the "animal" Oannes] passed down to humankind; and since that time no one has invented anything more.[12]

Although in their inscriptions the Babylonian and Assyrian kings proudly invoked the names of their predecessors who had ruled the land thousands of years before, they also sought to close the gap in time that existed between themselves and the beginning of all things. Elegant testimony to this desideratum was the annual New Year's celebration. In this important state ritual the king presented himself hand in hand with the World-God (i.e., with the image of this god which was usually worshiped in the temple) so as to recreate in a ritual performance the primordial struggle of this god with the powers of chaos, the eventual triumph over this adversary by the forces of order, and the ensuing creation of the world. Mesopotamian rulers legitimized themselves not only by tracing their descent from an "eternal seed,"[13] from a "precious seed from the time before the Flood,"[14] and from "families from the beginning of time,"[15] but, according to a well-known myth from the Neo-Babylonian period, the gods created "the King" immediately after creating humankind in order that he might "lead [them] righteously."[16] The god-willed task of a king consisted in preserving, defending and renewing the world as had been ordered in the act of creation. Thus reforms in Mesopotamia were fundamentally seen as the restoration of this order, which had flagged over time. The developments witnessed by Mesopotamian society in the course of centuries and millennia—developments that were of enormous importance for human history

—were virtually never described as "progress" but mostly as restoration. The ideal society and state for the Mesopotamians—their utopia, as it were—always had its settled place in a long-ago age and never in the future. It is therefore hardly surprising that in the first millennia B.C. Assyrian kings, through subtle allusions, portrayed their campaigns against the enemies of the empire as the ever recurring primeval battle of the World-God against the forces of chaos, ending with the triumph of world order in the work of creation.

It was not only this mythical period which provided the Babylonians and Assyrians with a paradigm for ordering history. In the cultural memory of Mesopotamia those kings of "distant yore" who had achieved military conquest and expanded their power sphere far beyond the Fertile Crescent were seen as outstanding figures and even "Savior Kings" whose works were the expression of an ideal monarchy it behooved one to emulate. Among these was Sargon of Akkade, who was the first to unite the Mesopotamian city-states into an empire (c. 2350 B.C.). Also personifying the kingly ideal was Hammurabi, who in the eighteenth century B.C. once more united a Mesopotamia that had lapsed into a farrago of petty states. It was chiefly in the first millennium B.C. that the methods were passed down by which these kings succeeded in maintaining harmonious relations with the world order as established by the gods in the act of creation—thus fostering the kings' memorable successes, along with an extensive epic literature on the "Savior Kings" and the original inscriptions of these rulers (which were always being scrupulously collected). In medical texts, for example, a certain medicine is extolled as having helped Hammurabi.[17] To preserve the health of their monarch, royal scholars of the first millennium B.C. compiled lists of such things as what amulet Sargon of Akkade wore into battle, or what stones were contained in the amulet chain fastened to Hammurabi's bed.[18] Also conscientiously collated were omens which portended the victories or successes of these kings. Posterity was not concerned with the personality of a "Savior King" but rather with his relationship to the gods, which was revealed in his successes, his proper performance of the rituals and in those cosmic phenomena interpreted as favorable signs. Royal successors of the "Savior Kings" wished to partake of the divine grace bestowed upon their historical models and thus sought to emulate them. For Mesopotamian society the past already contained (pre-formed) all possibilities for the future, and hence its preoccupation with bygone mythical or historical epochs was simultaneously a preoccupation with the future. No text illustrates this better than a fictitious autobiographical account of the deeds of Sargon of Akkade.[19] This document probably originated in the late eighth century B.C. at the court of the Assyrian King Sargon II, who, upon ascending the throne, likely chose his name so as to invest his reign with the glorious aura of the great Old-Akkadian king:

> ...I exercised kingship for [5]4 years.
> I mastered and re[igned over] the black-headed people.[20]
> (Through) the rockiest mountains [I] he[wed (a path)] with bronze pickaxes.
> I repeatedly climbed the highest mountains [(...)].
> I repeatedly crossed all the low mountain ranges.
> The lands of the sea I circled three times.
> [I] s[ubjugated] Dilmun.[21]
> I [climbed] the great wall of Heaven and Earth.[21]
> I did remove [its sto]nes [ ].
> Irrespective of which king comes after me,
> [May he exercise kingship for 54 years].
> May he mas[ter and reign over] the black-headed people.
> (Through) the rockiest mountains may he [hew] (a path) with bronze pickaxes.
> May he repeatedly climb the highest mountains.
> [May he repeatedly cross all the low mountain ranges].
> May he circle three times the lands of the sea.
> [May he subjugate Dilmun].
> May he climb the great wall of Heaven and Earth.
> [May he remove its stones].

Notes

[1] Akkadian is the oldest known Semitic language, coming down to us in the form of cuneiform documents from the period ca. 2800 B.C. to the first century A.D. Akkadian has two dialects: Babylonian and Assyrian.

[2] Sumerian is an agglutinate tongue unrelated to any other known language family. Sumerian itself is known to us through cuneiform documents ranging from the late fourth millennium B.C. to the first century A.D. As a spoken language it died out in the early second millennium B.C.

[3] See G. Goosens, "Les recherches historiques à l'époque néo-babylonienne," *Revue d'assyriologie et d'archéologie orientale* 42 (1948): 49–59; and Paul-Alain Beaulieu, *The Reign of Nabonidus, King of Babylon, 556–539 B.C.* (New Haven: Yale University Press 1989).

[4] See W. G. Lambert, "A New Source for the Reign of Nabonidus," in *Archiv für Orientforschung* vol. 22 (1968–69): 1–8.

[5] Kurigalzu reigned in the fourteenth century B.C.

[6] Nebuchadnezzar reigned from 604–562 B.C.

[7] W. G. Lambert, "A New Source," p. 5, line 24; and also see W. von Soden, *Akkadisches Handwörterbuch* 1399a.

[8] ki – bi – shè gi_4 (Sumerian) = *ana ashrīshu turru* (Akkadian).

[9] See, for example, J. J. A. van Dijk, "Inanna raubt den 'grossen Himmel': Ein Mythos," in *Festschrift für Rykle Borger zu seinem 65. Geburtstag am 24. Mai 1994: tikip santakki mala bashmu* ed. Stefan M. Maul, (Cuneiform Monographs 10: Groningen 1998): 9–38.

[10] See Paul Schnabel, *Berossos und die babylonisch-hellenistische Literatur* (Leipzig: B.G. Teubner, 1923) (reprint: Hildesheim 1968); and the translation of Stanley Mayer Burstein, *The Babyloniaca of Berossus: Sources from the Ancient Near East* 1/5 (Malibu: Undena Publications, 1978) 143–81(= 1–39).

[11] For Oannes in the cuneiform literature, see W. W. Hallo, *Journal of the American Oriental Society* 83 (1963): 176, fn. 79; W. G. Lambert, *Journal of Cuneiform Studies* vol. 16 (1972): 74; and R. Borger, *Journal of Near Eastern Studies* 33 (1974): 183–96; as well as A. R. George, *Babylonian Topographical Texts* (Leuven: Peeters, 1992), 269.

[12] See Schnabel, *Berossos und die babylonisch-hellenistische Literatur* 253.

[13] See A. L. Oppenheim et al., *The Assyrian Dictionary of the University of Chicago*, vol. Z, 95f. s.v. *zēru* 4b.

[14] See Grant Frame, *Rulers of Babylonia: From the Second Dynasty of Isin to the End of Assyrian Domination (1157–612 B.C.)—The Royal Inscriptions of Mesopotamia, Babylonian Periods*, vol. 2 (Toronto: University of Toronto Press, 1995) 25, Nebuchadnezzar I B.2.4.8, line 8: *zēru naṣru sha lām abūbi*.

[15] The Assyrian King Asarhaddon (680–669 B.C.) designated himself and the Assyrian royal dynasty as *zēr sharrūti kisitti ṣâti* ("Seed of the King-

dom, Eternal Family Tree"). See Riekele Borger, *Die Inschriften Asarhaddons, Königs von Assyrien, Archiv für Orientforschung*, supplement 9 (Graz 1956): 32, Brs. A, line 17 (here translated as "königlicher Same, Uradliger").

[16] See W. R. Mayer, "Ein Mythos von der Erschaffung des Menschen und des Königs," *Orientalia Nova Series* 56 (1987): 55–68.

[17] See, for example, E. von Weiher, *Spätbabylonische Texte aus Uruk*, part II (Berlin 1983), 194ff., Text No. 50.

[18] See, for example, E. von Weiher, *Uruk: Spätbabylonische Texte aus dem Planquadrat U 18*, part IV (Berlin 1993) 28ff., Text No. 129.

[19] See Joan Goodnick Westenholz, *Legends of the Kings of Akkade* (Winona Lake, Indiana: Eisenbrauns, 1997), 38–49.

[20] "Black-headed people" is an appellation for the people of Mesopotamia.

[21] Dilmun was the name of the island known today as Bahrain.

[22] This reading is uncertain. See the commentary of Westenholz, *Legends of the Kings of Akkade* 42f.

Epic Remains

Seeing and Time in the *Odyssey*

Karen Bassi

I. What the Singer Sees

When we think about the various visible objects that comprise the landscape of the ancient Greek epic, the first to come to mind are the Shield of Achilles, the scepter of the Achaians, and the bed shared by Penelope and Odysseus. These objects—whether unique or of a general class—are like snapshots in the epic narrative. Individually, they have been the focus of extensive scholarship on the poems, principally in terms of their metonymic relationship to the narrative at large (*ecphrasis*), their expression of vividness (*enargeia*), or their function in the exposition of character and social status. But the question of how visual perception is the basis of "evidence" about the past in the epics remains open. The fact that the appeal to "evidence" in this context is obviously anachronistic exemplifies the issues that frame the discussion. For in what sense do descriptions of visible phenomena in archaic poetry prefigure the criteria by which historical and archaeological evidence come to be defined as such?[1] In more general terms, what is the relationship between visible phenomena and temporal categories in the epics and how does this relationship contribute to our understanding of the dominance of visible (i.e., empirical) "evidence" in Western thought?

The idea that the past is a culturally shared phenomenon is an operating principle in the Homeric poems. Scholarship on this aspect of their content has focused primarily on the ways in which the poet gives *enargeia* (vividness) to the events he narrates, i.e., on the linguistic and formal strategies by which he makes the past "come alive" in the present of the oral performance. Egbert Bakker, for example, describes the past in the epics as "not so much an event referred to as a state of mind in the present."[2] And he goes on to make the case that epic performance challenges and collapses the distinction between the past time of the events narrated and the present time of their verbal enunciation:

> With regard to this experience of the past as something-to-be-performed, our usual notion of past tense, geared as it is to reference, the correspondence between language and facts in the past, is particularly inappropriate. If the past is something that is remembered, it does not exist in recorded form but owes its existence to the verbalizing, introverted consciousness of the performer that draws it into the present. The past, in fact, becomes "present," both in a temporal and a spatial sense: it is turned from "then" and "there" into "now" and "here" within the context of a special social event and through the actions of a special, authoritative speaker. The natural and frequent consequence of this situation is that the performer adopts the stance of an eyewitness or sportscast reporter, one who verbalizes things seen, staging the participants in the performance as spectators of the epic events from the past. Epic discourse strives to overcome the gap between the natural, visual icon and the arbitrary sign that is language, producing a sustained attempt at *enargeia* (vividness).[3]

Bakker's arguments are suggestive and persuasive, especially his discussion of the relationship between the conceptualization of time in the epic and the development of the past temporal augment in the Greek language. It may be true that the common absence of the augment in Homeric Greek is evidence for Bakker's conclusions that "Epic discourse … is to a certain degree, tenseless" and that "temporal reference is in fact irrelevant for the epic singer."[4] It may also be the case that oral performance demands a kind of "vividness" that is produced out of bringing past events into the present moment of the performance. But it is also the case that this reading comes out of the rhetorical tradition on the epic. We can ask, for example, to what extent its explanatory power answers to a desire (in the present) for an original epic performance that can transform modern *readers* into "spectators of the epic events." In other words, is this appeal to vividness simply a conditioned response to the predicament of confronting the oral epic as a literary artifact?[5] And to what extent—or in what sense—is a "visual icon" any less arbitrary than a linguistic sign, especially when the icon is itself a product of narrative description?[6] In sum, these questions bring us once again to the concept of visible "data" as evidence for past events, including (in this case) visualization as a feature of the poem's "original" oral performance.[7]

As we've seen, Bakker describes the epic performer as an "eyewitness," and he refers to the "transformation of visual evidence into language" where "visual evidence" refers to the cognitive images in the mind or memory of the poet, images that precede their description in poetic language.[8] Epic poetry obviously refers to events that exist both inside and outside the time and place delimited by the narrative frame. But in doing so it does not exactly transform visible evidence into language; the notion that visual images in the mind of the poet precede their description in language is both too literal and too tied to a strictly chronological priority.[9] It seems more accurate to say that the epic narrative creates the category of visible evidence (about past events) as a linguistic phenomenon. I use the term "evidence" here, as does Bakker, to refer to the truth-claims connected with a visual source, but with the caveat that this appeal to "evidence" is again anachronistic. As Bakker suggests, these claims are implicit in the epic poet's role as an extradiegetic eye-witness to the events about which he sings and explicit in the intradiegetic eye-witnessing of characters.[10] And in both cases the focus is on the visualized description of objects and events that constitute the past. But if we reject arguments based on what was in the epic poet's mind (or in his "mind's eye") prior to composing his poems—as I think we should—then how do we begin to understand the epic past as a product of visualization?[11]

This question has less to do with the formal descriptive properties of visual phenomena in the epics than with how these phenomena are linked to epistemological and temporal contingencies. At a more general level, these contingencies work against the idea of "timelessness" as an inherent property of Greek and, by extension, European literature.[12] We begin then with the general relationship between visualization, temporalization, and narration in the epic or, more to the point, with the idea of the past as a product of visual experience. In Bakker's analysis, this experience is figured through the "introverted consciousness" of the epic performer in real time who, by the use of vivid language, encourages or even requires his audience (actual or implied) to visualize the objects and events about which he sings. Bakker is right to insist that the performer or singer is a critical figure for understanding the ways in which epic language comments on the variables that inform its own production. But at the same time, conclusions or inferences based on our access to the consciousness or intent of an actual performer or singer raise more questions than they answer. What we want to know, rather, is how the persona of the singer is a defining feature of this commentary and, more to the point, how visual perception is a defining feature of this persona. The aim, however, is to integrate this analysis into the broader perspective of the poem's evidentiary and epistemological discourses.

If the *Odyssey*—as opposed to the *Iliad*—is "aédocentric," i.e., if the divinely inspired singer or *aoidos* is at the center of the poem's conception, then we can ask both why this is the case and what its effects are.[13] Obviously, this conclusion is based first of all on the fact that two bards, Phemius and Demodocus, are featured characters in the poem. And their "centrality" can be explained by the simple fact that the plot provides opportunities for them to perform at social gatherings in Phaiacia and Ithaca.[14] But it is also obvious that this is circular reasoning. It may also be overstating the case to say that singers play a central role in the *Odyssey*. But insofar as we can talk about an overt self-consciousness in the epic, it is to be found in those passages where the role of the singer and his effect on an audience are brought to the reader's attention. Here too, however, the common conclusion that the poet of the *Odyssey* praises himself through the characterization of his singers is not an adequate answer to the questions posed above; this is a point to which I will return. I want to suggest that the "self-conscious" presence of singers in the *Odyssey* is neither a function of the domestic locations of the plot nor a means by which the Homeric bard praises himself. Rather it is an essential feature of an epistemological predicament in the poem; we could even call it a crisis.

This predicament is the basis of Odysseus's remarks on the accuracy and effectiveness of Demodocus's song about the events in Troy in the *Odyssey* 8. The paradigmatic value of these remarks lies in the fact that they constitute a succinct commentary on how past events are to be represented in speech or song, complicated by the fact that they appeal to two seemingly contradictory evidentiary models: the divine teaching of the Muses or Apollo on the one hand and the first-hand experience of the singer as their human agent on the other (*Odyssey* 8.487–491; cf. 3.94, 4.324):[15]

> "Demodocus, I praise you beyond all mortals;
> surely either the Muse, child of Zeus, or Apollo taught you.

> For you sing about the fate of the Achaeans in a very ordered fashion,
> all the things the Achaeans did and endured and all they suffered
> as if you yourself were somehow present or heard it from another
> [who was present]."

The irony I refer to above lies in the fact that Demodocus is blind, as the poet tells us earlier: "The Muse loved him [Demodocus] and gave him both a good thing and a bad thing; on the one hand she deprived him of his eyes and on the other hand she gave him sweet song" (*Odyssey* 8.63–64). Stanford notes that, "This is the first appearance in European literature of that typical figure of the Blind Bard."[16] What is "typical," however, is especially in need of a critical understanding. In the absence of any clear ancient evidence, the common explanation that the singer's blindness is an expression of his "inner" sight or that his calling is linked to his physical incapacity for fighting is less than satisfactory. Demodocus' blindness is a bad thing, a *kakon*, and there is no indication that this *kakon* is compensatory or that it is the precondition for any extraordinary perceptual sense; moreover, not all singers are blind.[17] In short, blindness is linked with the "good" gift of song, but not in any obvious way. And the conclusion that the later tradition about the blindness of Homer was "doubtless promoted by this passage" (i.e., *Odyssey* 8.63–64) is only a footnote to the history of biographical criticism.[18] At the same time, it is not by accident or happenstance that the singer is blind in the Odyssey.

It must be admitted, of course, that in praising Demodocus "as if" he had been present at the events about which he sings, Odysseus does not say that he sings as if he had seen the sufferings of the Achaeans; this would be an intolerable inconsistency in the narrative and an implicit affront to the singer. And yet, visual perception is clearly implied in the reference to Demodocus having been present at the events about which he sings or of having heard about them from someone else who was present (*Odyssey* 8.491).[19] This understanding of the passage is corroborated in *Odyssey* 3.93–95 and 4.323–325 where Telemachus asks Nestor and Menelaus, respectively, about the death of Odysseus; either they saw the event with their "own eyes" (*ophthalmoi*) or they heard the story from some other wanderer. Here seeing with one's own eyes is clearly the principal vehicle of (empirical) proof about the past and hearing is a sort of second-order seeing. At the same time, hearing what happened from someone else implies hearing from someone who had seen the event with his own eyes; this is the effect of the reference to this other person as a "wanderer." These passages thus set out more explicitly the interplay of perceptual and evidentiary modes (hearing and seeing) that informs Odysseus's "as if" praise of Demodocus in Book 8.[20]

II. The Anti-Aoidos

Keeping in mind that the epic singer is the conventionally authorized preserver of a past worth telling, the equivocation in Odysseus's remarks calls for an explanation. As I have suggested above, this equivocation is expressed first of all in an unresolved conflict between divine agency ("The Muse or Apollo taught you") and human agency ("You were present at the events yourself"). It may be explained in part by the fact that Odysseus was present at the events recounted in Demodocus's first and last songs, or that Odysseus's own presence will testify to the truth or accuracy of the adventures he will begin to narrate in Book 9. In both cases—and by an act of displacement—Odysseus's physical presence at the events in question establishes the evidentiary criterion for relating significant past events. This distinction between human and divine agency—expressed in the distinction between sight and blindness—is central to the project of theorizing the conception of the past in the Greek epic.

We get a better understanding of this conceptual system in several passages in the *Odyssey* where Odysseus is himself compared to a singer, i.e., where by an "as if" statement he is judged in comparison with a divinely inspired *aoidos*. The most pertinent of these occurs in the midst of Odysseus's account of his trip to the underworld where Alcinoos comments that Odysseus has told his story "as a singer" would (*Odyssey* 11.362–369):

> And then Alcinoos spoke in return:
> "O Odysseus, as we look at you we do not think you are like
> a deceptive or thieving man, the sort that the black earth
> nourishes in numbers, men who are scattered abroad and
> make up lies, from which no one could learn [see] anything.[21]
> But the shape of words is upon you, and your thoughts within
> are noble,

> And you have told your story skillfully, as a singer would,
> Your own grievous cares and those of all the Argives."

De Jong comments that in comparing Odysseus to a singer "the Odyssean narrator pays himself [a compliment]."[22] But if this is so, then the "compliment" is simultaneously compromised by the negative comparison of men who tell lies and singers who (presumably) tell the truth, with the implication that Odysseus may in fact be one of the former. And in retrospect we know that Odysseus "can make many false things seem like true things when he speaks" (*Odyssey* 19.203). So, although there is no direct indication in the narrative that Odysseus's journey to Hades is any less credible than the other tales he tells, Alcinoos's comment at least hints at an implied incredulity with respect to this particular story or *muthos*.[23] In short, the simile raises more questions about Odysseus's truth-telling than it answers.[24]

Of course, the simile also refers implicitly to the *aoidos* of the epic itself (cf. *Odyssey* 1.1.) and in this respect Odysseus is like an *aoidos* insofar as the poem itself (the *Odyssey*) is the vehicle of his story or *muthos*.[25] But this conclusion becomes more complicated when we look at how Odysseus begins his first-person narrative in Book 9 where he clearly contrasts what he is about to do with the song of the *aoidos*. It is a good thing, he says to Alcinoos, to listen to the *aoidos* who is "like the gods in his voice" (*Odyssey* 9.4) and to partake of the festivities as he sings (*Odyssey* 9.1–11).[26] "But" (*de*) he continues, "your heart (*thumos*) is bent on asking me about my own grievous cares, with the result that I will groan all the more for my hardships" (*Odyssey* 9.12-13).[27] Odysseus's words mark a clear distinction between the singer's activity and his own; the former brings joy to the "whole people" (*demon hapanta*, *Odyssey* 9.6), the latter brings greater pain to himself. Moreover, unlike an *aoidos* who calls on the Muse for inspiration, Odysseus asks, "What shall I tell you first of all, what shall I tell later?" (*Odyssey* 9.14).[28] And although the gods have given him many cares (*Odyssey* 9.15) they have not given him the gift of song (cf. *Odyssey* 8.63). Finally, instead of telling someone else's story, he tells his own, beginning with the statement "I am Odysseus, the son of Laertes" (*Odyssey* 9.19). Seen in the light of these expressed differences in both form and content, Alcinoos's simile cannot be explained as a compliment the poet/narrator pays himself. Rather, it calls attention to the fact that by every criterion established in the epic, Odysseus is *not* like a singer (*aoidos*).

The significance of this conclusion becomes clearer when we look at Alcinoos's simile in greater detail, beginning with the fact that it is based on the skill (*epistamenōs*) with which Odysseus tells his story or *muthos* (*Odyssey* 11.368). In the epics, the semantic range of *epistamai* is restricted almost exclusively to practical (i.e., physical) skills or accomplishments; these include fighting, dancing, preparing a sacrifice or meal, carpentry, weaving, and medical treatment.[29] It is used far less frequently to describe skill in speaking as it is at 11.368 (cf. *Iliad* 4.404, 15.282, 19.80; *Odyssey* 8.240); and there is only one other instance in the epics where it is (indirectly) predicated of a story or *muthos*.[30] At the same time, *katalego* is routinely used in passages where the truth or accuracy of what is being said by a human speaker is being tested or emphasized.[31] This is the verb used by Odysseus at 9.14 when he asks, "What shall I tell you first of all, what shall I tell later?" and, more immediately, it is the verb used by Alcinoos when he asks Odysseus to tell him "truly" whether he saw any of his comrades in the underworld (*Odyssey* 11.369).[32] The fact that this pairing of *katalego* and *epistamai* at 11.368 occurs only here in the Homeric corpus suggests that the semantic range of both verbs is being stretched to accommodate the simile.[33] More to the point, the emphasis on accuracy and skill in terms of human experience and speech (and not divine inspiration) suggests once again the extent to which Odysseus's account of past events is not like a singer's.

In order to test this hypothesis, we can turn to the only other time in Homer where *epistamai* is used in a simile. Interestingly, in this instance too it is applied to Odysseus who, when he strings the great bow in the contest that precedes his victory over the suitors, is compared to "a man who is skilled at singing," in other words, to an *aoidos* (*Odyssey* 21.404–407):[34]

> But Odysseus of the many wiles
> as soon as he had taken up the great bow and looked it all over,
> like a man who is skilled (*epistamenos*) at [playing] the phorminx
> and singing,[35]
> he easily stretched the gut around the new peg, etc.

De Jong insists that "the primary function of the simile is to emphasize the ease with which Odysseus strings the bow...," and she compares it to the simile in *Iliad* 15.361–366 where the ease with which Apollo overturns the Greek embankment is described in comparison to a child toppling a sandcastle.[36] It is obvious that both similes are based on the shared ease of performing a physical task. But a more complex set of associations emerges when, based on shared linguistic features rather than on a thematic link (the "ease" of performing a task), we compare this simile with Alcinoos's at 11.368–369. What does the use of *epistamai* tell us about the activity of being "like a singer" in the poem? And, to return to the main argument: What do these similes tell us about the discursive and ideological distinctions between blind singers and sighted speakers and, more generally, about temporal relations in the epic?

In Alcinoos's simile, *epistamai* (in its adverbial form) refers to the content of Odysseus's account, i.e., to his ability to relate "with skill" his own cares and those of the Achaeans. In this respect, being like a singer seems to be based on the shared objects of divinely inspired song and human speech, i.e., "grievous cares" (*kêdea lugra*). But this perceived similarity only conceals a more complex set of differences. For, as I have tried to show, the appeal to practical or human skill in Alcinoos's simile at 11.368 implies what then becomes more explicit, namely, that Odysseus's tale is about events that he has experienced first-hand (unlike a singer).[37] In the simile of the bow, moreover, *epistamai* is found within its more customary semantic range of performing a practical or physical task, with the added indication that such a task requires visual knowledge: Odysseus is said by the suitors to be "one who looks (or knows how to look) at bows or who steals them" (*Odyssey* 21.397) and the narrator remarks (in the passage quoted above) that Odysseus "looked at [the bow] all over" (*ide pantê*).[38] Here skill is first measured in terms of visual perception, is then uneasily equated with "playing the phorminx and singing," but is finally and more plausibly qualified as the practical ability to string a bow.

Thus, while Odysseus is "like a singer" insofar as stringing a bow is like stringing a lyre, the overall effect of the simile is to emphasize that singing and fighting are in fact very different occupations.[39] One critical aspect of this difference is that fighting requires sharp visual acumen (exemplified in stringing the bow and, by extension, shooting an arrow); here we can also recall that the herald must help the blind Demodocus find his phorminx in Book 8 (*Odyssey* 8.66ff.). In other words, both similes express a significant dissimilarity which, enhanced by the presence of *epistamai* in each case, brings us back to the proposition that the *Odyssey* presents competing and contradictory epistemological and evidentiary discourses. This is obviously a complicated situation, since it also implies a conflict between the poet (represented most vividly by his "surrogate" Demodocus) and his main character. My point is that the divinely inspired authority of the former is challenged by the immediate presence, physical skill, and visual perception of the latter.

If we return to the "as if" statement in Odysseus' comments to Demodocus in *Odyssey* 8.487–491, we can see how it works to define the epic singer by his lack of empirical knowledge, i.e., by his physical and temporal distance from the events he recounts; this definition will become a point of departure for Socrates' criticism of the rhapsode in Plato's *Ion*. To put the matter more succinctly, Odysseus's comment that Demodocus sings (*aeidô*) about the sufferings of the Achaeans as if he himself had been there is the ironic counterpart to Alcinoos's comment that Odysseus tells (*katalegô*) his story as if he were a singer (*aoidos*). In the context of praise, the irony lies in the incongruity between the verbs of telling and singing and the evidentiary criteria to which they are attached: Demodocus sings his narrative as if he were present and Odysseus tells his as if he were a singer. Read together then, these similes create a chiastic displacement in which what is contrary to fact (i.e., that Demodocus was present at the events about which he sings or that Odysseus is a singer about the events at which he was present) demonstrates the instability and complexity of the poem's evidentiary discourse(s). Here too, we can recognize the banality of reversing the similes; to say that Demodocus sings like a singer or that Odysseus recounts events as if he had been there is to make the similes pointless in the service of stating the obvious.[40] Thus, another effect of this chiastic displacement is an implied ambivalence about the presence of an eyewitness speaker (Odysseus) within a form that is ideally or conventionally authorized by a divinely inspired (and sometimes sightless) 'singer.'[41] In saying so, however, I am not arguing for a linear trajectory from "primitive" to rational thought.[42] Rather, I am arguing for a dialogic relationship between divine song and human speech in which blindness and sight are key terms.

Alcinoos begins his praise of Odysseus's storytelling by commenting on the effect of looking at Odysseus (*eiskomen eisoroôntes*, *Odyssey* 11.363). It may be easy to ignore this introductory gambit—as modern commentaries generally do—or to explain it as a straightforward allusion to the correspondence between Odysseus's external appearance and his internal disposition. But how does the statement that Odysseus tells his story "as if" he were a singer follow from the assertion that he does not look like a liar? The notion that Odysseus is telling the truth is of course only implied in the denial that he is a liar. And it is only by extending this implication that we come to the conclusion that singers do not lie about the past. But the circuitous route by which this conclusion is reached makes the situation more complicated. A fundamental feature of this complication is that the visual cue is transferred by *metalepsis* from Odysseus's physical appearance to his words; they have a "shape" (*morphe*) which—contrasted with his inner disposition or *phrenes*—is externalized and, by metaphorical extension, visualized.[43] In other words, the effect of Alcinoos's statements depends upon a correspondence between a man's external (visible) appearance and his words (*epea*) by virtue of which his inner (unseeable) disposition can be deduced.[44]

Richard Martin has argued that *epos* (as opposed to *muthos*) in the Homeric poems has a discrete physical quality.[45] And he goes on to comment that an *epos* is an act of private communication that refers to "an item of exchange that is at the same time a physical object, like a weapon."[46] Even this too brief overview of Martin's argument suggests that the distinction he identifies is epitomized at *Odyssey* 11.362–369 where Alcinoos says that Odysseus's *epea* have shape (*morphe*) and where, in contrast, *muthos* refers to the whole of the speech which Odysseus has spoken "skillfully, as if [he] were a singer." And although Martin points out that the plural *epea* can be a substitute for the singular *muthos* in the *Iliad*, the simile here seems to rely on a clearly implied difference between the two.[47] The basis of this difference is specified in terms of speaking subjects: Odysseus produces *epea* while the anonymous singer is the producer of a *muthos* or, in Martin's terms, an account enacted in performance and prefigured by generic conventions.[48] Unlike this singer with whom—as I have suggested—he is so unsuccessfully compared, Odysseus's *epea* have a shape or *morphe* that by an act of transference refers to his own physical presence at the events about which he speaks.[49] In other words, *morphe* here emphasizes the physical and immediate quality of *epea* as an expression of the physical or immediate presence of their speaker; his words have shape because he has seen what he is talking about. In more general terms, the "shape" of Odysseus's words (*epea*) creates a link between visual perception as an evidentiary criterion and language as its medium.

III. Conclusion: The Past "as if" …

Insofar as Odysseus's *epea* refer to his own past actions, moreover, they are the products of human as opposed to divine agency.[50] And, unlike the singer's *muthos*, they are based on what he has implicitly seen with his "own eyes." Thus if the *Odyssey* is "aédocentric," this is not simply due to the self-consciousness of the poet (i.e., Homer). Or, if we do attribute it to the self-consciousness of the poet, it is not simply a matter of his self-congratulation. Rather, it is hyperbolic and, as a consequence, compensatory. But what is it compensating for? The answer I am offering here is that the *Odyssey* embodies an uneasy coexistence between two evidentiary modes, represented by the divinely inspired singer on the one hand and the poem's human protagonist on the other. In this respect, the poem's focus on the socially and culturally authorized role of the former compensates for the bold imposition of the latter. The formal expression of this compensatory gesture is the chiastic displacement I have described above, in which "as if" statements ("as if you were somehow present" and "as if you were a singer") demonstrate its agonistic potential.

More to the point, however, the epistemological hinge of this displacement is the appeal to visual or visually available proof. Conventionally and ideally figured as an ethical equation between a man's external appearance and his inner (unseen) disposition, this metaphoric equation is then extended to visual perception as the basis for making general truth claims about the past. In this respect, Odysseus's praise of Demodocus for telling the story of the Trojan War "as if he were present" can be productively read against Ranke's famous appeal to knowing the past "as it actually [or essentially] happened" (*wie es eigentlich gewesen*).[51] Ranke's phrase is of less interest here, however, than Ricoeur's commentary on it:[52]

> The *wie*—which, paradoxically, acts to balance the *eigentlich* —thus assumes the tropological value of "such as"—interpret-

> ed as metaphor, metonymy, synecdoche, and irony. What Hayden White terms the "representative" function of the historical imagination once again borders on the act of providing "ourselves a figure of..." by which the imagination manifests its ocular dimension. The past is what I would have seen, what I would have witnessed if I had been there...

What is striking is that Ricoeur's use of "such as" and "as if" statements to allude to the founding principles of nineteenth-century historicism are similar to those used in Homer, with the notable switch from the second person in Odysseus's remarks to Demodocus ("as if you yourself had been there") to the first person in the passage from Ricoeur ("if I had been there").[53] One might object that this is no more than a mere coincidence. But if we take seriously the possibility that such statements are a clue to the historicist project (as Ricoeur's references to the work of White also recommend) then we are urged to consider their theoretical significance, including their use in Homer. This does not mean taking the epic poems as "source" texts, nor does it call for compiling a data-set of similar "as if" statements.[54] It does, however, require taking Ricoeur's statements as symptomatic rather than idiosyncratic.

Doing so gives us insight into the rhetorical effect of the "as if" statement in the ancient epic and, in particular, to its "ocular dimension." Odysseus's address to Demodocus is a form of displacement that, together with the complementary (not complimentary!) assertion that Odysseus tells his story "as a singer would," points to the instability of the poem's evidentiary relationship to the past. To extend this conclusion, the "as if" statement implies that the time of the divinely inspired (but blind) singer is passing by or, in other terms, that the epic singer is on his way to becoming the repudiated "other" of the writers of history.[55] In this respect, when Odysseus praises Demodocus for singing as if he had been there, it is more accurate to say that he —rather than the poet—is indirectly praising himself. But as the source of this indirectness, the poet of the *Odyssey* alerts us to the fact that the human desire to see or to be present in a past that is not one's own is simultaneously the proof of its impossibility. It is this interplay of indirectness and impossibility that, it seems to me, Ricoeur is responding to when, by way of illustrating the methodological error of the historicist position, he says that "The past is what I would have seen, what I would have witnessed if I had been there."[56]

Ricoeur's statement thus illustrates how the persona of the historian develops as a response to the persistent illusion of seeing the past.

Notes

[1] This question is usually posed the other way around: Do actual Mycenaean artifacts constitute the material culture of the epic? See, for example, the "Introduction" to Hope Simpson and J.F. Lazenby, *The Catalogue of Ships in Homer's* Iliad (Oxford: Clarendon Press, 1970), 1: "Because our knowledge of the Greek world in the Late Bronze Age comes almost entirely from archaeological evidence, the most striking similarities between it and the heroic world are its material things." This somewhat circular formulation then leads to the question of whether the archaeological evidence is proof of anachronisms in the epic narrative. Of course, the project of "matching" artifacts to narrative description is a tricky one. In the case of a good match (i.e., the famous "boar's tusk" helmet of *Iliad* 10.261–265) does the object in question contribute to the truth-value of the poem or does the poem prove the significance of the artifact? Cf. J. Svenbro, *La parole et le marbre: aux origines de la poétique grecque* (Lund: University of Lund, 1976), 27–28 (hereafter cited as *La parole*). Also, Heidegger's discussion in *Being and Time* of ancient objects as "belonging" to past time; Martin Heidegger, *Being and Time, a Translation of Sein und Zeit.* Trans. Joan Stambaugh (Albany, NY: State University of New York Press, 1996), 348 [380 in the Niemeyer Edition]: "'The antiquities' preserved in museums (for example, household things) belong to a 'time past,' [*vergangenen Zeit*] and are yet still objectively present in the 'present' [*Gegenwart*]."

[2] Egbert Bakker, "Storytelling in the Future: Truth, Time, and Tense in Homeric Epic," in *Written Voices, Spoken Signs: Tradition, Performance, and the Epic Text*, ed. Egbert Bakker and Ahuvia Kahane (Cambridge, Mass: Harvard University Press, 1997), 12, emphasis in the original (hereafter cited as "Storytelling"). See also Egbert Bakker, "Discourse and Performance: Involvement, Visualization, and 'Presence' in Homeric Poetry," *Classical Antiquity* 12 (1993): 1–29 (hereafter cited as "Discourse and Performance") and Andrew Ford, *Homer, The Poetry of the Past* (Ithaca: Cornell University Press, 1992), 54–55 (hereafter cited as *Poetry of the Past*). Ford states, somewhat enigmatically, that the Homeric poems "make the invisible past appear to its hearers" (55).

[3] Bakker, "Storytelling," 15; emphasis in the original. Cf. Deborah Boedeker, "Presenting the Past in Fifth-Century Athens." *Democracy, Empire, and the Arts in Fifth-Century Athens.* Eds. Deborah Boedeker and Kurt Raaflaub (Cambridge, Mass.: Harvard University Press, 1998), 189, on the juxtaposition of legendary and recent events in Athenian public art as a "way of seeing the present as analogous to the [heroic] past..." Also, François Lissarrague, "The World of the Warrior," trans. Deborah

Lyons. *A City of Images: Iconography and Society in Ancient Greece.* Ed. Claude Bérard and others (Princeton: Princeton University Press, 1989) 46, on the depiction of contemporary warriors as epic heroes on fifth-century vases.

4 Bakker, "Storytelling," 28, emphasis in the original.

5 Cf. Bakker, "Storytelling," 20–23, for the notion of a "past *consciousness* that is recognized as past but that *still* comes into the present" in the minds of epic characters. Although Bakker makes some very suggestive arguments in this article, those based on the consciousness of characters or on what happens in the mind of a character are tentative at best.

6 Cf. Ernst Robert Curtius, *European Literature and the Latin Middle Ages*, trans. Willard R. Trask (Princeton: Princeton University Press, 1991) 13–16, for similar appeals to the "timeless present" of European literature (beginning with Homer, of course) and to the notion that understanding visual objects (i.e., the Parthenon frieze) requires little "mental effort."

7 See Gregory Nagy, *Poetry as Performance, Homer and Beyond* (Cambridge: Cambridge University Press, 1996) on performance and mimesis (hereafter cited as *Poetry as Performance*).

8 Bakker, "Discourse and Performance," 15–17: "Epic narrators in performance, too, are interpreters, not of visual evidence in their physical here and now, but of visual evidence provided by their memory" (17).

9 Cf. Richard T. Neer, *Style and Politics in Athenian Vase Painting: The Craft of Democracy, Ca. 530–460 B.C.E.* (New York: Cambridge University Press, 2002) 30, on empiricist and idealist notions that "the meaning of a text or image … precede[s] its 'realization' in form."

10 Bakker, "Discourse and Performance,"17 on *Iliad* 20.344–48.

11 Cf. Karl Bühler, *Theory of Language, The Representational Function of Language*. Trans. Donald Fraser Goodwin (Amsterdam: John Benjamins Publishing Company, 1990), 156, on the "displacement" between what is seen by the "mind's eye" and what is seen by the "bodily eye." Also, Hal Foster, *Vision and Visuality* (Seattle: Bay Press, 1988) ix, on the distinction between "vision and visuality," or between "the datum of vision and its discursive determinations." Foster's distinction is discussed by Robert S. Nelson, ed., *Visuality before and Beyond the Renaissance, Seeing as Others Saw* (Cambridge, 2000), 2.

12 Cf. O. Andersen, "Myth, Paradigm and 'Spatial Form' in the *Iliad*," in *Homer: Beyond Oral Poetry, Recent Trends in Homeric Interpretation*, eds. J.M. Bremer, I.J.F. de Jong and J. Kalff (Amsterdam: B.R. Grüner Publishing Company, 1987), 1–14, on the "timelessness" of mythological paradigms in Homer.

13 Svenbro, *La parole*, 25, with note 68 credits the neologism to a forthcoming article by Françoise Frontisi-Ducroux titled "Homère et le temps retrouvé." To date, I have been unable to find this article.

14 Agamemnon leaves a third, anonymous singer to watch over Clytemnestra, as mentioned by Nestor at 3.267ff. Another anonymous singer at the

court of Menelaus in Sparta is mentioned in passing at 4.17. On the figure of the singer in his social context in Homer, see Svenbro, *La parole*, Section 1.1.

[15] See Pietro Pucci, *Odysseus Polutropos: Intertextual Readings in the Odyssey and Iliad* (Ithaca: Cornell University Press, 1987), 228–235, on the "indecision about the specific responsibility of the Muse" in the *Odyssey*. Also, Kathryn Stoddard, *The Narrative Voice in the Theogony of Hesiod* (Leiden: Brill, 2004), 96–97, who argues for the Hesiodic narrator's "independence and separation from the Muses" and concludes that the *Theogony* differs from the Homeric poems in that the Muses inspire Hesiod to "sing his own words." Odysseus's tales may be said to represent another level of independence from the Muses: he tells his own story in his own words.

[16] W. B. Stanford, *The Odyssey of Homer* (New York: St. Martins Press, 1959) ad loc. (hereafter cited as *Odyssey*). See also Alfred Heubeck, Stephanie West, and J.B. Hainsworth, *A Commentary on Homer's* Odyssey, vol. 1 (Oxford: Clarendon Press, 1988) ad loc. (hereafter cited as *Commentary*).

[17] Cf. the so-called Palinode of Stesichorus, which I discuss in "Helen and the Discourse of Denial in Stesichorus's Palinode" *Arethusa* 26 (1993): 51–76. In his story about the Palinode in Plato's *Phaedrus*, Socrates says that Stesichorus's blindness is a punishment for telling a false story about Helen. We might conclude that the poet must be blind in order to tell the true story. Cf. *Iliad* 2.594ff. where Thamyris is physically punished (*pêros*) for trying to surpass the Muses in singing. G. S. Kirk, *The Iliad: A Commentary*, vol. 1 (New York: Cambridge University Press, 1985) ad loc. (hereafter cited as *Commentary*) concludes that *pêros* "cannot mean 'blind' here, since that was traditionally no handicap to a singer." But this conclusion relies on a too literal reading of the evidence. On the tradition of blind singers in general, see C. M. Bowra, *Heroic Poetry* (London: Macmillan and Company, 1964), 420–422.

[18] Heubeck, West and Hainsworth, *Commentary*, vol. 1 on 8.64.

[19] It may be pertinent that the expression "To see the light of the sun" (*horan phaos hêelioio*) means to "be alive," as (with no little irony) at *Odyssey* 10.498. Cf. *Odyssey* 4.539–540; Mimnermus 1.8. The metaphor is based on the belief that being (present) is a function of visual perception.

[20] The appeal to Apollo and the Muses at *Odyssey* 8.487–489 also seems to have less to do with the *content* of what Demodocus sings than with its *form*; the divinities taught the bard how to sing *kata kosmon*. Conversely, the *content* of what he sings is more immediately verified in the simile of his being present. In other words, the distinction between divine inspiration and human knowledge is implicitly expressed in terms of a distinction between form and content. This distinction is confronted in Plato's *Ion* (530b10-c1) where Socrates distinguishes between the rhap-

sode's knowledge of the "verses only" (*monon ta epê*) of Homer and his "thought" (*dianoia*).

21 Stanford, *Odyssey*, ad loc., translates this line as "from sources which ... no one could see [i.e., test] for himself'." Cf. Ford, *Poetry of the Past*, 124, note 53. Carol Dougherty, *The Raft of Odysseus, The Ethnographic Imagination of Homer's Odyssey* (New York: Oxford University Press, 2001), 43–53 (hereafter cited as *Raft of Odysseus*) argues that the image of the "deceptive and thieving man" in this passage is a means of equating pirates with bad poets.

22 Irene J. F. de Jong, *A Narratological Commentary on the Odyssey* (Cambridge: Cambridge University Press, 2001), 286 (hereafter cited as *Narratological Commentary*). Cf. Ford, *Poetry of the Past*, 124–125 and John Peradotto, *Man in the Middle Voice: Name and Narration in the Odyssey* (Princeton: Princeton University Press, 1990), 92–93.

23 Odysseus does say to Circe that "No man has ever yet gone to Hades in a black ship" (*Odyssey* 10.502). But that such a journey is unprecedented does not indicate that it is less credible. Cf. Plato, *Rep.* 614b2 where Socrates introduces the story of Er's sojourn in Hades: "What I am going to tell you won't be like the tale (*apologon*) told to Alcinoos..." On Plato's use of myth, see Penelope Murray, "What Is a *Mythos* for Plato." From *Myth to Reason? Studies in the Development of Greek Thought*. Ed. Richard Buxton (Oxford: Oxford University Press, 1999), 251–62, esp. p. 257.

24 On the difference between the so-called "lying" tales that Odysseus tells on Ithaca and the "true" adventure stories that he tells to the Phaiacians in Books 9–12, see Dougherty, *Raft of Odysseus*, 73–77. In general, the *aoidos* is an ambiguous figure in the poem, exemplified by the anonymous *aoidos* whom Agamemnon leaves to watch over Clytemnestra (*Odyssey* 3.267ff.). Stanford, *Odyssey*, remarks on line 267 that the singer plays a "virtuous part"...as propaganda for H's own profession. He does not appear in either Pindar's version of the story in *Pythian* 11 or in Aeschylus's *Agamemnon*. See also, de Jong, *Narratological Commentary,* 82–83. But the fact that this singer fails to fulfill his obligation is hardly evidence of his "virtuous part." Alex Hardie, "The Ancient Etymology of AOIDOS," *Philologus* 144 (2000): 166 (hereafter cited as "Aoidos") mentions the suggestion in the Homeric scholia that the singer (*aoidos*) left in charge of Clytemnestra had no genitals (*aidoia*), i.e., "he was a eunuch." Hardie also reviews the evidence for an etymological connection between *oida* (the verb meaning "to know") and *aoidos* (both alpha intensive and alpha privative) which, he argues, had its origins in Homer. And in this context he refers to Alcinoos's praise of Odysseus's "transparent truthfulness as a narrator" at *Odyssey* 11.368 (p.165). In addition to begging the question of "transparency," however, Hardie's evidence only proves that the question of whether or not singers "know" anything was fraught with uncertainty in antiquity. In the first

reference to song in the *Odyssey*, Telemachus complains that the suitors care too much (it seems) about the *kithara* and song (*Odyssey* 1.159). And although he tells Penelope not to blame the singers for the content of their song (*Odyssey* 1.345ff.) and even tells the suitors to be quiet so that they might listen to Phemius (*Odyssey* 1.369ff.), Telemachus's initial observation weakens the conclusion that the poet (!) is paying himself a compliment. In short—and in spite of the perceived interest in etymology among later authors—the singer in the *Odyssey* is less a foil for self-congratulation than (keeping with the biographical fiction) a figure of self-doubt. Cf. Nagy, *Poetry as Performance*, 112–113 on the definition of *aoidos*. On the difference between the singers Phemius and Demodocus in the *Odyssey*, see Svenbro, *La parole*, 21–24.

25 Alcinoos refers to the content of Odysseus's story as *theskela* deeds (*Odyssey* 11.374). On the uncertain meaning of *theskela* ("divinely inspired"?) see Albert Heubeck and Arie Hoekstra, eds., *A Commentary on Homer's Odyssey*, Vol. 2) Oxford: Clarendon Press, 1988), hereafter cited as Commentary, ad loc.

26 The same phrase, i.e., "like the gods in his voice," is used by Telemachus to describe the singer at *Odyssey*. 1.370–371.

27 Huebeck and Hoekstra, *Commentary*, vol. 2, ad loc. explain the sequence of ideas as follows: "Odysseus means: although it is indeed lovely to hear a bard perform (3–4), if your heart is now moved (*epetrapeto*), Alcinous, to ask me to sing of my own misfortune (viii 572ff.), it may bring you joy, but it will increase my sorrow still further." But, of course, Odysseus does not refer to what he is about to say as "singing." Nor is there any indication that his account may bring Alcinoos joy. Rather, it seems to me that the two activities, i.e., that of the singer and that of Odysseus, are clearly differentiated. The question remains why the commentators prefer to equate the two.

28 The chronological anxiety or self-consciousness expressed in the question "What shall I tell you first, etc." appears to be unique to Odysseus and to the *Odyssey*. The closest parallel is found in the *Iliad* where the narrator/singer asks, "Who was the first to be slain, and who last…?" (*Iliad* 5.703; 11.299; cf. 16.692). Odysseus's question is unique both because it is directed to himself and because it is concerned with the chronology of telling a story in general rather than with the chronology of similar actions (i.e., the order of battle casualties). At the beginning of the *Odyssey*, the singer calls upon the Muse to tell Odysseus's story "from some point, from whatever point you will" (*Odyssey* 1.10). On this phrase, see Heubeck and Hoekstra, *Commentary*, vol. 1. At the beginning of the *Iliad*, the singer calls upon the Muse to sing (it seems) from the point at which Agamemnon and Achilles first quarreled (*Iliad* 1.6), on which see Kirk, *Commentary*, vol. 1. Thus, chronology is clearly a feature of the epic proem, with the Muses as the arbiters of where to begin in both poems. But the proem to the *Odyssey* is also striking for its

rather cavalier attitude which is in stark contrast to Odysseus's pointed interest in ordering the events of his own story. In this respect, there is a difference in what we might call the methodological approach of the singer of the *Odyssey* and its protagonist.

29 See Dougherty, *Raft of Odysseus,* 32–35 on Odysseus's construction of his raft at *Odyssey* 5.244–57: "Odysseus's work on the raft is knowledgeable" (*epistamenõs*, 245). And she goes on to comment that: "[T]he goddess Calypso represents the divine or traditional component of raft-building; she provides the raw materials that all rafts are made of ..., but Odysseus himself must figure out how to build and steer the raft in order to get home, drawing on his skills, knowledge, and familiarity with tools" (35). Dougherty argues that the building of the raft "corresponds to the techniques of producing oral poetry."

30 At *Iliad* 14.90–94 Odysseus criticizes Agamemnon's story or *muthos* because it does not conform to that of a man who would "know how" (*hos tis epistaito*) to speak what is right.

31 So, for example, in the first appearance of the verb in the *Iliad*, Agamemnon tells Nestor that he has not lied about his "madness" (*Iliad* 9.115; cf. 19.186). In the *Odyssey*, Odysseus is often the first-person subject of the verb. At *Odyssey* 7.297 he tells Alcinoos, "All I have told you is true" (*alêtheiên katelexa*); at *Odyssey* 10.16 (=12.35) Odysseus says to Alcinoos that he told Aiolos "all the tale [of the Trojan expedition] according to what is proportionate" (*kata moiran katelexa*). The verb is also used by Odysseus when he asks Demodocus to sing of the Trojan horse at 8.496–498: "If you tell me these things according to what is proportionate (*kata moiran katalexês*), I shall immediately tell all men how the god willingly granted you divine song." See Richard Martin, *The Language of Heroes, Speech and Performance in the Iliad* (Ithaca and London: Cornell University Press, 1989), 97 and 102 (hereafter cited as *Language of Heroes*) on the meaning of *kata moiran* as "proportionate;" cf. *kata kosmon, Odyssey* 8.489. Also relevant are 17.108 and 122 where Telemachus insists to Penelope that he will tell her the "truth" and that he had told Menelaus the "truth" (*alêtheiên katalexô* and *alêtheiên katelexa*). At 23.225–226, Penelope comments that Odysseus has given "very clear signs (*ariphradea katelexas*) describing our bed." At *Odyssey* 3.97 and 4.327, Telemachus asks Nestor and Menelaus, respectively, to "tell me (*katalexon*) well how you got sight of him." On this last passage, see Heubeck, West and Hainsworth, *Commentary*, vol. 1. There are variations, of course: Telemachus asks Nestor to tell the "true story" (*alêthes enispes*) about the death of Agamemnon at *Od.* 3.247 and Nestor replies that he "will tell all the truth" (*alêthea pant' agoreusô*, 3.254). Cf. H. Fournier, *Les verbes "dire" en grec ancien* (Paris: C. Klincksieck, 1946), 117; Dougherty, *Raft of Odysseus*, 65; Bakker, "Storytelling," 60.

[32] Huebeck and Hoekstra, *Commentary*, vol. 2, on *Odyssey* 9.14–15 note that "*katalegein* suggests a methodically ordered account."

[33] See Doreen Innes, "Metaphor, Simile, and Allegory as Ornaments of Style," in *Metaphor, Allegory and the Classical Tradition: Ancient Thought and Modern Revisions,* ed. G.R. Boys-Stones (Oxford: Oxford University Press, 2003) 11 on the notion of "semantic stretch."

[34] See Joseph Russo, Manuel Fernandez-Galiano and Alfred Heübeck, *A Commentary on Homer's Odyssey,* vol. III (Oxford: Oxford University Press, 1992) ad loc. (hereafter cited as *Commentary*): "This is the only occurrence of *epistamenos* with the gen., as if it were a substantive."

[35] The *phorminx* is explicitly said to be the instrument of the *aoidos* at *Odyssey.* 4.17–18.

[36] De Jong, *Narratological Commentary*, ad loc. She reiterates the claim that these similes are the means by which the narrator "suggests his own importance."

[37] Cf. W. Moran, "Mimneskomai and 'Remembering' Epic Stories in Homer and the *Hymns*," *Quaderni Urbinati di Cultura Classica* 20 (1975) 195–211 for a discussion of the formula "don't you remember when?" in Homer. This phrase occurs only once in the *Odyssey* (24.114–117) where the respondent, Amphimedon, says that he does remember the events alluded to by Agamemnon. Moran comments that Amphimedon "knows the story from his personal experience of the events, and he knows how to act like a bard" (207). But this only raises the question of whether a bard "knows" or "remembers" events because of his "personal experience."" The answer I propose is that the bard/singer necessarily sings about events he does *not* know from "personal experience," i.e., events at which he could not have been present, and that this is an inherent feature of bardic discourse. To put the matter another way, Amphimedon is not "in effect, a bard" (Moran, 207) but an eyewitness. Cf. Svenbro, *La parole*, 22, who notes that both Demodocus and Homer are singers who know how to sing things about which they have "no personal knowledge" ("c'est seulement grâce au 'don de la parole' que l'aède sait chanter les choses dont il n'a aucune connaissance personnelle"). Hardie, "Aiodos," discusses the relationship between singing and knowing in authors later than Homer.

[38] Cf. Russo, Fernandez-Galiano, and Heubeck, *Commentary*, vol. 3, ad loc. and their note on 21.405 regarding the emphasis on Odysseus's "careful scrutiny of the bow."

[39] Cf. the speech of Polydamas at *Iliad* 13.729–734.

[40] Cf. Helene Foley, "'Reverse Similes' and Sex Roles in the *Odyssey*," *Arethusa* 11 (1978): 7–26.

[41] See the discussion of Dougherty, *Raft of Odysseus*, 66–78, who points to the contrast between the kind of fixed songs that a bard like Demodocus sings and the "wandering" songs of Odysseus, the eyewitness: "Odysseus… is a poet on the move, singing new songs for new audiences at new

places, and his songs are equally mobile. Never the same twice, his songs represent a flexible and mercurial notion of poetic truth, one that is authorized not by the Muses but rather by the actual travels of the poet himself" (70–71).

42 On the history and prevalence of this "from…to" trajectory in Classical (and Western) scholarship, see Buxton's excellent "Introduction" to Richard Buxton, ed., *From Myth to Reason? Studies in the Development of Greek Thought* (Oxford: Oxford University Press, 1999), 1–21. Buxton notes "the extent to which our reading of the myth/reason issue must depend on how we assess the narrative strategies at work."

43 Heubeck and Hoekstra, *Commentary*, vol. 2, ad loc. take *morphe* to refer to Odysseus's "charm." This reading is plausible, but fails to take account of the more literal meaning of *morphe* as a visible shape. Dougherty, *Raft of Odysseus*, 57 argues that *morphe epeon* is part of the "language appropriate to the Demodocus model of a poet who sings as part of a closed aristocratic social context." Cf. the phrase *krisis morphes in* Euripides's *Helen*; I discuss this phrase in "The Somatics of the Past: Helen and the Body of Tragedy," in *Acting on the Past, Historical Performance Across the Disciplines*, eds. Mark Franko and Annette Richards (Hanover: Wesleyan University Press, 2000), 13–34 (hereafter cited as "Somatics of the Past").

44 In addition to Alcinoos's speech at *Odyssey* 11.367, the phrase "the shape of words" (*morphê epeôn*) occurs only one other time in the Homeric poems where again it is embedded in a comparison with a man's external appearance or *eidos* (*Odyssey* 8.169–177).

45 According to Martin, *Language of Heroes*, 12, *epos* is "an utterance, ideally short, accompanying a physical act, and focusing on message, as perceived by the addressee, rather than on performance as enacted by the speaker."

46 Martin, *Language of Heroes*, 30. *Epea* can also be deceptive or dangerous, of course, as when Nestor says that, while the Greeks were in Troy, Aegisthus "kept beguiling the wife of Agamemnon with *epea*" (*Odyssey* 3.264) or when Demodocus says that Achilles and Odysseus "strove with violent *epea*" (*Odyssey* 8.77). These examples illustrate what Martin calls the private context of *epea* (37–38).

47 Martin, *Language of Heroes*, 29–30.

48 Singers produce song (*aoidê*), of course, but they can also produce *epea*. Thus, the Phaiacians take delight in Demodocus's *epea* (*Odyssey* 8.91). But the implication here too seems to me to be that Odysseus is listening to the story of his own sufferings. On *muthos* as referring to "actual genres of discourse which are also poetic genres" in the *Iliad*, see Martin, *Language of Heroes*, 42 and Chapter 2.

49 This conceit reaches its high point when Aristophanes puts words on a scale in order to give empirical "weight" to the poetic contest between Aeschylus and Euripides in *Frogs* (1378–1410).

[50] This observation may give greater specificity to Martin's conclusion that *epea* in the epics are *private* utterances as opposed to performance-based speech-acts. In this context, however, "private" does not mean "everyday," nor is it the opposite of "public;" Odysseus obviously tells his story to a formally constituted audience in a public setting. Rather, I am using "private" to refer to the act of telling his *own* story, defined as such by his self-reported physical presence at the events he recounts.

[51] On Ranke's phrase, see Georg G. Iggers and Konrad von Moltke, eds. *The Theory and Practice of History, Leopold von Ranke* (Indianapolis and New York: Bobbs-Merrill Company, 1973), xix (hereafter cited as *Theory and Practice*) esp. on the meaning of *eigentlich* as "characteristic, essential" as opposed to "actually."

[52] Paul Ricoeur, *Time and Narrative*. Trans. Kathleen Blamey and David Pellauer. (Chicago: University of Chicago Press, 1984–1988), vol. 3, 185.

[53] On the complicated history and meaning of the term *historicism* and its adherents, see the "Introduction" to Iggers and Moltke, *Theory and Practice*, esp. xviii–xix.

[54] It is accurate, if reductive, to say that these approaches, i.e., source criticism and the accumulation of data, are the fundamental principles of traditional historicist research. See the "Introduction" to Iggers and Moltke, *Theory and Practice*.

[55] See Thucydides, *History* 1.21 (cf. 2.41); Aristotle, *Poetics* 1451a36–b19. Cf. Wilhelm von Humboldt's 1821 essay, "On the Historian's Task," translated in Iggers and Moltke, *Theory and Practice*, 6–10, where he compares the activities of the poet and the historian. I discuss Aristotle's attempt in the *Poetics* to distinguish between the work of the historian and that of the poet in "Somatics of the Past."

[56] Humboldt's remarks are of interest: "For events are only half understood or are distorted, if one stops with their superficial appearance; moreover, the common observer constantly imbues this appearance with errors and half-truths. These are dispelled only by the true form of events which reveals itself solely to the historian whose eyesight is naturally keen and has been sharpened by study and practice." "On the Historian's Task," in Iggers and Moltke, *Theory and Practice*, 10.

Fourier and the Saint-Simonians on the Shape of History

Jonathan Beecher

One of the main intellectual consequences of the French Revolution was to leave many Europeans with the sense that the optimistic, rationalistic and egalitarian ideology of the Enlightenment had exhausted itself and been discredited with the failure of the radical phase of the revolution. There was a sense on many sides that the Enlightenment had been "on trial" during the French Revolution and that the understanding of human nature and history offered by the Enlightenment had proved inadequate. The period that followed the French Revolution was therefore marked by an intellectual reaction leading in two directions. First, to the belief that future thinking about society and politics would have to take a much greater account of the affective needs of individuals and of the role that long-established beliefs and customs must play in the maintenance of any social and political order. Second, history had to be understood in a way that the eighteenth century had failed to understand it. Not simply as "philosophy teaching by example"— a storehouse of moral lessons useful at any time—but rather as a force or power greater than the will of any individual.

In this essay, I discuss two of the most interesting French efforts to make historical sense of the post-revolutionary period: efforts made in the late 1820s by a group of young social thinkers called the Saint-Simonians and efforts made, started a generation earlier, by the utopian thinker Charles Fourier (1772–1837). Both Fourier and the Saint-Simonians were inspired by the desire to understand and overcome what they saw as the disorder of post-revolutionary society and culture. While neither was sympathetic to the Catholic and royalist reaction, both were strong critics of the rationalism of the Enlightenment and the narrowness of its conceptions of human nature and progress. Both believed that understanding begins by putting the present in historical perspective.

The Saint-Simonians, whose thinking was loosely inspired by the social philosopher Henri Saint-Simon (1760–1825), were brilliant, but troubled young men and women who saw themselves living in a society in which traditional social bonds had collapsed. They were eager for a faith that would give them direction and purpose, and they regarded Saint-Simon as the prophet of a new world in which science and love might collaborate to bring about the material and moral regeneration of humanity. In 1829, they organized what amounted to a lecture series in Paris in which they attempted to spell out the main elements of their doctrine. The following year, having constituted themselves formally as a "church," they began to devote themselves to missionary activity, sending groups of "preachers" out into the provinces to expound their social gospel. It was in their 1829 lectures that they presented the fullest account of their reflections on history.

In these lectures on history the Saint-Simonians took issue with the tendency of earlier historians to focus on "great men" and their accomplishments. The true subject of history, they argued, was humanity as a whole—the human species considered as "a single body developing progressively according to invariable laws."[1] The first of these laws stated that history was marked by an alternation between two stages or epochs—the organic and the critical. The organic stage was one in which the goal of all social activity was clearly defined and widely understood. All efforts were focused on the accomplishment of this goal, and the different social roles that different individuals had to play were recognized and accepted. "Truly superior individuals," they wrote, "are naturally found in positions of power; there is legitimacy, sovereignty, authority in the true meaning of these words. Harmony reigns in social relations."[2] In organic periods, strong and weak were joined by common beliefs and by shared answers to life's largest questions. Such periods were periods of religious unity, the Saint-Simonians asserted. They were periods in which shared belief culminated naturally and spontaneously in shared reverence and shared worship. Looking over the history of the West, the Saint-Simonians saw two great organic periods, each marked by the flourishing of a religion: the first was the period of early Greek and Roman polytheism and the second was the period of medieval Christianity.

Critical epochs were epochs in which the old shared beliefs and relations came under attack and the social order disintegrated, leaving only "an agglomeration of isolated individuals, at odds with one another."[3] Such periods were marked by competition, rivalry, skepticism, and egoism, which the Saint-Simonians called "the hallmark of critical epochs" and "the deepest wound of modern societies."[4] In critical periods all beliefs were abolished, all common sentiments extinguished, and the goal of collective action disappeared from view. In critical periods power lost its legitimacy, and true capacities were no longer recognized. Egoism replaced dedication, and devotion gave way to atheism. Historically, it was Socrates who initiated the critical epoch that replaced the first organic age, and Martin Luther who initiated the modern critical age which completed its destructive work with the French Revolution.

If history for the Saint-Simonians was marked by an alternation of organic and critical epochs, this alternation was itself subsumed within a larger pattern. And this was the Saint-Simonians' second law of history: an unstoppable general movement from conflict and antagonism to association and love. Thus history for the Saint-Simonians was progressive, as it had been for the philosophical historians of the eighteenth century. But unlike their predecessors, the Saint-Simonians rejected the notion that progress consisted in the development of the rational faculties at the expense of man's passionate nature. For the Saint-Simonians, human beings had infinite possibilities; and the development of the capacity for feeling, for love, for religious faith, for the exercise of the creative imagination was just as important as the development of the rational faculties and control over nature. Indeed, they warned, the one-sided development of the rational faculties could lead to disaster.

For the Saint-Simonians the end of the historical process was the attainment of "universal association." By this they meant "the association of all men on the entire surface of the globe and in all spheres of their relationships."[5] In work but also in play. In intellectual life but also in the emotional sphere. This would be the result of a long, ongoing process. For in the course of human history the units of association had grown steadily larger—from the family to the city to the nation to the Catholic Church, and finally to a church embracing all humanity and all the modes of human existence.

History was thus seen by the Saint-Simonians in spatial terms as a process marked by the gradual contraction of the area of antagonism and the expansion of association, and of love, in an ever widening circle from family to tribe to nation to church to all humanity. Here the Saint-Simonians were adapting, to their own purposes, the spatial imagery that the philosophical historians Turgot and Condorcet had used to characterize the spread of enlightenment and the gradual extinction of ignorance and superstition. For the Saint-Simonians, as Frank Manuel has written, "a history of the expansion of love replaced that of the successive advances of mind."[6]

In its detail the Saint-Simonians' theory of history was more complex. They recognized that the old conflicts of families, tribes and cities persisted within the larger structures of association. Still they asserted that the general tendency was for antagonism to assume less lethal forms over time. In the earliest societies warriors ate their captives; later they merely killed them; finally they enslaved them. In modern society a new form of slavery emerged. Wage slavery was milder than chattel slavery. But it was still slavery in the sense that the worker, dependent entirely on the sale of his labor and unable to bargain effectively, had no choice but to accept the conditions imposed by the employer. The goal was to replace the exploitation of man by man with the peaceful exploitation of the globe by, and in the interest of, all humankind.[7]

In their own time, the Saint-Simonians believed, the last of the great critical epochs was coming to an end. The destruction of the old order was now complete, and the Saint-Simonians saw themselves as the prophets of a new organic era which would be different from its predecessor. Different because it would bring history in the traditional sense to an end.

> Humanity is progressing [they wrote] towards a final state which will dispense with these long and painful alternations, and in which progress will take place continuously, steadily, without interruption, without crises … in which religion and philosophy, art and worship, dogma and science will no longer be divided, [and] in which duty and interest, theory and practice, far from being at war, will lead to the same goal, the moral elevation of man.[8]

With this shining vision the Saint-Simonian cycles would give way to a future marked by uninterrupted community and love.

Not all the Saint-Simonians' readers were equally impressed by this vision of the end of history. Nonetheless in its essential elements the Saint-Simonian vision of history resonated deeply among a host of contemporaries—including J.S. Mill, Carlyle, Feuerbach and Marx. To cite just one example, Mill later wrote that the Saint-Simonian theory of history helped convince him of two important points. First, that "institutions which, considered in themselves, are incapable of producing anything but the most frightful evils (the Catholic Church for example)" could have been "absolutely indispensable … at some particular stage in the progress of the human mind."[9] Second, and more generally, the reading of the Saint-Simonians led Mill to the conclusion that the history of his own time might reflect the peculiarities of an age of transition and not, as he had supposed, the perpetual attributes of human nature. This enabled him to believe for the first time, in the progress of the human species, and to "look forward, through the present age of loud disputes but generally weak convictions, to a future which shall unite the best qualities of the critical with the best qualities of the organic periods."[10]

If the Saint-Simonians were social prophets, centrally concerned with the problem of history, Charles Fourier has commonly been regarded as a quintessential utopian far more concerned with the ideal than the historical. As Karl Marx put it, utopians like Fourier were too absorbed in painting "fantastic pictures of future society" to give adequate attention to the historical process by which the new world was to be brought into being.[11] It is true that Fourier's descriptions of his ideal society have a timeless, pastoral quality that links them to the static and generally agrarian utopias of the Old Regime. Fourier's conception of human nature was completely ahistorical. In his view human beings were moved by a set of unchanging passions; he never made a serious effort to analyze the development of the human faculties or of human knowledge as a historical process. Rousseau's attempt (in the *Second Discourse*) to interpret the development of man's self-consciousness as a response to developing technologies and modes of cooperation was foreign to him. Even more so was Marx's vision of history as the story of human self-creation through labor.

Yet Fourier clearly wanted to give his theory a historical dimension; and in his writings he generally attempted to subsume both his critique of the contemporary world and his description of the ideal society within the framework of a panoramic account of the whole course of human history. According to this account, which ran from "the first infection of the seas by astral fluid" to the moment when the earth would cease to spin on its axis, the contemporary period—which he referred to pejoratively as "civilization"—was just the fifth, and one of the worst, of 32 distinct historical periods.

In its fullest outline, which was contained in his first work, the *Théorie des quatre mouvements*, Fourier's 32-stage scheme of historical development was divided into four phases. First there was a phase of "infancy" or "ascending incoherence," which was meant to last approximately five thousand years. It all began in good biblical fashion with a brief period that Fourier named "foreshadowing of happiness" or simply "Edenism." This was followed by four "unfortunate periods," which Fourier collectively described as "ages of perfidy, injustice, constraint, poverty, revolutions, and bodily weakness."[12] The first of these was a period of economic backwardness and psychic inertia to which Fourier gave the name of Savagery. Then came the even more lugubrious periods of Patriarchate, Barbarism, and Civilization. Finally, Fourier claimed, the childhood of human history would end with two transitional periods, in which certain reforms in economic or sexual mores would prepare the way for the ideal society.

Fourier's historical timetable thus called for an initial phase of roughly five thousand years of chaos and suffering. It was to be followed, however, by a second phase of "growth" or "ascending harmony" which would last for a full 35,000 years. According to the plan, this second phase of vigor, enlightenment, and passional fulfillment would be marked by eight periods and eight new "creations" culminating with a pivotal or "amphiharmonic" period of complete happiness. Thereafter life on the planet earth would gradually retrace its steps through a third phase of decline and a fourth of "decrepitude." Following the twenty-fifth period of its development, mankind would sink back from Harmony into Chaos, passing in reverse order through the initial seven periods. Finally, after just 80,000 years of existence, an end would come to animal and vegetable life on earth, and the planet would cease to rotate on its axis.[13]

Such are the broad outlines of Fourier's picture of history. It was of course an ambiguous picture in which the striving toward happiness was linked to a vision of final destruction. But Fourier never seriously explored the darker implications of this vision. He had much more to say about the youth than about the old age of the species. And when he asked himself how it was that a just and omnipotent God could tolerate the evils of civilization and the prior periods of injustice and perfidy, he had a quick and almost flippant answer. God had after all allotted to the harmonious periods of human history a span that was a full seven times longer than that given to the ages of constraint. That did not seem ungenerous.

In viewing human history as a recapitulation of the stages of individual growth from childhood to decrepitude, Fourier was appropriating one of the central metaphors in the great tradition of philosophical history running from Augustine to Henri Saint-Simon. But whereas Saint-Simon saw the principal social problem of the post-revolutionary period, the conflict of feudal and industrial civilizations, as a "crisis of puberty," according to Fourier's timetable humanity was still coping with the teething problems of infancy.[14] And whereas Saint-Simon believed that the motor of change was scientific or industrial activity, we find Fourier in his early writings groping toward a scheme of periodization in which each historical period is defined in terms of the status of women and attempting to work out a theory of social change in which sexual rather than economic relations played the decisive role. As he wrote in the *Quatre mouvements*: "Each historical period has an attribute that forms the *pivot of the mechanism* and whose absence or presence determines a change in period. This attribute is always drawn from love."[15]

Fourier argued that the period of Barbarism was marked by "the absolute servitude of women" and Civilization by "exclusive marriage and the civil liberties of the wife," and that the crucial feature of the ensuing period of Guarantism would be a form of "amorous corporation" within which women would enjoy considerable sexual freedom. If the barbarians had taken up the practice of monogamous marriage, Fourier wrote, "this innovation alone" would have made them civilized. And if the French revolutionaries had abolished the institution of marriage, the whole edifice of civilization would have rapidly collapsed.[16]

Neither historians nor Fourier's own contemporaries have known what to make of his theory of history. His early attempt to work out a scheme of periodization based on the premise that "social progress and changes of period are brought about by virtue of the progress of women towards liberty"[17] won him the admiration of Flora Tristan and Désirée Véret and other early French feminists. But the attempt to work out a theory of historical change centering on the changing status of women was not consistently pursued by Fourier. Apparently fearing that he would lose all hope of reaching an audience if he presented himself too explicitly as a prophet of sexual liberation, he focused in his later presentations of the theory on the priority of economic relations as catalysts of social change.

As for the general significance of Fourier's historical theory, it has often been asserted that he lacked both the interest and the insight of a Marx or of the Saint-Simonians with regard to the specifically historical dimension of human experience. It would seem that the thought that presided over the elaboration of Fourier's historical theory was less a desire to understand historical change than a penchant for symmetry. And it is surely not a coincidence that Fourier's account of the 32 historical periods was so organized as to dovetail exactly not only with his account of the various stages in the life of man but also with his conception of the "basic series" of 16 tribes and 32 choirs into which his ideal community was organized. By and large then, Fourier's elaborate theory of history might well be described as having played a decorative rather than an essential role in his doctrine.[18]

But surely more can be said about Fourier's way of conceptualizing history. First, Fourier's theory of history can be described as a thought-experiment, in which he was trying to imagine the emergence of a world consistent with human needs and expressive of human powers. In undertaking this thought-experiment, Fourier relied on a trusty guide which was what he called his "theory of universal analogy." This theory rests on two premises: the universe is a unified system, a web of hidden correspondences or hieroglyphs; and human beings are at the center of the system. According to Fourier, everything that transpired in the human world had some echo or correspondence in the world of nature. Each of the basic human passions was represented by its own color, musical note, geometrical form and celestial body. Similarly, every human personality trait, institution, or social relationship was faithfully depicted in some animal, vegetable or mineral species.

Since nature mirrored the conflict or harmony of the human passions, the system was anything but static. According to Fourier's cosmogony there were to be 18 (or 26) separate creations during the course of human history. The first creation had endowed the earth with a multitude of poisonous snakes, insects, sea-monsters, and other dangerous beasts. Each of these harmful creatures was a symbol or hieroglyph of some form of conflict or disharmony in the early stages of human history. For example, the 72 different "crimes of commerce" that Fourier enumerated were represented by 72 different types of poisonous snakes. But in the ideal world that Fourier called Harmony these snakes, along with a great many other dangerous animals, would disappear. New creations would replace them with a vast number of useful and gracious species (including anti-lions and anti-crocodiles) whose existence would reflect the new-found harmony of the human passions. Even in civilization one could detect portents of the future. The bee's hive and the beaver's dam were models of Harmony just as the wasp's nest was an image of civilization and the parrot (with its bright plumage and glib talk) was an image of the civilized philosopher.

The idea that the natural universe is a mirror of human concerns or of man himself is as old as civilization. When Plato wrote in the *Timaeus* (30B) that "this world is in truth a living creature endowed with a soul and with reason" he was apparently drawing on a substantial tradition of Pythagorean speculation on the macrocosm and the microcosm and the harmony of the spheres. There was also a long and influential tradition of Christian speculation about the invisible harmonies, likenesses and proportions that bound together the universe.[19] For centuries Christian philosophers had been finding allegories of the human condition in the flowers and the fields and even in the form of a nut. In the century of Fourier's birth there were still many who regarded the natural world not as a self-contained machine but rather as "a map and shadow of the spiritual estate of the soul of man."[20]

What was Fourier's relation to this whole tradition of thought? Obviously he did not share the Christian preoccupation with analogy as a means of leading the mind to God. Nor was he as concerned as classical thinkers with strictly political analogies—the notion of the body politic, the analogy between the faculties of the soul and the division of functions within the state. It is true that Fourier spoke of the human body as an "epitome of the movement of the universe" and that he determined the number of the passions and the size and structure of his ideal community on the basis of analogies relating to the number of bones, muscles, and even teeth in the human body.[21] It is also true that Fourier, who possessed a copy of Kepler's *Harmonice mundi*, was fascinated by Kepler's vision of the universe shaped and structured by music and mathematics.[22] But still there is one respect in which Fourier's thought departed radically from that of earlier thinkers who viewed the universe as a mirror of human concerns. For almost all the earlier thinkers the vision of the universe as a mirror of man served to provide reasons and justifications for the existing order of things. For the eighteenth-century illuminati, just as for the Christian cosmographers of the later Middle Ages, everything in the world was exactly as God had intended it to be: snakes, insects, disease, and sorrow were all expressions of God's plan, parts of a larger providential design. For Fourier, on the other hand, the natural and divinely ordained order of things was not to be identified with the existing order. It was not a given. Rather, it was an order that had to be brought into being. And analogy itself was a critical instrument that served to widen horizons and reveal possibilities.

Fourier was thus one of a number of thinkers of the romantic period who drew on the classical and Christian traditions of analogical thinking, but who saw analogy as pointing beyond the known world to a new world of restored harmonies and realized visions. He may have had little knowledge of Swedenborg and Saint-Martin, but he shared with some of their early nineteenth-century heirs the belief that since nature could be read emblematically, the physical world offered clues to man's future destinies.[23] For Fourier as for Blake the physical universe was not simply a mirror of a timeless human condition. It was also a rich source of symbols that could be used to evoke the promise of a more perfect order.

Thus Fourier's historical theory, with its portents of harmony and its multiple creations, can be seen as a myth that placed the human species at the center of the cosmos and turned the natural universe into a "forest of symbols" that could, if rightly understood, point beyond given and time-bound realities to a new mode of existence. It can be seen as a radical affirmation of the power of human beings to shape their own universe—as an assertion of the thoroughly utopian conviction that the only limit to our possibilities is our desire. But unlike the Saint-Simonian vision of history, which it in some ways resembles, Fourier's does not culminate in a final state beyond time. The shape of Fourier's history does not cease to be cyclical. And if he never seriously explored the darker implications of his cyclical vision, this prophet of passional gratification seems to have recognized that happiness, by its nature, is transitory.

In the historical thinking of Fourier and the Saint-Simonians we have then two attempts to organize and control time—past, present and future—each growing out of the post-revolutionary period and each beginning within a traditional time frame but exploding it. The Saint-Simonians adapted to their own purposes the spatial imagery of the *philosophes*, who saw history as the spread of reason and enlightenment. For the Saint-Simonians it was love, and not reason, that would spread in an ever widening circle. In Charles Fourier's theory of history there are echoes of even older frameworks and patterns. The initial stage of Edenism and the subsequent five thousand years of chaos and suffering evoke the Bible; and the view of human history as recapitulating the stages of individual growth and decline recalls one of the oldest metaphors in historical thinking. In Fourier's account of the four "unfortunate" periods of Savagery, Patriarchate, Barbarism and Civilization there are not only echoes but also explicit allusions to the "four stages theory" of socio-economic development that dominated European historical thought in the latter half of the eighteenth century.[24] But these shards and fragments of earlier ways of conceptualizing history were transformed by Fourier, with the help of his theory of universal analogy, into a whole which was very much his own. Neither decorative nor fundamentally derivative, Fourier's account of the four phases, 18 creations and 32 periods of life on the planet earth was, like much else in his thought, an affirmation of the underlying order and harmony in a world of apparent disorder and chaos.

Notes

[1] *Doctrine de Saint-Simon. Exposition. Première Année 1829*, eds. Celestin Bouglé and Elie Halévy (Paris: Riviére, 1924), 167.

[2] *Doctrine de Saint-Simon*, 196.

[3] *Doctrine de Saint-Simon*, 127.

[4] *Doctrine de Saint-Simon*, 146, 148–149.

[5] *Doctrine de Saint-Simon*, 203–204.

[6] Frank Manuel, *Shapes of Philosophical History* (Stanford: Stanford University Press, 1967), 105–106.

[7] *Doctrine de Saint-Simon*, 237–241; Manuel, *Shapes of Philosophical History*, 106–107.

[8] *Doctrine de Saint-Simon*, 200.

[9] John Stuart Mill, *Correspondance inédite avec Gustave d'Eichthal* ed. Eugène d'Eichthal (Paris, 1898), 27.

[10] John Stuart Mill, *Autobiography* (New York: Columbia University Press, 1966), 114–118.

[11] Karl Marx and Friedrich Engels, *The Communist Manifesto*, ed. Gareth Stedman Jones (London: Penguin, 2002), 255.

[12] "Tableau du cours du mouvement social," insert in *Oeuvres complètes de Charles Fourier,* 12 vols. (Paris: Anthropos, 1966–1968), I, 52.

[13] Fourier, *Oeuvres completes,* I, 33–41, 52–72. For a fuller commentary see Jonathan Beecher, *Charles Fourier: The Visionary and his World* (Berkeley: University of California Press, 1986), 318–331.

[14] Henri Saint-Simon, "Catéchisme des industriels. Deuxième cahier," in *Oeuvres de Saint-Simon et d'Enfantin*, 47 vols. (Paris: E. Dentu, 1865–1878), XXXVII, 96; and Fourier, *Oeuvres complètes*, I, 16.

[15] Fourier, *Oeuvres complètes*, I, 89.

[16] Fourier, *Oeuvres complètes*, I, 89 and XII, 622.

[17] Fourier, *Oeuvres complètes*, I, 132.

[18] These are positions that I have taken in Beecher, *Charles Fourier*, 327.

[19] Rudolf Allers, "Microcosmus from Anaximandros to Paracelsus," *Traditio*, II (1944): 319–406.

[20] John Cotton, cited in Perry Miller, *The New England Mind: The Seventeenth Century* (Boston: Beacon Press, 1961), 212.

[21] See for example Fourier, *Oeuvres complètes*, I, 289–290; IV, 180; and XII, 411.

[22] Simone Debout-Oleskiewicz, "Aperçus de l'influence de Kepler sur Fourier," *Revue internationale de philosophie*, XVI, 60 (1962): 195–199.

[23] Helpful guides to the varied romantic reformulations of classical and Christian analogical thinking include Auguste Viatte, *Les Sources occultes du romantisme*, 2 vols. (Paris: Honoré Champion, 1969), and various articles by Frank Paul Bowman, notably "Illuminism, Utopia, Mythology," in *The French Romantics*, 2 vols, ed. D.G. Charlton (Cambridge: Cambridge University Press, 1984), I, 76–112.

[24] Ronald L. Meek, *Social Science and the Ignoble Savage* (Cambridge: Cambridge University Press, 1976).

World History According to Katrina

Wai Chee Dimock

How does Hurricane Katrina change our understanding of the United States, the lengths and widths of its history, as well as its place in the history of the world? As a catastrophe that casts into doubt the efficacy and security of the nation-state, what alternatives does it suggest, what other forms of shelter, what other ways to organize human beings into meaningful groups? And how might these nonstandard groupings challenge American Studies as a discipline, given their deviations from the foundational norm, and the shape of the future they portend?

The nation-state seems "unbundled" by the hurricane in ways both large and small—not only as a system of defense but also as a psychological insurance, a political membership, an academic field. I want to use these unbundlings as an occasion to think about the scope of our work: both in terms of time frame, and in terms of geographical borders. And, on both fronts, it seems crucial to ask three interrelated questions. First, can the nation-state measure up to a phenomenon like Katrina? Second, is it completely sovereign, able to defend its borders on every platform, every mapping of the world? And, finally, can its geography and chronology give us a broad enough database, give us a body of evidence that can speak to the long-term relation between climatic events and human history?

It is instructive to begin with an essay on Katrina by Michael Ignatieff, published on September 25, 2005 in the *New York Times Magazine*. "When the levees broke, the contract of American citizenship failed," Ignatieff says. The breach is not just in the physical structures, or in New Orleans as a physical city, but in something even more consequential, namely, the integrity of the United States as a nation, its ability to *be* sovereign. According to Ignatieff, the most "basic term" of this sovereignty is "protection: helping citizens to protect their families and possessions from forces beyond their control."[1] And, just as the nation is defined by its power to protect, citizens are defined by their right to demand that protection. They are "entitled to this because they are Americans."[2] Nationality, in other words, ought to be synonymous with a guaranteed safety, an insulation from any harm that arises. It ought to be our bulwark against the storm. And the tragedy of Katrina is that it seems to have thrown that bulwark into question. Ignatieff summarizes the problem as follows:

> In America, a levee defends a foundational moral intuition: all lives are worth protecting and, because this is America, worth protecting at the highest standard. This principle was betrayed by the Army Corps of Engineers, by the state and local officials who knew the levees needed repair and did nothing, and by Congress, which allowed the president to cut appropriations for levee renewal.[3]

According to this analysis, the problem is that the sovereignty of the nation has not been sovereign enough. The United States ought to have been an invincible line of defense, and it has not been. The remedy, then, is also fairly simple: that line has to be firmed up, made invincible once again. The narrative that Ignatieff constructs begins and ends with the levees for this reason, because not only is the nation-state broken and then mended on their backs, but to see the problem as solely a problem of the levees is already to predetermine the solution, making Katrina an event internal to the United States, an engineering failure, something that can be fixed without changing our basic sense of what the sovereign nation amounts to, what it is equipped (or not equipped) to do, and the extent of protection it is able to offer its citizens.

Nonsovereign History

As must be clear, I find this approach unduly limiting. I'd like to explore a larger set of analytic coordinates than those suggested by Ignatieff, and to do so in a slightly roundabout fashion, by way of a debate that casts doubt on the sovereign claims of the nation, especially its adequacy as a unit of time, a debate that has galvanized historians no less than literary critics. James Sheehan, in his 2006 Presidential Address to the American Historical Association, specifically raises this as an issue. It would "be foolish to deny the importance of states," Sheehan says, "But the state was not and is not history's natural telos. The emergence of states was neither inevitable nor uniform nor irreversible."[4] Even though national chronology might look like the only chronology there is, a self-evident way of measuring time, we pay a steep price when we reify it and routinize it, allowing ourselves no other frame of reference. "Modern historiography is inextricably linked with the modern nation," Thomas Bender writes. "This has both given focus to historical inquiry and won for it a place in civic life. But it has also been disabling, silencing stories both smaller and larger than the nation."[5] As a unit of time, the nation tends to work as a pair of shutters, blocking out all those phenomena that do not match its schedule, reducing to non-events all those processes either too large or too small to fit its slots. Prasenjit Duara, historian of China—a country with a long record of just such disappearing acts—urges us to "rescue history from the nation" for just that reason.[6] To make sovereign borders the limits for relevant data is to make it a foregone conclusion that the form of the nation is the only form that matters.[7] It is to take that form and reproduce it in the form of the discipline, "naturalizing the nation-state as the skin that contains the experience of the past."[8]

That skin is very much the skin for those of us who call ourselves "Americanists." Much of the work done in the field is, to be sure, critical of the nation. Still, the very existence of an institutional entity called "American Studies" implicitly (and sometimes explicitly) reinforces the idea that the United States is a natural unit, a closed set, with clear dividing lines that separate it from other bodies of evidence. The field would not have been a field without this assumption. Territorial sovereignty is foundational to American Studies in this sense: not only does it generate a self-enclosed domain that legitimizes the field, it also institutes a cut-off line for what falls outside. As Anthony Giddens observes in a different context, "Sovereignty provides an ordering principle for what is 'internal' to states and what is 'external' to them."[9] The concept of "off limits" inversely defines the borders of a political jurisdiction; it also inversely defines the borders of a field of knowledge.[10]

This conflation of nation and field leads to a research agenda that is almost tautological: to study the United States, we need go no further than the United States. This makes things easier, though not everyone would agree that such a tautology is in fact valid, nor a good approximation of the forces that shape the world. Janice Radway, in her presidential address to the American Studies Association in 1998, makes a point of invoking this model—and rejecting it. In language strikingly similar to Duara's, she cautions us against any conception of the field as being like the territorial nation, lined with a skin:

> [F]ar from being conceived on the model of a container—that is, as a particular kind of hollowed out object with evident edges or skin enclosing certain organically uniform contents—territories and geographies need to be reconceived as spatially-situated and intricately intertwined networks of social relationships. [11]

Radway's challenge to the "container" model turns American Studies from an enclosed space to a fluid network, with no sovereign borders, nothing that will keep it defensibly separated from the rest of the world. What does this mean, in practice? Well, for one thing, we can't say, with any degree of finality, that anything is "extraneous," because extraneousness is not an attribute that is cut-and-dried, antecedently given. It is a provisional fact, a contextual variable, changing with the array of forces that happen to be in play, and with their different modes of interaction. This lack of intrinsic separation suggests that the analytic domain is always going to be heuristically stretched beyond any set of prescribed coordinates. The study of the United States can never be tautologically identical to the borders of the United States, because it can never keep the "outside" a permanent outside, externalized by defensible borders. The field then, according to Radway, can bear no resemblance to the territorial form of the nation. The nation is sovereign, or imagines itself to be. The field can have no such pretension.

What does it mean to write a history that is nonsovereign, with the seemingly extraneous being always ready, at a moment's notice, to morph into the unextraneous? I'd like to come back to Katrina as a test case, and explore two instances of this dynamic, when a seemingly secure jurisdiction suddenly bursts at the seams, becoming a kind of flooded container, flooded by an outside that refuses to stay out. To explore these two scenarios, I would like, first, to follow the unconventional coverage of Katrina by a local newspaper, the New Orleans *Times-Picayune*. The *Times-Picayune* received the Pulitzer

Prize for this report, so the importance of its work has certainly been recognized. But the history that it gives us *is* a nonsovereign history, not only because the initiative is coming from the ground up, from a local newspaper, not reflecting a national consensus, but also because this initiative produces a database that in no way matches the official borders of the United States. Nonsovereign history is off-beat, off-key, off-center. Its unorthodox paths jump from the micro to the macro and bypass the default center, going over and under the jurisdiction of the nation. Its scale is both smaller and larger: operating subnationally on the one hand, as a grassroots phenomenon, and transnationally on the other hand, as a cross-border phenomenon.

Cross-stitching Time

What the New Orleans *Times-Picayune* does, specifically, is to send its own staff writer, John McQuaid, to a different country—the Netherlands—in order to produce a corpus of evidence about Katrina not limited by nationality, but gathering information wherever relevant, tracing a series of zig-zags between two analytic poles. These zig-zags generate a cross-stitching of time, necessary, because the United States is not the only country in the world having to deal with storms and the flooding that comes with those storms. The Netherlands, throughout its history, has been facing this problem, and its collective decisions shed light on the U.S. flood for just that reason. What we will eventually see, in the robustness and thoroughness of the Dutch response, is an alternative time-line, the other side of the weave, as it were, a successful instance of democratic politics that remains only a utopian hope in the United States.

"The North Sea's furious winters can kick up storm surges more than 13 feet high—a lethal threat to a country where millions live below sea level, some as much as 22 feet down," John McQuaid notes.[12] On February 1, 1953, the Netherlands was hit by a North Sea storm that lasted for 33 hours. The storm surge (water pushed to the shore by the winds) was 150 inches higher than the normal sea level. The dikes collapsed in more than 450 places. Over 1,800 people died; some 4,000 buildings were swept away or badly damaged.[13] Out of a population of around 12 million, 100,000 had to evacuate. Twice as many people were killed by the flood as by the German bombing of Rotterdam in 1940.

The scale of the destruction is very much comparable to New Orleans, and the preceding circumstances are also quite similar. Simon Rozendaal, Dutch journalist writing in the *Wall Street Journal,* comments expressly on this. Rozendaal notes: "As in the American Gulf states, the Dutch levee system had been neglected. It was not long after World War II: the Netherlands had just lost its colony, Indonesia; and the Cold War diverted money and attention."[14] Local disasters are, in this sense, the almost predictable side effects of global geopolitics. They are part of a larger distributive pattern—a pattern of unequal protection that Ulrich Beck calls the global "risk society"—with the risk falling on the least privileged, and being maximized at just those points where the resources have been most depleted.[15] This was true of the Netherlands; it was true of New Orleans. In both cases, the military budget was funded at the expense of domestic infrastructures, paving the way for their eventual breakdown. The Lake Pontchartrain and Vicinity Hurricane Protection project, a public works project aimed at building up levees and protecting pumping stations on the east bank of the Mississippi in Orleans, St. Bernard, St. Charles, and Jefferson parishes, received less than 20 percent of the funding requested by the Army Corps of Engineers. This was not a secret; it was already public knowledge back in 2004. The *Philadelphia Inquirer* had run a story about this, reporting that it "appears that the money has been moved in the president's budget to handle homeland security and the war in Iraq."[16]

The *Philadelphia Inquirer* and the New Orleans *Times-Picayune* are helpless witnesses to a time-line waved aside, dismissed as unimportant. In the unfolding catastrophe, they have the status of a tragic chorus. They come bearing knowledge, and they go nowhere. They, along with various science magazines, have been writing reports for years—useless reports—about various warning signs: the erosion of the wetlands, the subsidence of the soil, and the presence of dangerous chemicals as well as dangerous artificial waterways such as the MRGO (the Mississippi River Gulf Outlet), which greatly increase the power of the storm surge. These warnings had absolutely no effect on government policies: this was true not only in the United States but also in the Netherlands. Six months before the 1953 disaster, the Dutch engineer Johan van Veen had calculated that the storm surge could rise up to 13 feet relative to the sinking coast. The Dutch meteorological service made the same prediction, but only 3 of the 1,000 water boards, which managed the dikes, had a subscription to this service.[17] In the case of New Orleans, the warnings had come from FEMA's own modeling of a hypothetical Hurricane Pam in 2004, and from dire forecasts appearing in *Scientific American*, *National Geographic*, *Popular Mechanics*, the *Times-Picayune*, the *Houston Chronicle*, the *New York Times*, as well as on the PBS science program, *Nova*.[18] But all this information came to nothing. It was not able to percolate to a higher level, not able to lead to the dismantling of the MRGO, for instance, or to the re-diversion of funds from Iraq back to these domestic projects. And it most certainly was not able to reverse the unequal protection endemic in this country.

Dutch Delta Works

So far, then, a cross-stitching of time seems to show only the same pattern: a common hazard, and a common failure of the democratic process itself, a kind of blockage between available information and government action. But here the symmetry ends. Flood protection in the Netherlands after 1953 diverges sharply from the United States, suggesting also that the Dutch democracy is now structurally very different from its American counterpart. It is this alternative thread of time that the New Orleans *Times-Picayune* tries to highlight by sending its staff reporter there.

Before 1953, the Dutch had tried to protect their settlements by canals lined with dikes, essentially the same as the levee system in south Louisiana. The 1953 flood revealed a major flaw in that strategy, a flaw that would now prove fatal for New Orleans. Levee-lined canals, it turns out, are fundamentally unsafe: during severe storms, they would themselves become deadly passageways, allowing the churning ocean to penetrate far inland. After Katrina, a team of Dutch engineers went to New Orleans to study the failed system, and they repeated their reservations about the over-reliance on levees. The Dutch engineer Jurgen Battjes points out,

> The region's levee-lined canals were conduits for Katrina's storm surge to pour into the heart of the city. From the east, water flowed into the Intercoastal Waterway and Industrial Canal, where floodwalls were topped and then collapsed, flooding the Lower Ninth Ward, St. Bernard Parish and eastern New Orleans. From Lake Pontchartrain, it flowed into the 17th Street

> and London Avenue drainage canals, which were breached, flooding central New Orleans.[19]

The Dutch Delta Works *(Deltawerken)*, begun shortly after 1953, adopted a different strategy. Rather than building higher and stronger dikes along the canals, as they had always done, the Dutch opted instead to construct giant barriers across all ocean inlets, sealing off the estuaries, and turning them into giant freshwater lakes. The first (in the Hollandse IJssel) went into operation in 1958. This was followed by the damming of the Veerse Gat and the Zandkreek in 1961, the Haringvliet and the Brouwershavensche Gat in 1971 and 1972. These closures blocked off the invading ocean; but they also destroyed the unique ecosystem of the estuaries, a unique mix of fresh water and sea water, and the breeding ground for many species of North Sea fish. Environmentalists as well as mussel and oyster fishermen fiercely opposed the plan for just that reason.[20]

From the 1970s on, the philosophy behind the Delta Works would undergo yet another shift, this time taking into account a twofold understanding of "protection," equalizing it across the entire habitat. The goal was not only to protect southwestern Holland against the storm surge of the North Sea, but also to protect the existing ecosystem of the river estuaries. The enormous Oosterscheldt Barrier was the result. One of the most spectacular feats of hydraulic engineering in the world, this barrier is 5.6 miles long, with 62 moveable flood gates, each the size of a twelve-story apartment building. This was followed by the equally immense Maeslant Barrier, which opened in 1997. Together, these massive public projects are designed to give the Netherlands a level of protection adequate to a flood that would come once every 10,000 years.[21]

The technology is certainly impressive, but even more so is the robustness of the democratic process that puts it to work. Flood protection in the Netherlands is accompanied by public debate every step of the way. It was this public debate that led to the change in direction in the 1970s. And it was this public debate that made it possible for this small nation to commit itself to these vast expenditures, and to plan ahead in terms of a statistical time frame of 10,000 years. Democracy, in the Dutch context, means three things: public information available to everyone; public debate having a direct impact on government policies; and a political will to limit vulnerability across the board, extending protection to populations both human and not human, currently voting and not yet born.

Against the robustness of that democratic culture, what happened in the United States must be called something else. To begin with, the New Orleans levees were designed to protect only against a storm that would come once every 50 years—in other words, only against a Category 3 hurricane. And even this modest level of protection was not always maintained, as Ivor van Heerden, Deputy Director of the Louisiana State University Hurricane Center, points out.[22] In a study released on June 1, 2006, the Army Corps of Engineers admits to this, accepting blame not only for the flawed design and construction of the levees but also for its underestimation of hurricane strength based on outdated standards.[23] This is a problem it has known for some time. "It's possible to protect New Orleans from a Category 5 hurricane," Al Naomi, senior project manager for the Corps, told the *Philadelphia Inquirer* on October 8, 2004. "But we've got to start. To do nothing is tantamount to negligence." The Corps submitted a proposal that year to Congress requesting $4 million to fund a preliminary study. Congress shelved the proposal, never bringing it on the floor, citing budgetary constraints resulting from the Iraq War.[24]

Unlike the active participation of the Dutch citizenry, policy decisions here were made—or not made—behind closed doors, by a legislative body acting only out of fiscal concerns, without ever opening up its reasoning to public scrutiny. Still, even if that public scrutiny had taken place, it is not clear whether the Dutch time-scale of the "10,000-year flood" would have been adopted. Long-term planning has never had much of a place on the federal, state, or municipal agenda: 10,000 years seem almost unimaginable. As the *Washington Post* reports:

> In 1982, the Orleans Levee District urged the Corps to 'lower its design standards to provide more realistic hurricane protection.' The levee district, stocked with political appointees, could spend freely on private investigators, riverboat gambling, and a $2.4 million Mardi Gras foundation. But it said it could not afford its share of protection from a 200-year storm, suggesting that 100-year protection would be fine. [25]

This strange sense of proportion might turn out to be one of the most destructive effects of the time scale of a young nation, one that allows neither a long past nor a long future to interfere with the short, but oversized centrality of the present. What does it mean never to think of time except in single and double digits? And how might these single and double digits affect a nation's ability to deal with events such as hurricanes, whose potential for harm outstrips those digits by many orders of magnitude? A nonsovereign history of Katrina shows that, beyond the broken levees, what needs to be mended is the democratic process itself, and its need for a frame of reference beyond the geography and chronology of the nation. The least we can do, to get started, is to move away from the United States as the default center, and to take stock of its dependent place in a larger field of forces.

The World's Water

That dependent place, in turn, radically changes the way we think about causality: the web that articulates it, the claims that can be pressed, and the responses needed as a result. The implications are far-reaching, because to draw a larger causal circle around the nation is also to draw a larger circle of accountability, to give a broad interpretation to the harm that it might have perpetrated at a distance, harm that might seem extraneous from one point of view. How, for instance, can we make a nation face up to the death and destruction that it is causing hundreds and thousands of civilians, thousands of miles away, on a different continent? Justice looks very different when it is framed in this way, seen as extended rather than encapsulated. Rather than being a problem of crime and punishment contained within a single nation, it becomes another instance of the flooded container: flooded, in this case, by the causal web that links it—against the illusion of sovereignty—to cross-currents affecting the entire planet, a seascape turbulent and borderless.

It is in this context that a "world history" can be said to be traceable from Katrina, for hurricanes are, of course, part of this seascape, this extended rather than encapsulated web of causality. They are indexes to the climate pattern of the world as a whole. Generated by air-sea interaction, this hydrology can be adequately studied only through "multi-basin indices," which is to say, by comparing data from the North Pacific, Indian, Southwest Pacific, and North Atlantic Oceans. Not only are hurricanes water-borne disasters, they are disasters unique to warm water: as long as the sea surface temperature remains below 26.5 degrees Celsius (80 degrees Fahrenheit), no hurricane will form. When oceans get heated up, on the other hand, they fuel a convection process that transforms cold-core tropical depressions into hot-core cyclones. Katrina itself strengthened to a Category 5 hurricane when it was passing over the Gulf of Mexico, where the surface waters were unusually warm, about 2 degrees Fahrenheit warmer than normal for that time of year.[26]

Sea surface temperature is the single most important factor in hurricane formation. And it was by looking at this data that the MIT climatologist, Kerry Emanuel, was able to predict what was to come. On July 31, 2005, one month before Katrina, Emanuel published his research in the online edition of the journal *Nature*. Tracking hurricanes by their "power dissipation index" (a combination of the lifetime of storms and their intensity), Emanuel shows that "this index has increased markedly since the mid-1970s," an upward trend strongly correlated with the rise in the sea surface temperature. Both the duration of hurricanes and their wind speeds have "doubled in the past 30 years" as the Pacific and the Atlantic have warmed by 1 degree Fahrenheit between 1970 and 2004. Since changing ocean temperatures are themselves indices to climate change, Emanuel sees the increasingly destructive hurricanes as "at least partly anthropogenic." He predicts "a substantial increase in hurricane-related losses in the twenty-first century."[27]

Emanuel's study was corroborated almost immediately in a parallel study by a team from the Georgia Institute of Technology, reported in *Science* on September 16, 2005. By looking at "the number of tropical cyclones and cyclone days as well as tropical cyclone intensity over the past 35 years, in an environment of increasing sea surface temperature," this study finds that "hurricanes in the strongest categories (4 + 5) have almost doubled in number... These changes occur in all of the ocean basins." How to explain this across-the-board jump? J. B. Webster, speaking for the Georgia Tech team, is even less ambiguous in seeing a strong correlation between the rising ocean temperatures and the rising concentrations of atmospheric carbon dioxide—chief of the greenhouse gases—though they concede that "attribution of 30-year trends to global warming would require a longer global data record and, especially, a deeper understanding of the role of hurricanes in the general circulation of the atmosphere and ocean."[28]

Whether or not hurricanes can be directly traced to global warming,[29] it seems clear that the database needs to be planetary in scope, studying all the oceans in conjunction. Studied in conjunction, they point to a changing world, becoming daily less hospitable, looking less and less like the planet that has supported our species and other species. We take it so much for granted that we never notice that its features have grown ominous. Of the weapons of mass destruction already lined up, the most deadly will probably come, not in the form of hurricanes, but as a hydrology simpler, less of a spectacle, though infinitely more catastrophic: namely, the rising sea levels due to the melting of the Arctic and Antarctic ice sheets.

In its 2001 report, the UN's Intergovernmental Panel on Climate Change (IPCC) predicted that sea-level rise in the twenty-first century will proceed "at an average rate of 2.2 to 4.4 times the rate over the twentieth century," while singling out the West Antarctic ice sheet as especially worrisome, since it "contains enough ice to raise sea level by 6 meters."[30] Meanwhile, seismic stations revealed a significant increase in "icequakes," caused by ice sheets breaking loose and lurching forward; the annual number of these icequakes registering 4.6 or greater on the Richter scale doubled from 7 to 14 in the late 1990s, it doubled again by 2005.[31] Satellite measurements of the earth's gravitational field showed a loss of 50 cubic miles of ice in Greenland in 2005, matched by a similar loss in West Antarctica.[32] The new IPCC report, issued in February 2007, stuck to a more conservative figure for the sea-level rise (7.8 inches to 2 feet by the century's end), but the human cost is staggering even at this rate.[33] The World Bank estimates, for instance, that even a three-foot rise in sea level would turn at least 60 million people into refugees.[34]

What would the United States look like then? The fate of New Orleans would have been sealed long before then, as would the fate of many other coastal cities. Al Gore, in *An Inconvenient Truth*, gives us a computer projection of what would be left of Florida if the sea level were to increase by 18–20 feet; it is a horrendous image. The century ahead will most certainly be dominated by this advancing seascape as the earth continues to heat up. Sovereign borders will be so diluted—literally—that they will be small comfort for U.S. citizens; even the world's largest military budget will not yield a credible line of defense. Yet the irony is that, while we are not capable of long-term protection, we are more than capable of action that has the potential for long-term harm. The balance between human history and nonhuman processes, always problematic, is now weighted more and more in the latter's direction, with a growing gap between the kind of habitat the human species has depended on and the kind of habitat the planet is becoming. The United States is ill prepared for this development, though there are signs now that the tide might be turning, that climate change might be re-educating all of us in the primacy of the planet over the sovereignty of any nation. At this critical moment, it is especially important for American Studies to rethink its space and time coordinates, to take up questions that might once have seem far removed—coming not only from hitherto extraneous fields such as Earth and Planetary Sciences, but also from hitherto extraneous populations, not traditionally included in the discipline.

Arctic Time-line

One such population is the Inuit living in the Arctic Circle. It is here that global warming is felt most directly and most severely, since the threshold for catastrophic change is much lower at the two poles: the difference of 1 or 2 degrees can have drastic consequences for the glaciers and the ice sheets. When it comes to climate change, the Arctic is ahead of the rest of the world: it has a time-line of its own. In December 1995, the IPCC issued a landmark report noting this uneven development. This was reaffirmed in 2004 by the eight-nation Arctic Climate Impact Assessment, which concluded that the Arctic is experiencing "some of the most rapid and severe climate change on earth."[35] In *An Inconvenient Truth* this is dramatized as the plight of the polar bear; and what makes the world unliveable for the polar bear also makes it unrecognizable for the Inuit. They have a word for it, *uggianaqtuq*, referring to the weather, a "familiar friend now behaving strangely."[36] In November 2000, the Inuit released a 45-minute video to document this fatal alienation. Entitled *Sila Alangotok: Inuit Observations on Climate Change,* it offers an extensive record of melting ice, eroding coastlines, and the appearance of wildlife never seen before, including the Pacific salmon and the robin.[37] It is this unrecognizability of the world, the unrecognizability of their habitat, that makes it necessary for the Inuit Circumpolar Conference (a federation made up 150,000 native peoples in Canada, Greenland, Russia, as well as the U.S.) to seek legal action against the world's foremost emitter of greenhouse-gases.

This is not easy to do. Currently, the infrastructure for transnational legal action is still very sketchy. Just as we don't have the legal instrumentalities to prosecute the United States for the harm that it is causing Iraqi civilians, we also don't have the legal instrumentalities to prosecute nations for the environmental harm that they are causing their distant neighbors. We do, of course, have courts that operate on a transnational level. There are four of these at the moment: the International Court of Justice at the Hague; the International Criminal Court, also at the Hague; the Court of Justice of the European Communities at Luxembourg; and the European Court of Human Rights at Strasbourg. The first of these, the International Court of Justice, created in 1945, will only hear cases brought before it by nation-states.[38] The other three courts do in fact hear cases brought by non-state applicants,[39] but the grievance of the Inuit does not rise to the level of the International Criminal Court, and, not being a member of the EU, its case also cannot be heard in the two European courts. However, with the help of environmental groups such as Earth Justice and the Center for International Environmental Law, the Inuit were able to file a petition against the Bush administration with the Inter-American Commission on Human Rights on December 7, 2005, "seeking relief from violations resulting from global warming caused by acts and omissions of the United States."[40]

Almost all the deteriorating conditions of the Arctic can be traced to climate change, not only changes in the "quality, quantity and timing of snowfall," but also the destruction of coastal communities through the increasingly erratic behavior of water in all its forms:

> Permafrost, which holds together unstable underground gravel and inhibits water drainage, is melting at an alarming rate, causing slumping, landslides, severe erosion and loss of ground moisture, wetlands and lakes. The loss of sea ice, which dampens the impact of storm on coastal areas, has resulted in increasingly violent storms hitting the coastline, exacerbating erosion and flooding. Erosion in turn exposes coastal permafrost to warmer air and water, resulting in faster permafrost melts. These transformations have had a devastating impact on some coastal communities, particularly in Alaska and the Canadian Beaufort Sea Region…

> Other factors have also affected water levels. Changes in precipitation and temperature have led to sudden spring thaws that release large amounts of water, flooding rivers and eroding their streambeds. Yet, after spring floods, rivers and lakes are left with unusually low levels of water, further diminished by increased evaporation during the longer summer. These changes affect the availability and quality of natural drinking water sources. The fish stocks upon which Inuit rely are profoundly affected by changing water levels. Fish sometimes cannot reach their spawning grounds, their eggs are exposed or washed ashore, or northward moving species compete with the native stocks for ecological niches.[41]

Violent storms, floods, soil erosion, loss of wetlands—these are problems we associate with New Orleans and the Gulf of Mexico. It should not come as too much of a surprise, though, to see these problems also played out, thousands of miles away, in the Arctic Ocean, since there is in fact no line separating these two bodies of water. This single, already-damaged hydrology makes it clear that climate, geology, and human and nonhuman life are all complexly interdependent, part of the same precarious web. The catastrophe, already writ large in this seemingly remote part of the world, is one we cannot ignore. Sheila Watt-Cloutier, Chair of the Inuit Circumpolar Conference, received the UN Lifetime Achievement Award for Human Development,[42] and, along with Al Gore, was nominated for the 2007 Nobel Peace Prize.[43] The Inter-American Commission on Human Rights began its hearings on climate change in March.[44] While the Commission has no power of enforcement, a finding in favor of the Inuit could be the basis for future lawsuits in U.S. federal courts. World history is being made in front of our eyes. According to the Inuit, this begins with local knowledge, with micro-evidence and nonstandard chronologies. If these space and time coordinates look unfamiliar, perhaps the study of the United States needs to become unfamiliar to itself in just this way.

Notes

[1] Michael Ignatieff, "The Broken Contract," *New York Times Magazine*, Sept. 25, 2005.

[2] Ibid.

[3] Ibid.

[4] James J. Sheehan, "Presidential Address: The Problem of Sovereignty in European History," delivered at the annual meeting of the American Historical Association, Philadelphia, January 3, 2006, reprinted in *American Historical Review* 111 (February 2006): 1–15, quotation from 1–2. For other well-known critiques of the nation-state, see Ernest Gellner, *Nations and Nationalism* (Oxford: Oxford UP, 1983); Eric Hobsbawn, *Nations and Nationalism since 1780* (Cambridge: Cambridge UP, 1990); Ernest Renan, "What is a Nation?" in *Nation and Narration*, ed. Homi K. Bhabha (New York: Routledge, 1990), 8–22.

[5] Thomas Bender, "Preface," in *Rethinking American History in a Global Age* (Berkeley: University of California Press, 2002), vii.

[6] Prasenjit Duara, *Rescuing History from the Nation: Questioning Narratives of Modern China* (Chicago: University of Chicago Press, 1995).

[7] For an important critique of the nation-form, see, Etienne Balibar, "Racism and Nationalism," and "The Nation Form: History and Ideology," both in Etienne Balibar and Immanuel Wallerstein, *Race, Nation, Class: Ambiguous Identities* (London: Verso, 1991), 37–68, 86–106.

[8] Prasenjit Duara, "Transnationalism and the Challenge to National Histories," in Bender, *Rethinking American History*, 25–46, quotation from 25.

[9] Anthony Giddens, *The Nation-State and Violence* (Berkeley: University of California Press, 1987), 281.

[10] For a sustained argument linking sovereignty to the conditions of knowledge, see Jens Bartelson, *A Genealogy of Sovereignty* (Cambridge: Cambridge UP, 1995).

[11] Janice Radway, "What's in a Name? Presidential Address to the American Studies Association," November 20, 1998, *American Quarterly* 51 (March 1999): 1–32, quotation from 15.

[12] John McQuaid, "Beating back the sea: How the Dutch fight to save their low-lying land," *The Times-Picayune*, November 13, 2005. Available online at the Pulitzer website, http://www.pulitzer.org/ear/2006/publicservice/works/neworleansps12.html.

[13] Molly Moore, "Rethinking Defenses against Sea's Power," *Washington Post*, September 8, 2005, A22. At http://www.washingtonpost.com/wpdyn/content/article/2005/09/07/AR200509072400.html.

[14] Simon Rozendaal, "Katrina, Juliana, and Wilhelmina," *Wall Street Journal*, September 7, 2005, A16. Rozendaal is science writer for the Dutch news-magazine, *Elsevier*.

[15] Ulrich Beck, *Risk Society: Toward a New Modernity*. See also *Unequal Protection: Environmental Justice and Communities of Color*, ed. Richard D. Bullard (San Francisco: Sierra Club Books, 1994); *Global Environmental Risk*, eds. Jeanne X. Kasperson and Roger Kasperson (Tokyo: United Nations University Press, 2001); Susan L. Cutter, *Hazards, Vulnerability, and Environmental Justice* (London: Earthscan, 2006).

[16] Michael Eric Dyson, *Come Hell or High Water: Hurricane Katrina and the Color of Disaster* (New York: Basic Civitas, 2006), 81.

[17] Rozendaal, "Katrina, Juliana, and Wilhelmina."

[18] Dyson, *Come Hell or High Water*, 77–86.

[19] McQuaid, "Beating Back the Sea."

[20] "The Dutch Struggle against the Waters," information supplied by the Dutch Ministerie van Buitelandse Zaken. Available online at http://www.hollandring.com

[21] Ibid. See also, "How the Netherlands Prepared for a 10,000-Year Disastrous Flood," at http://www.larouchepub.com/eiw/public/2005/2005_30-39.pdf/43-47_38_ecodutch.pdf.

[22] According to van Heerden, the levees could actually offer protection only against a Category 2 storm, with wind speeds of up to 110 miles an hour. See "Levees Rebuilt Just in Time, but Doubts Remain," *New York Times*, May 25, 2006. See also Ivor van Heerden, *The Storm: What Went Wrong during Hurricane Katrina: Inside Story from a Louisiana Scientist* (New York: Viking, 2006).

[23] John Schwartz, "Army Builders Accept Blame Over Flooding," *New York Times*, June 2, 2006.

[24] Paul Nussbaum, "New Orleans' growing danger," *Philadelphia Inquirer*, October 4, 2004. At http://hurricane.lus.edu/in_the_news/ phillyinquirer100804.htm.

[25] "The Slow Drowning of New Orleans," *Washington Post*, October 9, 2005, A01. At http://www.washingtonpost.com/wp-dyn/content/article/2005/10/08/AR2005100801458.

[26] "Was Katrina's Power a Product of Global Warming?" at http://pewclimate.org Katrina weakened to category 4 shortly before landfall in Louisiana and Mississippi.

[27] Kerry Emanuel, "Increasing destructiveness of tropical cyclones over the past 30 years," *Nature* advance online publication, 31 July 2005. At http://www.nature.com/nature/journal/vaop/ncurrent/full/nature03906.-html.

[28] P. J. Webster et al. "Changes in Tropical Cyclone Number, Duration, and Intensity in a Warming Environment," *Science*, 16 September 2005. At http://www.sciencemag.org/cgi/content/full/309/5742/1844.

[29] Scientists who disagree with Emanuel and Webster think that the more destructive hurricanes are caused, not by global warming, but by a natural cycle called "multi-decadal oscillations." For a summary and docu-

mentation of the debate, see the "Global Warming and Hurricanes" website of the Geophysical Fluid Dynamics Laboratory, at the National Oceanic and Atmospheric Administration. At http://www.gfdl.noaa.gov/~tk/glob_warm_hurr.html.

[30] *Climate Change 2001: The Scientific Basis: Contribution of Working Group to the Third Assessment Report of the Intergovernmental Panel on Climate Change*, eds. J. H. Houghton et al. (New York: Cambridge U Press, 2001) *642*.

[31] Jim Hansen, "The Threat to the Planet," *New York Review of Books*, July 13, 2006, 12–16, quotation from 13.

[32] Ibid., 13.

[33] *Climate Change 2007: The Physical Science Basis: Summary for Policy Makers. Contribution of Working Group I to the Fourth Assessment Report of the Intergovernmental Panel on Climate Change*. Available at http://www.usgcrp.gov/usgcrp/links/ipcc.htm#4wg1.

For a good summary of the report, see Bill McKibben, "Warning on Warming," *New York Review of Books*, March 15, 2007, 44–45.

[34] Juliet Eilperin, "Clues to Rising Seas are Hidden in Polar Ice," *Washington Post*, July 16, 2007.

[35] *Arctic Climate Assessment Report* (Cambridge: Cambridge UP, 2004).

[36] "The Inuit, First Witnesses of Climate Change," a UNESCO report. Available at http://portal.unesco.org.

[37] A BBC report gives a good account of the video. See "Climate Change in the Canadian Arctic," at http://bbc.co.uk/worldservice/sci_tech/ highlights/ 010510_canadianarctic.shtml. See also "Responding to the Global Climate Change: The Perspective of the Inuit Circumpolar Conference on the Arctic Climate Impact Assessment," at http://inuitcircumpolar.com/index.php?ID=267&Lang=En.

[38] The International Court of Justice (ICJ) was created in 1945 under the Charter of the United Nations. See http://www.icj-cij.org/icjwww/generalinformation/ibbook/Bbookframepage.htm

[39] The Court of Justice of the European Communities, the legal institution of the European Union, whose charge is to enforce "community law... separate from, yet superior to national law," was initially created under the Treaties of Paris and Rome in 1952. See http://europa.en.int/institutions/court/index_en.htm. http://europa. The European Court of Human Rights (ECHR), the judicial arm of the Council of Europe, started out as the Convention for the Protection of Human Rights and Fundamental Freedoms (1950), and became consolidated as a single, full-time court on 1 November 1998. The International Criminal Court (ICC) was established on 17 July 1998 when 120 states adopted the Statute of Rome. The Statute provides for its entry into force 60 days after 60 states have ratified it, which happened on 11 April 2002. Accordingly, the ICC went into effect on 1 July 2002.

[40] At http://www.inuitcircumpolar.com/index.php?ID=316&Lang=En.

[41] Ibid., 2–3.

[42] At http://www.msnbc.com/id/16920923.

[43] See, for instance, http://www.cbc.cn/north.story/2007/04/27/north-award.html.

[44] At http://www.earthjustice.org/news/press/007/inter-american-commission-on-human-rights-Hearing-on-global-warming.html.

Historical Figures: Mediations, Citations, Narrations

Intricate Temporalities

The Transfiguration of Proper and "Improper" Sounds from Christian to Jewish Environments[1]

Ruth HaCohen

Three modes of communication situate the background of this paper. They relate to three realms of experience and to the intersections and conflations of their respective configurations. These modes are music, narrative, and ritual. All three engage us in the real present—they call our attention at a certain moment, they demand from us a blank duration for the projection of their inhering experiences, and they launch us into "unreal" presents whose mental fabrication they stimulate.[2] When we hear a story, listen to a piece of music, or attend a ritual, we are often willingly interpreting their messages by assuming their referred stretches of virtual past, present, or future, as our temporary lived present.[3] It could be the story of Exodus activated by a certain ritual in which we find ourselves involved; the cosmological history of the planets rendered in a piece of music that wishes to usher us into its heavenly mode of existence; or a story—literary or cinematographic—of futuristic adventures that takes us into its fantastic realms.

In such operations we can discern at least three modalities of imaginary *present* tense: past-present, present-present, and future-present. Since the present is doubly presented in each of these modalities, both as the time in which they historically/biographically occur and as the time which characterizes the phenomenal features of its mental objects, these three modalities should be rather called: the past as present in the present, the present as present in the present, and the future as present in the present, or in short: the past-present-present, the present-present-present, and the future-present-present. While the experience of such modalities seems to depend on the willing imagination alone, encoded modes of communication—of the kind historical and literary narratives, musical works, or performative rituals deploy—facilitate their activation, rendering them communal. These three modes often interpenetrate each other (in opera, moving pictures, religious services and the like), enhancing or being enhanced by an imagined present the other mode conjures up. In the following, I will concentrate primarily on the temporal experience projected by certain musical works/events which call upon and utilize the two other modes of communication: narrative and ritual. These temporal experiences—directly or indirectly reported by real or imagined audiences—are implied in the performative events these works historically launched. Various beliefs and practices nourished such experiences, themselves often enlisted for promoting or reinforcing certain hegemonic or subversive ideologies. The following theoretical discussion regarding the nature of musical temporality will bring me to more specific theological–ideological environments related to certain Christian ritualistic music, and to subsequent philosophical and literary transfiguration of their inhering presents into assumed Jewish contexts.

In and Out the Musical Present

It was the American philosopher, Susanne Langer (1895–1985), who half a century ago coined the trenchant phrase "Music makes time audible and its form and continuity sensible," thus summarizing her theory of musical perception and expression.[4] Building on the Bergsonian contrast between spatialized, measured time that is objective, and existential–experiential duration (*la durée pure*) that is subjective, Langer's maxim implies the notion—following Ernst Cassirer's theory of symbolic forms—regarding the ability of musical configurations to isomorphically mold and qualify the internal dynamics and emotional undulations of "being in time."[5] It is the experience of passage in individual life, along its psychic tensions, building-ups and resolutions, which Langer claims, is the "model for the virtual time created in music." In the process of creating music as a symbol of subjective time "every kind of [psychic] tension transformed into musical tension, every qualitative content into musical quality, every extraneous factor replaced by musical elements."[6] Isomorphism, in this case, is achieved through a common factor of emotional life and musical unfolding time (like mass which is a common factor of both a sculptured body and an organic body), and a superimposed iconic morphology, musically configured. This effects music as a symbolically condensed modality, a "sonorous image of passage, abstracted from actuality to become free and plastic and entirely perceptible."[7]

Several interrelated objections to this argument can be raised without refuting its basic claim: what is the role of music's *real* passage of time, along its regular pulse and pace, and its derivative *measured* time, in the making of this symbolic structure? How does this measured—and thus objectified, or at least, intersubjectified time—stand in relation to the ordinary, spatial–mechanical time which Langer, following Bergson, categorically contrasted to the experiential, musical one? How, in turn, does the internal, subjective, musical experience she describes relate to the communal experience of music, to its public performative presence, which genealogically preceded artistic, individualized music, and has never ceased to exist alongside it?[8] After all, it is the conjunction of the subjective and the intersubjective, of private time and communal time that renders "the audibility of time" through music, of social, religious and historical significance.

The analysis of the musical concept of time by the Jewish–Viennese émigré, the musician, musicologist and philosopher Victor Zuckerkandl (1896–1965), stems precisely from these sorts of considerations.[9] Zuckerkandl's point of departure is the nature of time as a phenomenological category perceived mainly as an imaginary extrapolated axis rather than an ontological given one:

> Anyone who thought that he could arrive at time by thinking away everything that is *in* time, the moving things, the sense perceptions that crowd upon us, our feelings, all the concrete content of our consciousness, and retaining 'only the form', would be in for a rude awakening; with the things that are in time, time itself would have slipped through his fingers.[10]

The experience of music is different, he claims, for in it, the concrete content of the experience is time itself. "Through tone," argues Zuckerkandl, "time becomes concrete experiential content." The experience of musical rhythm, he stresses:

> is an experience of time made possible through tones. In the unique phenomenon of the musical rest, we have as it were the crucial experiment for our thesis; the rest shows us, with a clarity that leaves nothing to be desired, what happens if tones are not just thought away but actually left out: what remains is not

> abstract empty form but a highly concrete experience, the experience of rhythm.[11]

In music, Zuckerkandl further maintains, time is a dynamic field, the forces of which—meter and rhythm—are the effects of time, of duration (and the nature of duration) that elapses through tones. Time, as such, cannot be differentiated, as substance, from its qualities; its activity is experienced as waves of sorts, dynamically oscillating and accumulating. Zuckerkandl draws further similes from the water metaphor, only to sharply differentiate the "target domain" (musical time) from its "source domain" (water); in music, he tells us "we see tree trunks moved, bridge piles subjected to pressure, soil carried away and deposited—but there is no water, the stream bed is empty."[12]

What follows is that musical time is a sort of pastime, or past-time which contradicts the common notion of passing time, of a present which is consigned to non-existence (the past), or being conquered from the not yet existent (the future).[13] In a musical occurrence "the entire past is preserved in its present and given directly with it"[14] and so is the future which the present is pregnant with. Had it been otherwise, claims Zuckerkandl, the experience of musical time as oscillating and accumulating waves would have been impossible. This sort of unfolding, he further maintains, should not be viewed in terms of regular remembering and foreshadowing, for it does not involve a mental re-presenting. The present in music (of a given, heard piece) is within us as the music flows; past events and future ones are felt as if they are impelled in the moment, beyond, or better before, re-calling and fore-knowing what is more abstract and conceptual. This oneness of being renders the experience of musical time as an epitome of the experience of present, Zuckerkandl concludes, much above other presentational forms on the one hand, and the "hourglass concept of time" (the Bergsonian spatialized time), on the other. Musical time, in the language of cognitive psychology, is thus an extended "psychological present"; it is a continuous–simultaneous being "stored by time" and through it.[15]

Here Zuckerkandl chooses to directly oppose Bergson, whom he, like Langer, is otherwise inspired by. The dichotomy the French philosopher wished to establish—between internal and external, subjective and objective time—is untenable if considered from the point of view of meter and rhythm, cognitively shared parameters or schemata. Musical time, argues Zuckerkandl, does not "belong to the psyche," i.e., it is not merely subjective. While he keeps the dichotomy between physical, eventless time, and organic, biological time—"time that remembers… a living time" he opts to view them as the extreme parts of a continuum, occurring within scientific—that is conceivable, knowable—time. In other words, according to Zuckerkandl, music only accentuates being in time (and not necessarily symbolizes it, as Langer claims), through the very fact that it uses time as its basic quality. Following Heidegger (though not explicitly on this issue), such being in time is in and of itself communal. Meter and rhythm, Zuckerkandl could argue, are not only forces affected by the flow of time, but also, by the same token, forces that further mold its experience for those attending it or performing it as a shared qualified being. It synchronizes their actions (of singing, breathing, playing) or reaction (in bodily motion or in more passive apprehension of musical stimuli), rendering them audible. In fact, there is no other phenomenon which seems as effective in this respect—an insight which was not lost on Luther, Rousseau or Nietzsche, though each bestowed on it a different social–communal role. Influenced by different phases of modern aestheticism, however, neither Langer and Zuckerkandl pursued this direction.

Musical Present Fraught with its Pasts: Ritual and Notated Music

Thinkers who to begin with were more interested in the communal–anthropological impact of music were more sensitive to these aspects. Thus we find in Franz Rosenzweig's *Stern der Erlösung* the following interesting insight concerning the singing of the chorale and other forms of religious music which he—rebutting modern aestheticism—opposes to the disintegration of real time into ideal time in concert music that aspires to pure experience:[16]

> The chorale feels the space, sung by all together in a mighty unison; it is the real basis for the ecclesiastical employment of music. It still survives in Bach's oratorio, and the Catholic Church has also continued to foster it albeit the musical mass leads in another direction. In the chorale, silence descends over speech, which otherwise has its own particular word to say in every individual mouths. But it is not that silence which simply

> listens mutely to the recited word, but the silence of one's singularity in the unanimity of the chorus…
>
> Now music attunes the souls in preparation for this community of life, as it is then realized in the sacrament. At their entrance into the common space, the souls had only been pretuned for community in general, for potential community; in the common singing of the chorale they are pretuned for real community. Even the musical mass, although it is only heard and not sung in common, is at bottom a pretuning of all individuals for community, just as much as is the chorale. For listening to music is a different kind of listening altogether from listening to a recited text or to a sermon; it does not establish a community but it arouses the assembled ones each for himself, to the same feelings—each for himself as the sight of the audience in a concert will show at once. Thus to hear a musical mass is fully equivalent to singing a chorale in this respect. The common space absorbs the individual bodily; thereupon he is seized as discoursing individual in his soul; and as discourse is disciplined by rhythm and melody, the personality of the personal word of the individual learns to fall silent. He speaks, but what he says are not his words but the words set to music which are common to all. [17]

In his quest for the conditions—theological, architectonic, liturgical, literary—sufficient or necessary for the constitution of an ideal religious community, Rosenzweig, a German–Jewish philosopher (1886–1929), phenomenologically observes both Christian and Jewish rituals, practices and beliefs which he deems as partially complementary, partially contradictory. The central role he gives to the Lutheran Chorale—a musical genre characterized by a declamatory togetherness achieved by means of shared rhythmic pace, harmonically qualified[18]—stems precisely from these temporal qualities. It is the shared, felt "present" of the chorale which transforms the congregation into a well-tuned community—a *vorgestimmte Gemeinschaft*, an effect unachievable through other artistic and ritualistic modes or means. Though Rosenzweig chooses not to explicate it, this form of being together in time is missing in Jewish ceremonial practices.[19] The fact that Rosenzweig, in this book that hails the Jewish way to redemption, considers the Christian church as exemplary in this respect should not be underestimated, especially when considering the heavy charges of Christians against what they had perceived, since the Middle Ages, as "synagogal noise/nausea."[20] Nor should we overlook the fact that Rosenzweig sees the establishment of a unified religious community through a shared musical practice as a vehicle for reconnecting itself to its imagined (absent) pasts.

The chain of "musical absences," repeated performances, say, of a Bach Passion (the, not coincidental, example of Rosenzweig) which recalls the first performances in Bach's time, has a strong ritualistic effect.[21] The regular repetition of certain orders (in the double sense of the word) operates as imaginary linchpin, chaining the individual annual instances of a piece's performance, while combining and reinforcing them in the imaginaries of the religious community.[22] Obscuring linear historicity—the dimension of change and difference of lapsing time—it eternalizes, in a recurring form, that which it commemorates or anticipates, simultaneously experienced as present, future, and past. When historical narrative plays a role in such commemoration or anticipation, as it does in the case of the Passion, music helps substantiate its temporal objects. Intrinsic musical qualities can further animate personae from the past, enacted in the ritual, re-presenting their emotive plights or embodying their confrontations and reconciliations.

This brings us back to the notion of superimposed temporalities with which I started this essay, now imbued by phenomenological as well as ideological content: The flow of music bestows on the listener a temporal experience of a continuous present, sensitizing him or her to the quality of existential/biological/spiritual time, and to its intersubjective dimensions. It is also able, and sometimes aspires to, entertain temporal dimensions that are not apparently contained within the "pure" continuous temporal qualities of a given musical piece or event. These dimensions are not transparently audible; they call for special initiation or socialization. Once so attuned, historical past times, imaginary future ones, as well as myriad of virtual presents, encroach into the short span of the "real" musical duration of actual listeners, enriching or subverting its coherent flow. In other words, not only pure phenomenal time becomes audible through music—also historical, ritualistic and biographical times infiltrate into its ideal duration, whether desired or not, enhancing its communal impact. In this respect, it is not categorically different from other temporal media (narrative, film, dance). Nonetheless, grafting mixed external times on the accentuated musical duration (in the sense pointed out by both Langer and Zuckerkandl) has special effects and consequences. It is because music is basically a social–communal experience of being in time that the grafting of other, shared, temporalities is so effective.

All this is implied in the above quotation from Rosenzweig, which, in turn, draws unwittingly on a basic trait of Western music: its reliance on a strict notational system that guarantees "the authoritative identification of a work from performance to performance," as Nelson Goodman put it, enabling its faithful preservation over time.[23] This formidable apparatus evolved over the course of several centuries of trial and error, and was urged by, and contributed to, constantly developing musical techniques.[24] As such it has tremendous historical, cultural, and aesthetic implications. In the terms used in the present paper this apparatus enabled a normative re-presentation of recognized musical presents (otherwise termed as "musical works"), allowing the reactivation of their basic temporal-dynamic qualities.[25] Musical works thus conceived could constitute well-defined musical presents, launched into unlimited future presents, which in turn would retrospectively accumulate behind each actual present as a trail of past presentations (performances).[26]

The potential of this unique cultural mechanism is too rich to be fully accounted for here. A not very typical, however powerful, example for this mechanism is a memorial concert that repeats a program once performed. A concert of this kind took place in Yad Vashem (The Holocaust Martyrs' and Heroes' Remembrance Authority, Jerusalem) in December 1995; it consisted of the same program of a concert played in Terezin concentration camp probably during 1943 including three string quartets by Beethoven, Brahms, and Gideon Klein (see figure 1.)[27]

Apparently, using the above nomenclature, this sort of commemoration involved a present-present-past modality (the experience at present of a present that was so experienced in the past); a shift of the virtuality of musical time to its occurrence at a certain historical moment, which, in turn, becomes relived, revived in the actual present. Such mutual shifts undermined the idealized present that works such as Beethoven's and Brahms' quartets had ideologically professed for previous audiences. Past-present-present and present-present-past became intermingled, disclosing new flows, new narratives in the heard music, suggesting new listening strategies, unheard con-

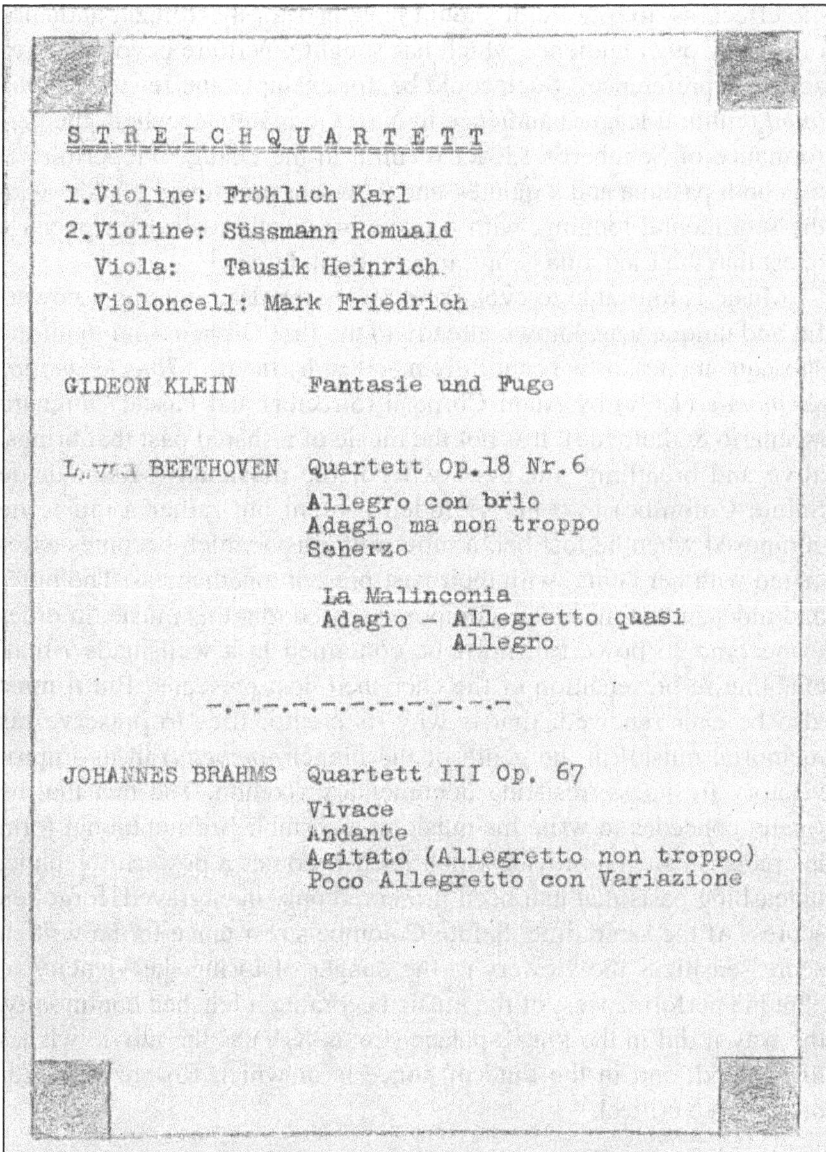

S T R E I C H Q U A R T E T T

1.Violine: Fröhlich Karl
2.Violine: Süssmann Romuald
Viola: Tausik Heinrich
Violoncell: Mark Friedrich

GIDEON KLEIN Fantasie und Fuge

L. van BEETHOVEN Quartett Op.18 Nr.6
Allegro con brio
Adagio ma non troppo
Scherzo

La Malinconia
Adagio – Allegretto quasi Allegro

-.-.-.-.-.-.-.-.-.-.-

JOHANNES BRAHMS Quartett III Op. 67
Vivace
Andante
Agitato (Allegretto non troppo)
Poco Allegretto con Variazione

Figure 1: *The original ad of the concert played in Terezin concentration camp*

tinuities and discontinuities. This past-present-present—the presentation of the past through musical experience—carries a strong nostalgic effect, as, to be sure, it should have had for the Terezin audience or for any other audience which has sought repertoire beyond "pure" aesthetic preferences. Such could be, for example, the Jewish *Kulturbund* (cultural league) audience in Nazi Germany, for whom the performance of Schubert's Lieder (central in the Leauge's repertoire[28]) was both pastime and a quintessential present past-time, redolent with the sentimental longing, with a quest for what was lost for ever—a quest that the Lied, qua genre, inherently indulges.[29]

Music is thus able to evoke and revive the dead in a most powerful and unique way, known already to the first Orpheus and to all the subsequent ones, as is beautifully manifest by the film *Tous les matins du monde* (1991) by Alain Corneau (director) and Pascal Quignard (scenario & dialogue). It is not the music of a shared past that brings, alive and breathing, the dead wife of the musician (Monsieur de Sainte Colombe) into his secluded present but rather a music he composed when he lost her, a mournful music which becomes associated with her being, with their past-present togetherness. The bitter and independent musician–composer sensed that this music, in order to become so powerful, must be contained in a well-made ritual, enabling re-presentation of the cherished, lost present.[30] But it must also be ever renewed; that is why its creator tries to preserve his memorial music (in the genre of the French *tombeau*) in its improvisatory freshness, resisting documentary fixation. The fact that he finally concedes to write his music in an established notational form is crucial in such a movie, which itself becomes a powerful vehicle, unleashing pasts that had been preserved only in engraved, forgotten scores. At the same time, Sainte Colombe's resistance to the written score sensitizes the viewers to the danger of losing the vivacity of genuine performances, of the music becoming a fetished commodity, the way it did in the king's palace (Louis XIV) as the movie wishes to contend, and in the kind of concerts of which Rosenzweig, for one, was so critical.[31]

Harmonizing the Catholic *Improperia*

Intersected modes of communication, though of a different kind, characterize the case I would like to discuss in the rest of this paper, a case, in which entangled pasts materialize into an apparently innocent musical present, which itself becomes a repetitive, continuous symbol of these pasts. At the same time, and because of its central place, this particular musical present, or work, turns into a commanding form, begetting new "presents" (works). One of the sonic results of this "musical present," carries the presents of bygone pasts in an even more entangled way, as we shall see below. At a certain moment, an unexpected mode of communication infiltrates into its imagined flow, undermining its avowed past-present-present, recalling a repressed past, which its model work (past-present) effectively concealed.

A Gregorian hymn consisting of series of reproaches addressed by an imagined Christ to fictional Jews—the Good Friday *Improperia*, is our point of departure. While its refrain quotes a verse from the Old Testament the *Popule meus* from *Micha* 6:3: *Popule meus quid feci tibi? aut quid molestus fui? Responde mihi!*" [O my people, what have I done to you? How have I wearied you? Answer me!] its many strophes appear as a retort to a well-known Jewish hymn, central in the Passover Seder, enumerating God's graces for his chosen people.[32] The original Jewish text praises God and Israel in words such as "For had He divided the sea for us, and had not caused us to pass over on dry land—it would have sufficed." The contrasting Catholic text rebuts (or, vice versa, had the Christian preceded the Jewish one): "I have opened the sea before you, and you have opened my body with your lance."[33]

This structure, alternating with the Greek Sanctus—the *Treisagion*—makes possible, in its plainchant rendition, the musical and emotional involvement of the choir as a representative of the congregation. Characterized by a melodic lament, (recited on the first mode), the main refrain adorns the monotony of the long series of reproaches. (Example 1a and b) Characteristically, the hymn heightens an incantation effect, contracting the duration of this long text into a concentrated instance of its archetypal strophe—a thick musicalized presence. This effect, centrally situated in the grim Good Friday's ceremonies of the Church, conveys the ancient antagonism to the Jews, so basic to this High Holiday, through repeated presents of shared vocal experience.

Example 1a

Example 1b

(Transcribed from *Liber Usualis*, p. 704, 706.)

Palestrina set the *Improperia* to music in 1560, when the Council of Trent, aiming at an effective reaction to the ascending Lutheran faith, was approaching the conclusion of many years of discussions. It was Palestrina who would be later recommended by this venerable council or their successors to provide the proper music for the contra-reformation enterprise, meaning, music both harmonious and well-balanced, avoiding secular connotations, while achieving intelligibility and communicative force. His *Improperia* already exemplifies this new aesthetics. It resembles the style and musical flow of the Lutheran chorale, which was already flourishing at the time. Palestrina gave these sentences "such an appropriate and beautiful musical setting," according to the Catholic Encyclopedia of 1910, that Pius IV ordered it to be annually used in the Sistine Chapel. Palestrina's setting, the Encyclopedia further eulogizes, is "unsurpassed in simple beauty, dramatic feeling, and depth of impressiveness."[34] The music is relatively simple and monotonous. Foregoing polyphonic complexity, in which Palestrina otherwise excels, it highlights certain chordal progressions (in fauxbourdon style) allowing for few embellishments (Example 2). Incantation effects were thus not missing, though they were now more harmoniously wrapped.

Example 2

Could the experience of repetitive, exclusive harmoniousness, of such "Proper Sounds"—a counterstatement to the Jewish "*Im*propriety," to which the text refers, subliminally relate to the Jews' categorization as unmusical and inharmonious? Was it meant, as such, to imprint on the souls of innocent believers an implicit juxtaposition of opposite aesthetic and theological mental representations of the Christian and Jewish communities? I leave this question open, while calling attention to the fact that exclusive harmoniousness achieved by musical means did not start with this music and would still go a long way, well into modern times.[35] It relies on a long tradition of Christian attribution of proper and improper sounds to Christians and Jews respectively. Unwittingly, however, it also points beyond that history, towards what could be, and perhaps was, for a short while, a shared or reciprocal liturgical duration of the two communities. For in its simple, antiphonal rendition, it highlighted a structure—to be found in many psalms and hymns, borrowed, at least in part, from the Ancient Jewish synagogue. Concomitantly, through its repetitive structure and annual performance, it became associated with the Sistine Chapel and the Holy Week, rendering it into a model for subsequent music composed for the holiest days of the Catholic year celebrated in this shrine.

Collapsed Musical Presents: Jewish Psalms as a Catholic *Miserere*

We should bear these associations in mind, when we consider the next example, a musical work similarly composed for the Sistine Chapel, seventy-eight years later. It is Gregorio Allegri's famous *Miserere* for the *Tenebrae* services held during the Holy week. Even for non-expert ears, it is rather obvious that Allegri's and Palestrina's work belong to same style, and since the annually performed Palestrina's *Improperia* was the most famous work of its kind the assumption that Allegri viewed it as a model is highly probable. Thus *Miserere*'s sonoric structure could deliver—for the chosen community of listeners—the exclusive import of Palestrina's work.

In the spirit of the emotionalism of the new age (the early "Baroque"), Allegri further elaborated the model, adding to it soaring soprano parts. Those who were listening to the work in its future successive presents could associate it, whether knowingly or unknowingly, with a rather intricate discriminating appeal: *Miserere mei Deus*, have mercy on me, O God, *miserere*, those entitled to Your mercy; and yet *Miserere*, for I know my transgression, because I, the sinner, need God's wisdom and love. In short: *Miserere* the individual, rather than the entire congregation;[36] a prayer of the person well aware of the worthlessness of his being. These new qualities are sonically iconized through the soaring, improvising voices, searching for their individual expression and redemption, using free counterpoint—the *super librum* technique in which the chapel's castrati singers excelled.[37] The chordal progressions of the later work (in comparison to that of Palestrina), deploying rudimentary tonal harmony, enhances directionality.[38] Thus past and present times, communitarian vs. individual values, strictly written parts together with free improvisatory sections, render this piece into a unique sort of musical present, overcoming the trap of fixity that entirely inscribed pieces can fall into.

Cherishing the uniqueness of this performance practice, and its associated religious–aesthetic values, popes and cardinals had a strict prohibition on copying, writing down, or transmitting Allegri's *Miserere* to anyone outside the holy chapel. These peculiar circumstances have shrouded the work with a particular legendary fog regarding its reception and dissemination. Nevertheless, towards the end of the eighteenth century, the work was brought to England by Sir Edmund

Burney who *could* have received it from a 14-year old boy that wrote it down from memory—a certain W. A. Mozart. Whether this "smuggling" took place in this way or another, the work was enthusiastically received by subsequent British audience who cherished it in their musical pantheon. Becoming a model for devotion and sacredness, many of them, even Burney himself, continuously distinguished such music from what they conceived as Jewish anti-musicality.[39]

At a certain fictional moment in the year 1866, a young English man, not indifferent to prejudices and discriminating views against the Jews and their sonoric worlds, enters a traditional, non-reformed Jewish synagogue—in Frankfurt—at Friday Night (*Leil Shabbat*) service, in an attempt to get an impression of the service's own sonoric "propriety." Suspicious of the ensuing ceremony, he nevertheless succumbs to the unfamiliar experience, as reported by the narrator:

> …he immediately found an open prayer-book pushed towards him and had to bow his thanks. However, the congregation had mustered, the reader had mounted to the *almemor* or platform and the service began. Deronda, having looked enough at the German translation of the Hebrew in the book before him to know that he was chiefly hearing Psalms and Old Testament passages or phrases, gave himself up to that strongest effect of chanted liturgies which is independent of detailed verbal meaning—like the effect of an Allegri's *Miserere* or a Palestrina's *Magnificat*. The most powerful movement of feeling with a liturgy is the prayer which seeks for nothing special, but is a yearning to escape from the limitations of our own weakness and an invocation of all Good to enter and abide with us; or else a self-oblivious lifting up of gladness, a *Gloria in excelsis* that such Good exists; both the yearning and the exultation gathering their utmost force from the sense of communion in a form which has expressed them both, for long generations of struggling fellow-men. The Hebrew liturgy, like others, has its transitions of litany, lyric, proclamation, dry statement and blessing; but this evening all were one for Deronda: the chant of the *Chazan's* or Reader's grand wide-ranging voice with its passage from monotony to sudden cries, the outburst of sweet boys' voices from the little quire, the devotional swaying of men's bodies backwards and forwards, the very commonness of the building and shabbiness of the scene where a national faith, which had penetrated the thinking of half the world, and

> moulded the splendid forms of the world's religion, was finding a remote, obscure echo—all were blent for him as one expression of a binding history, tragic and yet glorious. He wondered at the strength of his own feeling; it seemed beyond the occasion—what one might imagine to be a divine influx in the darkness, before there was any vision to interpret. The whole scene was a coherent strain, its burthen a passionate regret, which, if he had known the liturgy for the Day of Reconciliation, he might have clad in its antithetic burthen: "Happy the eye which saw all these things; but verily to hear only of them afflicts our soul. Happy the eye that saw our temple and the joy of our congregation; but verily to hear only of them afflicts our soul. Happy the eye that saw the singers when tuning every kind of song; but verily to hear only of them afflicts our soul."[40]

Internal stream of consciousness of the visitor, still considering himself an Englishman—George Eliot's Daniel Deronda, intermingles here with the author's voice and view which encourage the gentile readers (G. Eliot's embedded audience) to overcome their resistance to the experience they become now privy of. Allegri's music (together with that of Palestrina) is suggested by the author (echoing, perhaps, Daniel's own train of associations) as a sympathetic familiar model for the soundscape of a contemporary Jewish synagogue. This music is thus entrusted with the solemn role of transformation of attitude towards the holy music of the despised Other. This music's special "touches of sweet harmony" (to quote Shakespeare's *Merchant of Venice*) become paradoxically wide enough to include "a national faith, which had penetrated the thinking of half the world, and molded the splendid forms of the world's religion." Its soaring improvisatory elements stands for a past-present-past echo (the 1638 Allegri music for Catholicism's Holy of Holies) of a present-present-present sound (the contemporary Frankfurt Jewish Synagogue) bringing, at the same time, further remote musical presents into life (the music of ancient Israel, which is the past-present-present of the contemporary Jewish community)—the origin of the psalmodic structure of the *Miserere*.[41] This is beautifully exemplified by the quotation from the ancient Jewish *Piyyut* (poem) which Eliot brings, skipping Daniel's thrilled consciousness, in a *tour de force* of counterfactual association. This enables her to deepen the penetration into the possible moving consciousness of the host Jewish community:

> The whole scene was a coherent strain, its burthen a passionate regret, which, if he had known the liturgy for the Day of Reconciliation [Yom Kippur, or Day of Atonement][42], he might have clad in its antithetic burthen: "Happy the eye which saw all these things; but verily to hear only of them afflicts our soul. Happy the eye that saw our temple and the joy of our congregation; but verily to hear only of them afflicts our soul. Happy the eye that saw the singers when tuning every kind of song; but verily to hear only of them afflicts our soul."

This *Piyyut*, sung in the highly elevated Yom Kippur service right after the celebratory "Order of the Work," which powerfully narrates and commemorates in past-present-present the High Priest's ritual of the Holiest Day in the Holy of Holies, brings the congregation, once it is over, back to the grim present-present; deprived of temple and High Priest, it found itself back in a shabby synagogue and an unredeemed existence. These complicated imagined vocalities and intricate temporalities accentuate what Eliot elsewhere declares:

> All the great religions of the world historically considered, are rightly the objects of deep reverence and sympathy—they are the record of spiritual struggles which are the types of our own. This is to me pre-eminently true of Hebrewism and Christianity, on which my own youth was nourished.[43]

In between, as a repressed memory, simultaneously resounding with these past-presents and presents of the past, and without ever becoming an explicit message, one could hear glimpses of the dispelling *Improperia*, dressed in Palestrina's harmonious attire. Would these sounds, themselves becoming repressed through this sympathetic gesture, return, at a later point, to the foreground of these continuously presented pasts? Certainly they do, though luckily not so much in England. A huge, unbelievable attempt would be made to erase any signs of a possible musical mutuality these communities, and their auditorial past-presents share. This was to undermine the attempts of the Rosenzweigs and the Eliots to variously co-opt vexed traditions for the sake of sympathetic reconciliation and reverberation. Yet presents enshrined in their evoked pasts have the potential to sympathetically reactivate what seemed forever buried and lost.

Notes

[1] The main part of this paper was researched and written during my stay as a fellow at the *Wissenschaftskolleg zu Berlin* during the year 2004–5.

[2] The term mode deployed here follows, to a certain extent, the Aristotelian concept of "mode" or "manner of imitation" (*Poetics*, chapter 1), which Aristotle distinguishes from "medium" and "object of imitation." While "medium" focuses on the artistic métier, and "object"—on the reality represented, "mode" or "manner" refers to the presentation being engaged. Taking music as mode, rather than artistic medium, emphasizes its phenomenological, rather than formal qualities. "Modes" is close, but not identical with the concept of "modality" used below, which is inspired by Ernst Cassirer's theory of symbolic forms, see n. 4. The idea of a mental screen on which illusory images are projected is central to E. H. Gombrich's theory of artistic illusion, see *Art and Illusion* (London: Phaidon Press, 1960).

[3] I follow the tradition of Samuel Coleridge who, in discussing the condition for poetic illusion in the so-called fantastic literature, famously argued, that the poet's aim is "to transfer from our inward nature a human interest and a semblance of truth sufficient to procure for these shadows of imagination that *willing suspension of disbelief* for the moment, which constitutes poetic faith." See Samuel Taylor Coleridge, *Biographia Literaria* (Oxford: Oxford University Press [1907] 1947) 5–6. An exemplary modern theory that goes in a similar direction is Kendall L. Walton's *Mimesis as Make-Believe: On the Foundations of the Representational Arts* (Cambridge: Harvard University Press, 1990).

[4] Susanne Langer, *Feeling and Form* (New York: Charles Scribner's Sons, 1953), 110.

[5] Henri Bergson's ideas are developed in *Essai sur le données immédiates de conscience* (1889) translated as *Time and Free Will: An Essay on the Immediate Data of Consciousness* by F. L. Pogson (London: George Allen and Unwin, 1910); and later in *Matiére et mémoire* (Paris: Presses Universitaires, [1896] 1946), trans. as *Matter and Memory* by N. M. Paul and W. S. Palmer (New York: Macmillan Col., 1911). Ernst Cassirer's distinction between "quality" and "modality" is most pertinent to this discussion. "Quality" for Cassirer is "the particular type of combination by means of which [a given relation] creates series within the whole of consciousness, the arrangement of whose members is subject to a special law." Temporal succession, causation, and spatial relations, are such universal qualities. "Modality," on the other hand, is the inner transformation the basic quality undergoes when occurring "within a different *formal context.*" See *The Philosophy of Symbolic Forms* vol. 1, trans. Ralph Manheim (New Haven: Yale University Press, 1955) 95, emphasis in the original. Cassirer exemplifies his point by juxtaposing physical time and musical time. Obviously, "being in time" reverberates Heidegger; see below.

[6] Langer, *Feeling and Form,* 113.

[7] Ibid.

[8] Another caveat can be raised about Langer's claim, only indirectly related to the present discussion. It concerns the role of space in the total musical experience. Aesthetic tradition, associated with Lessing's *Laocoön* (1766), maintains that visual qualities determines spatial mental categories whereas auditorial (linear) qualities (or their derivatives, in language, drama and dance) effect temporal mental categories. The mixture of categories and mental dimensions is arguably phenomenologically weaker, hard-pressed and borrowed. Though Langer herself acknowledged the centrality of movement to music's basic imaginary configuration, she tended to underestimate space in the experience of music.

[9] *Sound and Symbol; Music and the External World* (New York 1956/*R*, 2/1973; Ger. orig., Zürich, 1963, as *Die Wirklichkeit der Musik: Der musikalische Begriff der Aussenwelt*, 2/1973).

[10] Ibid., 202.

[11] Ibid., 203.

[12] Ibid., 208. Zuckerkandl's phenomenological differentiation of source and target in this metaphor can imply that also in music, time is not concrete, thus contradicting his basic claim (above), unless resorting to the physical forces metaphor which implies that such forces (magnetism, gravitation and the like) are known only through their effects. In music, Zuckerkandl could claim, these effects are more direct than in temporal unfolding of other qualities.

[13] Ibid., 224.

[14] Ibid., 225.

[15] A major question arises here as to the boundaries of such extended presence. I fully agree with Raymond Monelle that it is limited to short musical stretches and cannot be applied to the dimensions of a Bruckner symphony (even more so, a Mahler symphony). There is definitely a point where such presents, especially in works of the classical romantic period, are grafted onto a more extended narrative unfolding, which calls for different sort of retention and memory. However, since I am dealing here with pieces of relatively short span, I can fully embrace Zuckerkandl's claim. For Monelle's insightful discussion of the concept of musical time developed in the wake of Bergson's distinction see *The Sense of Music, Semiotic Essays* (Princeton: Princeton University Press, 2000), 86–92. For an extreme view on the limitedness of cognitive simultaneity in music see Jerald Levinson, "Gurney and the Appreciation of Music" *Iyyun*, Jan. 1993, 181–205, and my reaction "Beyond the Musical Moment: Gurney in Context", ibid., 207–213.

[16] "Das Freventliche an der Musik sind die idealen Zeiten, mit denen sie die wirkliche Zeit zersetzt," see Franz Rosenzweig, *Der Stern der Erlösung* (Frankfurt a/M.: Suhrkamp, 1988) 401. Translated as *The Star of Redemption*, by William W. Hallo (Boston: Beacon Press, 1964), 360.

[17] Ibid., 362–3 (English). Der den Raum füllende, von allen gemeinsam in mächtiger Einstimmigkeit gesungene Choral ist die eigentliche Grundlage der kirchlichen Anwendung der Musik; noch in den Bachschen Passionen lebt er fort, und auch die römische Kirche hat ihn wiedergepflegt, wenn auch die musikalische Messe von ihm wegführt. Im Choral ist die Sprache, die sonst aus jedes Einzelnen Mund ihr eigenes und besonderes Wort zu reden hat, zum Schweigen gebracht. Nicht zu jenem Schweigen, das einfach stumm dem verlesenen Wort zuhört, sondern zum Schweigen seiner Eigenheit in der Einmütigkeit des Chors…

Zu dieser Gemeinsamkeit des Lebens, wie sie dann im Sakrament sich verwirklicht, stimmt nun die Musik die Seelen vor. Die beim Eintritt in den gemeinsamen Raum nur zur Gemeinschaft überhaupt, zur Möglichkeit von Gemeinschaft vorgestimmten Seelen werden im gemeinsamen Singen des Chorals vorgestimmt zur wirklichen Gemeinschaft. Auch die musikalische Messe ist ja, obwohl bloß mitgehört, nicht mitgesungen, doch im Grunde ebenso sehr wie der Choral eine Vorstimmung aller Einzelnen zur Gemeinschaft; denn Musik Hören ist ein ganz anderes Hören als das Hören eines vorgelesenen Texts oder einer Predigt: es gründet keine Gemeinschaft, sondern es erregt die Versammelten, einen jeden für sich, zu den gleichen Gefühlen,—einen jeden für sich, wie der Anblick eines Konzertpublikums unmittelbar zeigt. So ist das Anhören der musikalischen Messe in dieser Beziehung völlig gleichwertig dem Gesang des Chorals. Der Einzelne wird, nachdem ihn leiblich der gemeinsame Raum aufgenommen hat, nun in seiner Seele als redender Einzelner ergriffen, *und indem die Rede in die Zucht von Rhythmus und Melodie genommen wird*, lernt die Eigenheit des eigenen Wortes des Einzelnen das Verstummen. Er spricht, aber was er spricht, sind nicht seine Worte, *sondern die allen gemeinsamen Worte zur Musik*. Rosenzweig, *Der Stern der Erlösung*, 402–3. [My italics.]

[18] Harmony, in this connection, is a technical term related to the kind of the chords resulting from the rhythmically simultaneous, and melodically different parts (soprano, alt, tenor and bass) and their sequential progression.

[19] In *Keeping Together in Time, Dance and Drill in Human History* (Cambridge Mass.: Harvard University Press, 1995), William H. McNeill stresses the effectiveness of "moving together" of voices and muscles for consolidating group solidarity but devotes his discussion more to dance and bodily movement than to music.

[20] I deal with this accusation in "Between Noise and Harmony: The Oratorical Moment in the Musical Entanglements of Jews and Christians," *Critical Inquiry* vol. 32, no. 2 (Winter 2006), 250–277 and in my forthcoming book: *Between Noise and Harmony: The Jew as an Aesthetic Category in Vocal Fiction*. For the etymology of noise/nausea and its relation to the Jews see Leo Spitzer, "Noise" in *Word* II (1945), 260–276.

[21] This especially true when considering the history of the reception of *St. Matthew Passion*, discussed in my publications mentioned in n. 19. For recent study of its reception in 19th-century German culture see Celia Applegate, *Bach in Berlin: Nation and Culture in Mendelssohn's Revival of the "St. Matthew Passion,"* (Cornell: Cornell University Press, 2005).

[22] The first "imaginary" in the above sentence deploys the regular adjectival sense of the word, whereas the second—its sense as a noun, what Yaron Ezrahi conceives as "composites of facts and fictions that may acquire a degree of reality when they are widely treated as unquestionable givens." Imaginaries are thus distinguished from images which are concrete and usually discrete representations of entities and objects (see *Necessary Fiction: Imagining Democracy between Modernity and Post Modernity*, forthcoming). See also Charles Taylor, *Modern Social Imaginaries* (Durham: Duke University Press, 2004).

[23] *Languages of Art* (Indianapolis: The Bobbs-Merrill Company, 1968), 28 and passim. Goodman discerns three syntactic requirements and two semantic ones at the logical basis of such systems.

[24] The historical development of this notational system in the Middle Ages was analyzed from this point of view by Ruth Katz in "Musical Notation in the Light of Goodman's Requirements" in *How Classification Works: Nelson Goodman Among the Social Sciences*, eds. M. Douglas and D. Hull (Edinburgh: Edinburgh University Press, 1993).

[25] On the concept of work and its genealogy in Western music see Reinhard Strohm, "Looking back at ourselves: the problem with the musical work-concept," in *The Musical Work: Reality or Invention?* ed. M. Talbot (Liverpool University Press, 2000), 128–52 and idem, "'Opus': An Aspect of the Early History of the Musical Work-concept," in *Complexus effectuum musicologiae. Studia Miroslavo Perz septuagenario dedicata*, ed. Tomasz Jeż (Kraków: Rabid, 2003), 309–19.

[26] This idea in somewhat different terms is developed in Eero Tarasti, *Existential Semiotics* (Bloomington: Indiana University Press, 2000), especially 17–36.

[27] I attended this unique concert, powerfully experienced by its audience. I would like to thank Ms. Anita Tarsi, the head of Beit Theresienstadt (Theresiendtadt Martyrs Remembrance Association) for helping to locate the original ad of this concert, and for providing me with a photocopy of it. *Kunst und Kultur im Theresienstadt. Eine Dokumentation in Bildern*, ed. Rudolf M. Wlaschek (Gerlingen: Bleicher Verlag, 2001), 34. Karl Fröhlich, the director, founder (and first violinist) of this quartet, survived the camp and later immigrated to the USA.

[28] Lily E. Hirsch, *Imagining "Jewish Music": Der Jüdische Kulturbund and Musical Politics in Nazi Germany 1933–1941*, PhD Dissertation, Duke University, 2006. Hirsch gives variety of reasons for the centrality of Schubert in this repertoire.

29 The nostalgia associated with this musical poetic genre, stems from the combination of certain (romantic) texts and the symbolic dramatic setting in which they are embedded (consisting of expressive-narrating voice, "past tensing" or "presenting-past" events and emotion, and further voicing embedded protagonists, and of an atmospheric [pian]) accompaniment, painting the surrounding and supplying internal motives to the expressive surface). It also relates to certain embedded cultural practices, as discussed by Lawrence Kramer in *Classical Music and Postmodern Knowledge* (Berkeley: University of California Press, 1995), 143–173.

30 The ritual consists of a "communion" of sorts—including wafers and wine, which, in the possible world the film launches, is being realistically copied by a nearby painter, Lubin Baugin (whose historical equivalent (1610–1666) actually painted exactly such a painting, named *Las Dessert de Gaufrettes*, part of the Louvre collection today) becoming then an objectified present—a vehicle for reactivation of the séance.

31 The fact that I mention Franz Rosenzweig in this connection and not a more typical critic (also Frankfurtian) of such an "aberration" of classical music—Theodor Adorno—is because the former, as we have seen above, addresses his criticism precisely to the handling of time in such music.

32 Eric Werner, relying on Egon Wellesz, believes that the series of reproaches directly reflect an old rabbinic text, though with a "sharp anti-Jewish twist." Werner even goes so far to claim that the old tune of the Jewish poem "it would have sufficed" דיינו likewise wandered from the Jewish Seder table to the Good Friday anti-Jewish vocal utterances, but this seems rather far fetched. See *The Sacred Bridge,* vol. 1 (New York: Columbia University Press, 1959), 224 and 453; whether the direction of rebuttal was opposite—from the *Improperia* to *Dayenu*, is a question which is not directly relevant to my argument; in any event, their confrontational relation is self evident. Werner further elaborated on the theme in the second volume of the book (*The Sacred Bridge,* vol. 2 New York: Ktav Publishing House, 1984); his analysis was criticized, e.g. by Peter Jeffery in "Werner's *The Sacred Bridge,* Volume 2: A Review Essay," *The Jewish Quarterly Review,* vol. 77, no. 4 (April 1987), 289–291.

33 *Ego ante te aperui mare, et tu aperuisti lancea latum meum.*

34 P. J. Morrisroe, transcribed by Christine J. Murray, *The Catholic Encyclopedia,* vol. VII, 1910 by Robert Appleton Company, Online Edition 2003 by K. Knight; http:/newadvent.org/cathen/07703a.htm.

35 See Ruth HaCohen, "Between Noise and Harmony" and the related book referred to in n. 20 above.

36 This psalm entered the Catholic service rather late (its first polyphonic renditions are dated to the early sixteenth century).

37 For excellent and accessible information see Dave Lampson, "Gregorio Allegri Miserere", http:/www.classical.net/music/comp.lst/works/allegri/

miserere.html; see also Fredrick Hammond, *Music and Spectacle in Baroque Rome* (New Haven: Yale University Press, 1994) and J. Amann, *Allegris Miserere und die Aufführungspraxis in der Sixtina nach Reiseberichten und Musikhandschriften* (Regensburg, 1935).

[38] See n. 15 above; the new directionality contributed to a more compelling musical unfolding.

[39] It is the same Burney who included in his travelogue a negative, however typical, report of his visit into a synagogue in Holland. See Charles Burney, *An Eighteenth-Century Musical Tour in Central Europe and the Netherlands* 2nd ed. P. A. Scholes (London [1775] 1952) vol. II, 229–30.

[40] George Eliot, *Daniel Deronda,* Terence Cave, ed. (Penguin: [1876] 2003), Ch. 32, 367–8. For a more extensive discussion of this excerpt in relation to Eliot's theory of sympathy and in comparison to Wagner's notion of *Mitleid* (compassion) and his conception of the Jewish synagogue see my "Ritual Music and the Compassion Controversy: The Jew as an Aesthetic Category in Richard Wagner and George Eliot," *Theory and Criticism [Te'oria U-vikoret]: An Israeli Forum* 17, Autumn 2000: 35–74 (in Hebrew) and my forthcoming book.

[41] There are significant commonalities in the performative structure of both kinds of music, as discussed in my "Ritual Music" (see preceding note), even in the less improvisatory version practiced in Eliot's England. See Jerome Roche/Noel O'regan, "Allegri, Gregorio," *Grove Music Online* ed. L. Macy (Accessed 21.11.05), <http://www.grovemusic.com>.

[42] It is clear that Eliot translates here the German "Versöhnungstag," which she probably had learned from the German Jewish scholars with whom she corresponded, or from related sources.

[43] George Eliot's letter of 20 October 1873 to John Walter Cross. See Eliot, George, *The George Eliot Letters*, ed. Gordon S. Haight, 9 vols. (New Haven: Yale University Press, 1954–78).

Quoting from the Past or Dealing with Temporality[1]

Britta Duelke

In her famous essay on Walter Benjamin, Hannah Arendt summarized his theoretical reflections on history, tradition and authority in the following formula: "Insofar as the past has been transmitted as tradition, it possesses authority; insofar as authority presents itself historically, it becomes tradition."[2] We will never know whether Benjamin himself would have endorsed these lines, which clearly show the hand of Arendt. Yet the formula does have the appearance of something that comes close to Benjamin's very own style of thinking and writing, to such an extent that one could easily take it for a quotation of the "real" Benjamin.[3] Arendt's programmatic reading of Benjamin made reference to his own life experiences, to the breaks in tradition and the loss of authority which occurred in his actual lifetime—and also to his discovery that the transmissibility of the past relates to its capacity to be cited in the present: that is, its citability.[4] Benjamin was well aware that the original context of a quotation from tradition or history was far less important than the new context it helped to create. An isolated quotation does not necessarily make sense. The citability and meaning of a quotation—even if it is only a fragment of the original text—are determined by the frame in which it is used and from which it gains its meaning and potential authority.[5]

I use Arendt's "Benjaminian formula" as a starting point here for two reasons. It addresses some interesting questions with regard to quotations, not only in written traditions—of the sort Benjamin might be thought to be addressing—but also in relation to oral traditions of the sort that are central to my Australian case study. Moreover, it blurs in a very peculiar way the conceptual distinction between tradition and history by emphasizing their difference.[6]

Variations

Australian Aboriginal societies have traditionally been seen as unchanging and ahistorical—the very prototypes of a *classic* "traditional society." More recently, questions regarding the stability and historicity of traditions in Aboriginal Australia, or more precisely questions regarding the representations of the past in relation to present conditions and the interpretations offered of the "now," have attracted considerable interest, and not only among academic circles. In the context of Australian land rights and native-title legislations, the ideas of tradition and history, and the meanings attached to them, have acquired a wide-ranging political and social significance and have also exercised a material effect on people's lives.[7]

The area I focus on is in the northern, tropical region of the Northern Territory. Until now, people in this part of Australia have had little experience of the Native Title Act, but have been substantially affected by the Northern Territory Land Rights Act.[8] In 1976, land rights legislation was introduced in the Northern Territory (where Aborigines now represent over 28 percent of the population). This entitled Aboriginal people to claim ownership of traditional land, provided that the land in question was unalienated Crown land, and provided that claimants could prove that they belonged to a "local descent group" representing the "traditional owners."[9] Claimants were required to prove traditional descent and foraging rights as well as to present traditional knowledge and ceremonies demonstrating "primary spiritual responsibility" for the land in question. Now that the majority of land claims in the Northern Territory have been settled and the scheduled areas (the so-called *Schedule 1 Land*, that is, mainly ex-reserve land) have been transferred more or less automatically to Aboriginal ownership, nearly half of the land in the Northern Territory has become Aboriginal Land under the Act.

Religious, ritual and land-related traditions have always been central to Australian Aboriginal anthropology.[10] The constitutive basis for the anthropological examination of these traditions has been the exploration and analysis of traditional knowledge and practice which have long been accepted as significant for identity and *Weltanschauung*. Both the theoretical considerations and the conventional understanding of tradition have relied heavily on notions of stability, continuity and duration.[11] Consequently, the Land Rights

Act was underpinned by anthropological, legal and political notions of unchanging tradition that reflected not only the apparently dominant Aboriginal ideology of non-change and immutability, but also mirrored the pervasive anthropological view of traditional Aboriginal society as a "cold," unchanging and, therefore, *traditionally ahistorical* society.[12] This stress on tradition led to at least a partial obliteration of history and change both on the formal (i.e., the legal) as well as the ideational level.[13] It also created a political arena in which the concepts of tradition and traditional knowledge had to be interpreted and negotiated. This not only applied in relation to the Act and the institutions associated with it, but also had an impact at the local level, on social movements, power structures, the economy, the management of knowledge, and people's relationships with the land itself. The practical implementation of the Land Rights Act, however, indicated that long-held ideas about tradition and history in Aboriginal society needed to be reconsidered,[14] since the experiences made in these processes demonstrated that traditions were not just continuity-bearing, atemporal or detemporalized *re*-productions of the past but the foundations of and medium for particular lived social realities.

The granting of Aboriginal land through a legislative act should be viewed against the backdrop of a history of other externally applied policies and directives, including "colonization," "separation," and "assimilation."[15] For Aboriginal people, these policies have resulted not simply in an undifferentiated (social) "stock of knowledge" that has left the realm of tradition unaffected.[16] In many cases these historical legacies have led, on the contrary, to profound stratifications and dichotomies in people's perception and understanding of "traditionality," and the values assigned to it.[17] Consequently, the different and occasionally even conflicting meanings of what constitutes the realm of tradition for various Aboriginal groups and individuals must be understood in the historical context of shifting intra- and inter-social structures, values and the political power deriving from them.[18]

Novel circumstances and modifications in social, territorial and cultural relations require adaptations which affect the contents and valuations of knowledge and traditions as well as the role, composition and recruitment of knowledge-bearing groups and individuals.[19] To a certain degree innovations tend to favor the reproduction, revival, and creation of tradition, and the development of notions of traditionality that mesh with the people's individual experiences.[20]

Land grants in the Northern Territory have affected communities on different levels. The fact that some descent groups now have titles to land while others do not has had a profound impact on the dynamics of intra- and inter-ethnic structures. On the local level some of the new realities brought about by land grants require new explanatory frameworks. In turn, these have led to new "reality-creating practices" and new social roles, which are nevertheless still based essentially on a recourse to *the past*, that is, to tradition *and* history. The frequent negotiations and re-interpretations of the meaning of tradition have certainly caused local conflicts and initiated rapid processes of change; or rather, the conflicts and negotiations over present "realities"—involving an interplay between the actual state of affairs and the historical account(s) offered of them—have resulted in the creation of new accounts. However, the very fact that those quasi-ahistorical and apparently unchanging traditions have been negotiated, transformed, and declared to be "new traditions," simply serves to indicate once again that all social reality is a reconstruction of previous realities.[21] Since knowledge and tradition are tied to frames of meaning, the existence of different frames raises the question of authenticity. The question here, however, is not so much whether a particular claim to authenticity is "true" (or whether a particular tradition is "old"), but rather of how claims to authenticity become socially and legally acceptable.[22]

Contexts

The following case study examines the ways in which notions of "the past" in relation to land were negotiated at a former Catholic Mission Station in the so-called Top End (i.e., the northernmost half) of the Northern Territory. To focus on this particular community may appear paradoxical, since the people living on the Mission are migrants from areas (*Schedule 1 Land*) already deemed Aboriginal land under the Land Rights Act.[23] Yet this situation provides an opportunity to analyze the way in which the past is viewed, not as a monochronic reference point but as a negotiable polychronic framework which allows for a dominant ideology of stability and immutability to be maintained, in order to legitimate and to explain a present built upon the persistence of tradition, while, at the same time, serving to accommodate within itself the inevitability of change and dynamism in practice and ideology, not to mention historical circumstances.

The Mission people chose *not* to live on their own land but preferred to live on land that still belongs to the Roman Catholic Church which holds a freehold title to it. Being "alienated land," the Mission land has therefore been excluded from a traditional land claim lodged on behalf of another group that has continued to live on its land and who I will now call the Original Owners. The decision of the Mission people to live not on their own, but on other people's land, had many reasons one of which was a strong desire to maintain a particular life style, without, however, giving up the relations to their own traditional homelands.

The waves of migration into the wider Mission area during the first three decades of the twentieth century are still part of the living historical memory of the Mission people. Even today the older members of the Mission community recall how they, their parents and grandparents left their traditional homelands, which lie to the south-west of the Mission. According to their own accounts, people of different language groups and backgrounds walked north-east to the small white settlement, excited by the news of the new arrivals—the Euro-Australian and Chinese miners, farmers, crocodile hunters, and other entrepreneurs. Their desire for "whitefella goods," mainly tea, sugar and tobacco became, in their own words, "too strong": "Them oldfellas, they bin sit down now, settle down." The accounts of elderly people indicate that, from the early 1930s onwards, the traditional homelands that the Mission people had abandoned were left "empty," since nobody was living there permanently. However, this did not mean that they "forgot" their homelands. Their work shifts on cattle stations and farms, the mustering trips during the dry season (April–October) and the so-called "walkabouts" during the wet season (November–March) provided them with opportunities to return to their homelands on a more or less regular basis, so that they never lost touch.

During the 1930s the Original Aboriginal Owners, of what is now Mission land, had not only experienced sixty years of colonization, including thirteen years of Jesuit missionary presence (1886–1899), but had also been forced to accept that their land had become the target of migrating Aboriginal groups. For both the Original Owners and the migrants, memories of that time are riddled with stories of fights and disputes. Nevertheless, the migrants stayed, working at cattle stations and other agricultural enterprises and only occasionally returning to their home territories. In 1955, when the current Mission was opened by Missionaries of the Sacred Heart, the migrants re-located to it. Even today they represent the vast majority of the approximately 400 Aboriginal people living on the Mission.[24]

Twenty years later, after the passing of the Land Rights Act in 1976, and in keeping with developments in many parts of Australia, Mission people began, rather slowly, to move back closer to their old homelands. In this particular case, they began by returning to a recently established outstation[25] which I will call the Home Community, located about 100 km south-west of the Mission. Almost no one from the Mission could resist the initial attraction of this Arcadian site, where people were meant to live free from Euro-Australian domination and in strict accordance with traditional rules. Traditions and ceremonies were reinstated. The Home Community flourished and became one of the most important—though highly disputed—ritual places of the area. Funds began to flow and there was a revival of traditional Law.[26] From about the mid-1980s onwards this Home Community gained an increasingly negative reputation in regional land-rights politics. Disputes regarding boundaries were numerous, and conflicts increased when the Home Community began an active campaign to block the establishment of other outstations in the vicinity. For various reasons Mission people did not stay in the Home Community for long. Only a few felt obliged to endure the somewhat despotic life style there, and many—in accordance with what appeared to be a traditional way of resolving conflicts—simply wandered off. Ritual disputes and, finally, an escalation of violence in the early 1990s were the official reasons given for the virtual complete return of the population from the Home Community back to the Mission.

Mission people had, and still have, strong but ambivalent feelings towards both places: the Mission and the Home Community. No one wants to give up the traditional homelands—but neither do they want to live on them. With the introduction of land rights, however, the questions of where one lives and one's relationship towards the land have become questions not simply of taste or comfort, but of identity and status. Traditional relationships towards the homelands imply authenticity, legality and legitimation, and carry—for both Euro-Australians and Aborigines—notions of stability, credibility, and naturalness.

Thus, although the migrants saw the area around the Mission as a home with which they had long-term historic associations, they nevertheless felt ambivalent and felt that they ought not to live there, since this appeared to go against a certain notion of tradition. Their decision to remain on the Mission, in other words, bore the taint of inauthenticity and wrong-doing since merely *historical* relationships imply time and change, categories that are alien to the prevailing—or "orthodox"—concept of tradition.

Yet, for the Mission people, it was not the Home Community but the land in and around the Mission that was associated with their actual lives and experience, representing what Koselleck has called the "lived experiential space."[27] Over time, moreover, their relationship with the Mission land gained importance even from a "traditional" perspective. Local sites became sites of historical events[28]—most notably, the so-called conception,[29] birth, and death places of relatives, that is, sites often given prominence within a traditional perspective. Equally, the common traditional practice of using the name of an initiation site as a personal name for the person initiated there indicates a willingness to acknowledge historical events and to incorporate them within a given and established framework. This practice not only serves as an aid to remembering the event, but also as a means of establishing identity by combining notions of body, event, time and place. For the Mission people this meant that because of their long residence on, and identification with, the Mission land, they have been able to claim, over the years, certain rights to land which, although derived "historically," could well be legitimated "traditionally."[30]

Nevertheless, Mission people found themselves in serious conflict when it came to clarifying their relationship with the Mission—their "actual life-space"—on the one hand and their traditional homelands on the other. This conflict increased when, in 1990, the claim by the Original Owners to the land around the Mission was decided and parts of the land were returned to them. Complicating matters further, the Mission itself, and the land immediately around it, which were still owned by the Catholic Church (i.e., the areas were alienated and not unalienated Crown land), had been excluded from the

Original Owners' land claim, but now the Church offered to hand over the Mission's freehold title to a land trust set up by Mission people—that is, not to the Original or Traditional Owners but to the people whose families had originally built the Mission in the mid 1950s.[31]

Most of the Mission people supported the idea of a land trust, immediately engaging in debates about who should be nominated as a trustee and on what basis. Only a very small group based their claims on rather idiosyncratic interpretations of the Land Rights Act,[32] while the majority referred neither to the Act nor to "eternal traditions," invoking instead their long use and occupancy of the Mission land and their "rights" accrued from the past. In other words, they used the options available for making historical claims to land within a traditional framework.

In order properly to legitimate their claims, however, the Mission people needed to prove, both to themselves and to others, that they were "traditional," that they were "real" Aborigines. According to a widely-held view, any such claim had to have a basis in original and traditional land ownership. Now that they had defined themselves in opposition to traditionalism, the Mission people's claim to being "real Aborigines" had come under threat. Their situation demanded legitimation on all sides—in opposition to the Original Owners, to other Aboriginal groups, to the Church, and to the State, but above all in relation to their perception of themselves as "traditional" and therefore "real" Aborigines. Under these circumstances, the Mission people had to reconsider their historical and traditional involvement with the Mission land (and its people) as well as their own homelands. This prompted them to undertake detailed communicative reconstructions of, and frequent quotations from, the past in order to create a model for explaining, determining and legitimating the present in *polychronic* terms, that is, within both a traditional and a historical framework.

There was no need for the Mission people to justify their relationship with their own traditional homelands. This relationship was part of living memory, stably anchored in time, safeguarded by myth, and based on traditional criteria of land ownership. The relationship to the Mission land, on the other hand, was based on the concrete fact that the mission people had lived there for many years; it too had considerable "temporal depth" and was a central part of the people's living memory. But this relationship appeared to be too recent (that is, not "past" and not "bygone" enough) to be accepted as traditional, and it was not directly justified by myth. Thus, one relationship to land required explanation, the other did not.[33] In confronting this difference, the Mission people had to forge a new link between history and tradition.

People had to come to terms with a past consisting of antagonistic dimensions. The situation created a "critical tension" which required what Nancy Munn in a different context calls a "double-movement" or "bi-directional structure."[34] Here the double movement meant employing and mediating symbols drawn from varying frames—that is, from historical, traditional, temporal, local, discursive, explanatory and rhetorical frameworks.

Mission people strongly emphasized their close ties to their traditional lands and gave expression to these connections by, for example, negotiating the rights to recount (or paint) certain land-related myths, by recalling dream-travels and memories of the "olden times" when the "old people [were] still on their own land," by presenting themselves as traditional land owners whose rights were acknowledged under the Act, and last but not least by repeatedly expressing their desire to return and to set up an outstation on their own land. By referring to (or quoting from) their continuing relationship with their own homelands, the Mission people placed themselves in a complex and indisputable context of traditionality which also served as a guarantee of authenticity. This anchoring in a quasi-"holistic" traditional context also enabled them to legitimate their "new" or "historical" relationships to the Mission land. In fact, it was only by invoking "tradition" that they were able to stake their claims to specific historically important sites (that is, on "conception," birth and death places).

The double-movement as it was employed by the Mission people revealed the truth of the hermeneutical rule that one can only comprehend the whole by looking at the individual parts, and the individual parts by looking at the whole. To push the matter further, we might take up Hans-Georg Gadamer's notion of the circle and the hermeneutical consequences it unleashes, accentuating the concepts and procedures which he calls "*den Vorgriff auf Vollkommenheit*," (the anticipation of completeness [or perfection]). For Gadamer this "anticipation" represents a formal—and immanent—requirement of any capacity to construct a "complete unity of sense/reason/meaning" (*eine vollkommene Einheit von Sinn*) as well as a "transcendent expectation of sense/reason/meaning" (*eine transzendente Sinnerwartung*).[35] For the practices in question one could—within the limits of the circle[36] and by employing some interpretative liberty—go beyond Gadamer and state both an "anticipation" as well as a "recourse" to completeness (or perfection) (*einen "Rückgriff" auf Vollkommenheit*). In other words, "exceptions," "special cases" and "modifications" which are founded and legitimated by a *complex* of tradition *can* by means of a "recourse to persisting traditions" create a new traditional "completeness."[37]

Over the years the Mission people have never lost the connection with their own land, and the "authenticity" which derived from this traditional connection lent a certain legitimacy to their claims to the Mission lands. On this basis, they have succeeded in winning acknowledgement for *some* of their claims from those who continue to see themselves as the Original and Traditional Owners of the Mission lands, although the latter have expressed considerable misgivings and—naturally—remain highly critical of the "migrants."[38]

For the Mission people the attempts to secure a legal title to the Mission land were not based on a fake or *ad hoc* construction of land-related traditionality. They see themselves as having developed a close relationship with the Mission land, and strong social ties with those they have lived with for decades; together they constitute "the Mission mob," a group with a localized identity. They share a common and distinctive life style, or, as it is commonly put, "that style, we got'im from here." They uphold traditions and traditional etiquette—"no matter [regardless of] that modern world." They have continued performing ceremonies, and indeed in recent times these activities have actually increased. Although they are important, these ceremonies and other "traditional representations" (paintings and so on), which often serve as official or publicly-acknowledged indicators of traditionality and authenticity, are not "essential" to everyday Mission life or "that style."[39] Rather, it is the tacit, yet strict observance of social rules such as kinship and marriage prescriptions, as well as social etiquette, which continues to provide people not only with a stable social universe, but also with a sense of continuity and "traditional orientation."

All social and communicative interactions on the mission, including seating arrangements in motor vehicles and rooms, the exchange of gifts and food, ritual trade, ceremonial life, even drinking sessions and a long list of other social engagements follow these long established rules. The local kinship system, of the "Kariera-type," systematizes and stabilizes the Mission's entire social world. People continually point to the traditional kinship order and to rules that were established in mythic times. They insist that their youngsters marry "straight," that is, give preference to classificatory cross-cousin-marriage; they maintain avoidance and joking relationships, and carefully observe traditional forms of address and related aspects of etiquette; in other words, they *live* these traditions, or, as the Mission people say: "We follow the Law, that culture now we got'im, no more whitefella … proper blackfella way." Violations of these rules are the main cause of social conflict in everyday life or the vehicle for expressing the sources of these conflicts.[40]

Kinship remains the key cultural category for the Mission people and as such represents the "connective structure"[41] which links the social and the temporal dimension of their lives. It also creates the social world, which stands for a "symbolic universe,"[42] in which mutually-shared frames of experience, expectation, and action are created and reflected. The singular importance the Mission people attach to kinship and kinship relationships, as the apparently fixed and unchanging underpinnings of their world, goes beyond any simple notion of "construct"; it is the very expression of cultural stability, the realization of a symbolic universe which remains stable and enduring even if innovations in other life spheres—social, political and historical—demand integration within it. For the Mission people, daily life, conducted as it is, within the stable, consistent and long-standing order that kinship represents, is the true source of their sense of continuity and, hence their primary point of reference to tradition and traditionality. On the local level an "intact" kinship order implies "intact" traditions.

It is important to note that in their reconstructions of the past, the Mission people did not eliminate the historical memory of migration. Rather, they emphasized those elements of their traditional culture that could accommodate historical change.[43] The debates about the Mission land concentrated on the Mission people's historical attachment to it, and the latter saw it as right that they should gain title to the land on the grounds that they had created the Mission site. They had worked on the Mission all their lives and had built it from nothing, or, in their own words: "We bin build'im up from dust … nothing there before." They had transformed a formerly unnamed site into a newly named place of relevance, a place with a concrete meaning and a (his-)story. Moreover, while the Mission people insist that there is no sacred site (belonging to the Original Owners) on the Mission land, they refer to their own sites of historical importance: the so-called conception sites, and the birth and death places of kin. These sites are not only personally and socially meaningful as well as historically important to them, but they traditionally help to justify entitlements to land.

Mission people refer to *the past*, not just temporal history or atemporal tradition. They use the model of a polychronic past which represents a specific communicable and transmittable experience of their "life world."[44] By referring to rights derived from the past, they do not deny the development of the present; rather, they link the present with tradition and history, and project the present, that has been made meaningful by the past, into the future. In other words, they explain and legitimate the present in terms of the past, emphasizing both the persistence of tradition and the historical relationships which include change.[45] Here, reference to a polychronic past provides the essential framework necessary for orientation and legitimation, allowing for both notions of stability and agreed "eventfulness"; at the same time it provides the conceptual means to incorporate change. It thus serves as a basic argument in favor of the existing social reality.

As Friedman puts it: "The construction of a past […] is a project that selectively organizes events in a relation of continuity with a contemporary subject, thereby creating an appropriated representation of a life leading up to the present […],"[46] and into the future. The selective organization involved in this project should be viewed, as Ricœur points out, "against a background of a wider dialectic, that of historical consciousness, on the understanding [that] the epithet 'historical' does not qualify a particular science, namely historical science, but the human condition, or as is customary, its character of 'historicity'."[47] In that sense a polychronic past concerns a past which "acquires its double sense of having been and no longer being only in relation to the future."[48]

Contextual Variations

The example of the Mission makes apparent certain substantial historical changes in the local relationships to the land. This does not imply, however, that these changed relationships are necessarily "untraditional."

As Berger and Luckmann have shown, tradition can be seen as a component of social reality,[49] as a constantly self-renewing product of the interaction between institutional processes and legitimating symbolic worlds of meaning.[50] Precisely because they are permanently open or available, so to speak, as instruments for use, traditions have the capacity to integrate different, even antagonistic temporal horizons of the past into different horizons of meaning within the present. In this context it is pointless to question the age or authenticity of traditions; for what possible truth criteria could be invoked or applied, apart from those which similarly claim an opaque transhistorical validity? To apply such criteria would be to engage in an endlessly circular argument.

Everyday life on the mission tacitly follows rules and courses of action which elucidate the interplay between "tradition" and "modernity"—not as something exotic, but as ordinary and familiar elements of everyday life. Relating as it does to history and tradition, to stability and change, kinship serves best to represent forms and processes on an everyday level. The normative and meaning-generating "functions" of the kinship system guarantee a stability which prescribes and organizes the framework of daily life, both at a conceptual level and as a guide for action.[51] Kinship is a dominant component of social life and thus of the construction of the social universe, including life on the Mission—a *Dominantsetzung*, as we might say in German, a thing which has a prevailing force in the "fixing" or "setting" of social reality.[52] It is seen as having a permanent validity, as something eternally present—"from the start," "from always," as people might say. Consequently there is a continual investment in maintaining and "living" kinship as a constant, "intact" reality. Kinship belongs to what Berger and Luckmann refer to as "'first-order' objectivations" of everyday life[53]; it is quasi-institutionalized, objectively available and subjectively plausible. Since, on the Mission, kinship is valued as the *prime* example of the continuity of tradition, any change in any other sphere of life can be seen as secondary or of less relevance.[54] This does not necessarily give rise to contradictions—integrating change means absorbing it into the framework of an existing order without dissolving this order. Or to put it another way: as long as change is comprehended in terms of metaphors and symbols which stand for eternity and immutability, a given order can be maintained as eternal and immutable despite the changes which may objectively have occurred.[55]

In stating their grounds for taking over the Mission site, the Mission people make explicit reference to traditions that are at once stable and dynamic. "Spatial traditions" relating both to the homelands and to the Mission land are anchored and reflected in people's "culture of remembrance," that is, a culture always striving for "the whole" by being selective.[56] Not unlike quotations, traditions are tied to particular time frames and to particular frames of relevance; accordingly they are subject to particular criteria of changeability which are determined by the frames in which they are used and from which they gain their meaning and potential authority.[57] Nevertheless, in the process of legitimating the relationships with the Mission land, the various "spatial traditions" became intertwined in a "meandering" attempt to bring together meanings of disparate orders.[58]

The Mission people's explanations are based neither on the "society-culture-split" nor on "invented traditions,"[59] but on the accentuating and emphasizing of rules and orders which are very much part of the "traditional canon" but are now subject to different inflexions and different valorizations.[60] For them, their rights to the Mission land are guaranteed by a traditionally polychronic history of events (*eine traditionell polychrone Ereignisgeschichte*) which refers to the present, points into the future, and is tied to the land. These event-related (and thus historical) derivations are safeguarded and covered by traditions; they are indeed components of traditions. This is possible because traditions facilitate an interpretation of the world which is always new, always aimed at generating stability, and thus at taking "reality" outside the field of contestation.

It is well known that living traditions have the capacity to absorb change and to rework it into tradition. In the context of both a codified legal doctrine and the local everyday practice, this fundamental feature—its ability to be both changeable and stable—is certainly problematic. However, the notion of tradition is necessary for the construction of legitimating symbolic worlds of meaning within time. Traditions are the means to constantly deal with "the simultaneity of the non-simultaneous"[61] and they therefore differ fundamentally from the ideological petrifaction and to a certain extent monochronic positioning of absolutes found in traditionalism.[62]

Let me return to my starting point: Arendt's use of the "Benjaminian formula" and the multi-layered questions it raises concerning the historicity, authority, authenticity and citability of both, traditions and quotations. One possible approach to coming to terms with these questions might be a more systemic conceptualization of tradition such as the one employed by the poet T. S. Eliot—a contemporary of Benjamin and Arendt—who perceived tradition as a sort of mega-memory and who was also well aware of its selective qualities.[63] According to Aleida Assmann, Eliot saw both tradition and memory as based on a concept of "wholeness,"[64] however, this wholeness does not beseech an *apokatastasis* (lit. a restitution or restoration), no anything and everything, no total recall. Memory as well as tradition are completely quotable but with the important reservation that every present quotes them in a new and different way, or in Eliot's own words, "the *whole* existing order must be, if ever so slightly, altered."[65]

Notes

1 This chapter is a heavily modified version of "'No Matter That Modern World'—An Aboriginal Approach to a Polychronic Past: A Case Study From Northern Australia," in *Tradition and Agency: Tracing Cultural Continuity and Invention*, eds. Ton Otto and Poul Pedersen (Aarhus: Aarhus University Press, 2005), 267–290. An abridged version was published as "Knowing Traditions, Dealing with History? On Concepts, Strategies and Practices," in *The Power of Knowledge, the Resonance of Tradition*, eds. Luke Taylor et al. (Canberra: Aboriginal Studies Press, 2005), 199–213.

My exposition is based on the results of various periods of research in Northern Australia conducted between 1989 and 2001. I gratefully acknowledge two research grants (1989–1991 and 1992) by the Australian Institute of Aboriginal and Torres Strait Islander Studies, Canberra, as well as funding and travel grants by the Johann Wolfgang Goethe-University, Frankfurt am Main, (1989, 1992–1996), by the "Hochschulsonderprogramm III des Landes Hessen" (1997–2000) and, between 2000 and 2002, by the Deutsche Forschungsgemeinschaft and the Kulturwissenschaftliche Institut, Essen.

My thanks to John von Sturmer to whom I owe particular gratitude, and also to Maggie Brady, Mary Edmunds, Diana Olsberg, Bob Tonkinson, Ton Otto and Poul Pedersen for their helpful comments on earlier drafts.

2 Hannah Arendt, "Walter Benjamin, 1892–1940," in Arendt, *Men in Dark Times* (San Diego: Harcourt Brace and Company, 1983), 193.

3 Walter Benjamin, *Das Passagen-Werk,* in *Gesammelte Schriften,* vol. 5 no. 1, ed. Rolf Tiedemann (Frankfurt a/M: Suhrkamp, 1991), 609. See also Benjamin, *Ursprung des deutschen Trauerspiels,* in *Gesammelte Schriften,* vol. 1, no. 1, eds. Rolf Tiedemann and Hermann Schweppenhäuser (Frankfurt a/M: Suhrkamp, 1991), 207–238; and Benjamin, *Einbahnstrasse,* in *Gesammelte Schriften,* vol., 4 no. 1, ed. Tillman Rexroth (Frankfurt a/M: Suhrkamp, 1991), 138.

4 Arendt "Walter Benjamin, 1892–1940,"193; cf. Benjamin, *Illuminationen: Ausgewählte Schriften* 1, ed. Siegfried Unseld, revised edition (Frankfurt a/M: Suhrkamp, 1977), 136–139, 251–263, 380.

5 Martin Scharfe, "Erinnern und Vergessen: Zu einigen Prinzipien der Konstruktion von Kultur," in *Erinnern und Vergessen: Vorträge des 27. Deutschen Volkskundekongresses Göttingen 1989*, Brigitte Bönisch-Brednich, Rolf W. Brednich, and Helge Gerndt, eds. Beiträge zur Volkskunde in Niedersachsen 5. Schriftenreihe der Volkskundlichen Kommission für Niedersachsen e.V. 6 (Göttingen: Schmerse, 1991), 41; Siegfried Wiedenhofer, "Erinnerte Tradition und tradierte Erinnerung in Humanismus und Reformation," in *Mnemosyne: Formen und Funktio-*

nen der kulturellen Erinnerung, eds. Aleida Assmann and Dietrich Harth (Frankfurt a/M: Fischer, 1993), 309.

[6] In her seminal work on time and tradition, Aleida Assmann, a literary scholar, remarked that the common usage of the word *tradition*—being part not only of our everyday life but mainly of what in German is referred to as *Sonntagssprache*, or "Sunday language"—acts as a guarantee: Nobody using the term runs the risk of having to define it. This statement certainly carries some truth despite the growing academic and public attention paid to the categorical, conceptual, political and/or legal dimensions of traditions—whether they are perceived as "invented" or not.—A. Assmann, *Zeit und Tradition: Kulturelle Strategien der Dauer* (Cologne: Böhlau, 1999), 63. See also Eric Hobsbawm, "Introduction: Inventing Traditions" in *The Invention of Tradition*, eds. Eric Hobsbawm and Terence Ranger (Cambridge: Cambridge University Press, 1984), 3; and Alain Babadzan, "Anthropology, Nationalism and 'the Invention of Tradition'," *Anthropological Forum,* vol. 10, no. 2 (2000): 134, 151. Like most people using the word tradition, Arendt's above mentioned formula does not provide a definition, instead, she states and describes. In other words, Arendt relies on what Assmann calls a "classifying concept of tradition" (63) by constructing a line or connection in order to describe "facts" as they appear to the retrospective gaze of the observer. This notion of tradition develops solely in the eye of the observer and does not approach questions of intentionality and relation between the single elements of the line. For example, whoever proves the existence of a certain traditional line of thought or motif, let us say a line connecting Augustine, Pascal, Husserl and Derrida, is in no need of affirming that the authors themselves sought an intentional relationship with each other. To talk about traditions in this classifying sense simply means to make a *post hoc* statement, a statement made once, a continuity of motifs, ideas and *topoi* that have been (or has believed to have been) detected (ibid.). In contrast to what Assmann (ibid.) calls an "emphatic concept of tradition" (to which I shall return in my case study), the impact of a classifying concept of tradition is *per se* of limited range and of a non-obligatory or non-compulsory nature, it stands for an retrospective observation and leaves the normative, the binding and the future-related aspects out of sight. This non-obligatory and non-compulsory nature of a classification or "ordering"—no matter how arbitrary or unnecessary it may appear to be—can, however, become just the opposite when fixed as "traditionalism."—See Britta Duelke, "*...Same but Different...*" – *Vom Umgang mit Vergangenheit: Tradition und Geschichte im Alltag einer nordaustralischen Aborigines-Kommune*, Studien zur Kulturkunde 108 (Cologne: Köppe, 1998), 232–235; 243–249; and Babadzan 142ff.

[7] Gaynor Macdonald, "'Recognition and Justice': The Traditional/Historical Contradiction in New South Wales," in *Fighting Over Country: Anthropological Perspectives*, eds. Diane Smith and Julie Finlayson. Research

Monographs 12 (Canberra: Australian National University, Center for Aboriginal Economic Policy Research, 1997), 65–82; Gillian Cowlishaw, "Did the Earth Move For You? The Anti-Mabo Debate," *The Australian Journal of Anthropology* vol. 6, no. 1–2 (1995): 43–63; Mary Edmunds, *Conflict in Native Title Claims*, Issues Paper No. 7 (Canberra: Australian Institute of Aboriginal and Torres Strait Islander Studies / Native Title Research Unit, 1995).

8 *Native Title* (Cth) *Act*, 1993; *Aboriginal Land Rights* (Northern Territory) *Act*, 1976. Reprint no. 2, reprinted 31 July 1989 (Canberra: Australian Government Publishing Service, 1989), in the following *A.L.R.A.*

9 "'[T]raditional Aboriginal owners,' in relation to land, means a local descent group of Aboriginals who (a) have common spiritual affiliations to a site on the land, being affiliations that place the group under a primary spiritual responsibility for that site and for the land; and (b) are entitled by Aboriginal tradition to forage as a right over that land."—*A.L.R.A.* § 3 [1].

10 Ronald M. Berndt, "Traditional Concepts of Aboriginal Land," in *Aboriginal Sites, Rights and Resource Development: Proceedings of the Fifth Academy of the Social Sciences in Australia Symposium*, Ronald M. Berndt, ed. (Perth: University of Western Australia Press, 1982), 7.

11 Edward Shils, *Tradition* (Chicago: University of Chicago Press, 1981); Karl Mannheim, *Strukturen des Denkens*, eds. David Kettler, Volker Meja, and Nico Stehr (Frankfurt a/M: Suhrkamp, 1980) 101ff.; Karl Mannheim, *Konservatismus: Ein Beitrag zur Soziologie des Wissens* [1927], eds. David Kettler, Volker Meja, and Nico Stehr (Frankfurt a/M: Suhrkamp, 1984), 92ff.

12 This traditional view of Aboriginal Australia still has a certain currency even if the perspectives and approaches have shifted in other respects—see, for example, Erich Kolig, "A Sense of History and the Reconstitution of Cosmology in Australian Aboriginal Society: The Case of Myth Versus History," *Anthropos* vol. 90, no. 1–3 (1995): 50–59; and Tony Swain, *A Place for Strangers: Towards a History of Australian Aboriginal Being* (Cambridge: Cambridge University Press, 1993). Burke and Ricoeur among others, have, in another context, pointed to the general epistemological and methodological weaknesses of this thesis—see Peter Burke, "Historians, Anthropologists, and Symbols," in *Culture Through Time: Anthropological Approaches*, ed. Emiko Ohnuki-Tierney (Stanford: Stanford University Press, 1990), 268–83; and Paul Ricoeur, "Memory-Forgetfulness-History," *Zentrum für interdisziplinäre Forschung der Universität Bielefeld-Mitteilungen* vol. 2 (1995): 3ff, and Ricoeur, *Zeit und Erzählung, 1: Zeit und Historische Erzählung*, trans. Rainer Rochlitz (Munich: Fink, 1988), 125–35, passim. Gellner concisely summarizes his critique as follows: "But how on earth can one say, almost in the same breath, that one does not know anything about the past of an illiterate tribal society (there being no records), *and* that one

knows it to have been the *same* in the past as it is in the present?" Ernest Gellner, *The Concept of Kinship and Other Essays on Anthropological Method and Explanation,* new edition (Oxford: Basil Blackwell, 1987), 115.

13 Cf. Aboriginal Land Commissioner, *Finniss River Land Claim*, Report by the Aboriginal Land Commissioner, Mr. Justice Toohey, to the Minister for Aboriginal Affairs and to the Administrator of the Northern Territory (Canberra: Australian Government Publishing Service, 1981) 20ff; and Francesca Merlan, "Entitlement and Need: Concepts Underlying and in Land Rights and Native Title Acts," in *Claims to Knowledge, Claims to Country: Native Title, Native Title Claims and the Role of the Anthropologist*, ed. Mary Edmunds (Canberra: Aboriginal Studies Press, 1994) 20ff. *Aboriginal tradition* is a defined term within the Land Rights Act meaning, "the body of traditions, observances, customs and beliefs of Aboriginals or of a community or group of Aboriginals, and includes those traditions, observances, customs and beliefs as applied in relation to particular persons, sites, areas of land, things or relationships." (*A.L.R.A. 1976*: § 3 [1].) As Merlan—"Entitlement and Need," 20—points out: "In and of itself, the [Land Rights Act's] definition does not seem to demand an anchoring in pre-colonial or pristine forms of Aboriginality." Consequently there have been numerous cases in land claim hearings in which the Land Commissioners were willing to accept that Aboriginal tradition was neither static nor immutable—see Aboriginal Land Commissioner *Finniss River Land Claim,* 22. However, the experiences gained from land claim processes have demonstrated that "underlyingly, even the most liberal acceptance of change in tradition cannot get away from the fact that the notion still, in the context of land claims, has the force of requirements of distinctiveness and of congruousness in the modes that experts have come to expect of Aboriginal land tenure systems."—Merlan "Entitlement and Need," 20.

14 Francesca Merlan, "The Limits of Cultural Constructionism: The Case of Coronation Hill," *Oceania* vol. 61, no. 4 (1991): 341–52; Merlan, "The Regimentation of Customary Practice: From Northern Territory Land Claims to Mabo," *The Australian Journal of Anthropology* vol. 6, no. 1–2 (1995): 64–82.

15 Duelke, *Same But Different*, 56–80.

16 Alfred Schütz and Thomas Luckmann, *Strukturen der Lebenswelt*, vol. I (Frankfurt a/M: Suhrkamp, 1979) 30ff, 50, 60, 133ff, pt. 4; vol. II (1984) 44–50.

17 Diane Smith, "'That Register Business': The Role of the Land Councils in Determining Traditional Aboriginal Owners," in *Aboriginal Landowners: Contemporary Issues in the Determination of Traditional Aboriginal Land Ownership*, ed. L.R. Hiatt, Oceania Monograph 27 (Sydney: University of Sydney Press, 1984), 84–103.

[18] Merlan, "The Limits of Cultural Constructionism"; Robert Tonkinson, "Anthropology and Aboriginal Tradition: The Hindmarsh Island Affair and the Politics of Interpretation," *Oceania* vol. 68, no. 1 (1997): 1–26.

[19] Rolf Gerritsen, "Thoughts on Camelot: From Herodians and Zealots to the Contemporary Politics of Remote Aboriginal Settlement in the Northern Territory," paper presented to the Australasian Political Studies Association 23rd Annual Conference: Canberra, 28–30 August 1981.

[20] Nicholas Thomas, "The Inversion of Tradition," *American Ethnologist* vol. 19, no. 2 (1992): 213–32.

[21] Thomas Luckmann, "Grundformen der gesellschaftlichen Vermittlung des Wissens: Kommunikative Gattungen," in *Kultur und Gesellschaft, eds.* F. Neidhardt, M.R. Lepsius, and J. Weiß, Kölner Zeitschrift für Soziologie und Sozialpsychologie, Special edition 27 (Opladen: Westdeutscher Verlag, 1986), 200.

[22] Janet Wall Hendricks, "Power and Knowledge: Discourse and Ideological Transformations among the Shua," *American Ethnologist* vol. 15, no. 2 (1988): 216–38, esp. 235; for an opposite view, see Brian D. Haley and Larry R. Wilcoxon, "Anthropology and the Making of Cumash Tradition," *Current Anthropology* vol. 38, no. 5 (1997): 761–94.

[23] "Schedule 1" of the Land Rights Act listed the areas of land (mainly reserve land) to be transferred to Aboriginal Land Trusts. The migrants' homelands which were part of a scheduled reserve were therefore automatically transferred, that is, the migrants became Traditional Aboriginal Landowners under the Act without having to go through a land claim process.

[24] Most of the Original Owners continue to live outside the Mission. In 1990, for example, only five adults from this group resided on the Mission itself while many others preferred to live in their own little community located a few kilometers away.

[25] Outstations are decentralized satellite-communities.

[26] In Aboriginal English the standardized term "Law" (spelled with a capital "L"), also commonly called "Dreamtime Law," refers to the rules and orders derived from the mythical past.

[27] Reinhart Koselleck's conceptual framework is based on the fundamental polarity between "experiential space" (*Erfahrungsraum*) and "expectation horizon" (*Erwartungshorizont*)—Koselleck, *Vergangene Zukunft: Zur Semantik geschichtlicher Zeiten*, 2nd edition (Frankfurt a/M: Suhrkamp, 1992), 349–75. As Ricoeur points out, "'Experiential space' implies *the totality* of what is inherited from the past, its sedimentary traces constituting as it were the soil in which desires, fears, predictions and projects take root—in short, every kind of *anticipation* which projects us *into the future*. But there is no experiential space except one, diametrically opposed to an expectation horizon, and the expectation horizon is irreducible to the experiential space: the dialectic between these

two poles ensures *the dynamic nature of historical consciousness.*"—Ricoeur "Memory-Forgetfulness-History," 3, my emphasis added.

28 Michel de Certeau has written extensively about the more general meaning and importance of events for notions of space, spatial syntax and spatial syntactics. His reflections—although mainly referring to western urban contexts—offer some interesting parallels to and variations on the above mentioned notions—see Michel de Certeau, *Kunst des Handelns*, Ronald Vouille, trans. (Berlin: Merve, 1988), 215ff.; cf. Walter Benjamin, *Das Passagen-Werk*, in *Gesammelte Schriften* vol. 5, no. 1, Rolf Tiedemann ed. (Frankfurt a/M: Suhrkamp, 1991), 133–78.

29 A so-called conception site usually refers to a site at which a person's mother becomes aware of her pregnancy. Conception affiliations can create links between a person and a particular site. Ideally this site is within the territory a person already holds rights in, notably, genealogically-derived rights. However, if the conception site is not located in an area of already established rights (and this happens quite frequently), conception affiliations may provide the means for developing rights in sites in somebody else's country. The fact that a person was conceived at a particular site, however, is not necessarily a prerequisite for integrating this person into the "local descent group" as defined in the Act—see Graeme Neate, *Aboriginal Land Rights Law in the Northern Territory* (Chippendale, N.S.W.: Alternative Publishing Co-Operative, 1989), 61ff.

30 Nicolas Peterson, "Rights, Residence and Process in Australian Territorial Organization," in *Aborigines, Land and Land Rights,* eds. Nicolas Peterson and Marcia Langton (Canberra: Australian Institute of Aboriginal Studies, 1983), 137ff.

31 This was a timely move on the part of the Church since the threat that the Mission land might be included in the land claim was eliminated. Moreover, the transfer of the title was also an effective way of getting rid of some of the Mission's notorious financial difficulties, since it has become increasingly difficult to acquire government funding for Aboriginal communities which are not located on Aboriginal Land. The "generous" act of the Church was also meant as a tacit statement that the Church supported Aboriginal autonomy and *indigenous* land ownership. Yet, the original offer stated that the Church would retain ownership and control of certain valuable Church buildings, including the hospital and the school—the two central communal institutions of the Mission—and would thus retain some considerable influence and power—cf. John R. von Sturmer, "The Politics of Residence," in *Aborigines and Uranium: Consolidated Report on the Social Impact of Uranium Mining on the Aborigines of the Northern Territory*, Australian Institute of Aboriginal Studies, ed. (Canberra: Australian Institute of Aboriginal Studies, 1984), 109. Originally the plan for the Mission Land Trust was designed for the migrants who represented the majority of the Mission people. The few

Original Owners also residing on the Mission were not to be included, but see above endnote 24 and below endnote 38.

32 For example, the notion of "descent group" as defined in the Act was subject to discussion, and distinctions between people with primary and secondary rights to the Mission land appeared to resemble those made in published accounts of land claim hearings (mainly from Central Australia) in which primary and secondary spiritual rights to land were discussed. Here, however, primary and secondary rights did not refer to the complex and complementary ritual relationships between agnatic and uterine kinsmen, or patrilineal or matrilineal affiliations to land which are not customary in this region, but to whether the people in question were among the founding members of the Mission (primary), or among those who moved in later (secondary)—see L. R. Hiatt, "Land Tenure and Contemporary Land Claims," in *Aboriginal Landowners: Contemporary Issues in the Determination of Traditional Aboriginal Land Ownership*, ed. L.R. Hiatt, Oceania Monograph, vol. 27 (Sydney: University of Sydney Press, 1984), 19ff.

33 Hans Blumenberg, *Arbeit am Mythos* [1979], 5th edition (Frankfurt a/M: Suhrkamp, 1996), 165–93.

34 Nancy Munn, "The Transformation of Subjects into Objects in Warlpiri and Pitjantjatjara Myth," in *Australian Aboriginal Anthropology*, ed. Ronald M. Berndt (Nedlands: University of Western Australia Press for the Australian Institute of Aboriginal Studies, 1970), 144.

35 Hans-Georg Gadamer, *Wahrheit und Methode: Grundzüge einer philosophischen Hermeneutik* [1960], *Gesammelte Werke* 1, 5th Revised Edition (Tübingen: Mohr, 1986), 299. In this context the German word *Sinn* comprises the English "sense," "reason," and "meaning."

36 On the problematic aspects of using the "circle metaphor" see, for example, Martin Heidegger, *Sein und Zeit* [1927] (Tübingen: Niemeyer, 1993), 152ff., 314ff.

37 Britta Duelke, "Allochronien? Von kulturellen Praxen im Umgang mit Dauer und Veränderung," in *Zeit deuten: Perspektiven–Epochen–Paradigmen*, ed. Jörn Rüsen (Bielefeld: Transcript, 2003), 186ff.

38 From the perspective of the Original Owners these entitlements were certainly limited to rights of residence, not ownership. In October 1996 the legal title of the Mission land was handed over to a land trust of twenty-seven trustees, two of whom belonged to the group of the Original Owners. The Mission people had had intensive discussions amongst themselves and with the Original Owners which lasted up until the very day the title was due to be transferred. They decided to pursue the title for a twelve-year lease—and not the freehold title as originally planned. By postponing the final decision regarding the freehold title, the Mission people and the Original Owners hoped to gain more time in order to seek satisfactory alternatives.

[39] The Mission people are well known for their traditional arts and crafts as well as for their expertise in traditional medicine and bush skills. The local artists frequently reflect on items that are now produced and sold as *traditional* arts and crafts, and they are well aware that (regardless of the quality of a given work of art) many of their customers prefer to buy items with a traditional "flavor": in their experience "whitefellas often only buy paintings when they got a story and a Dreaming." Dilly bags, for example, have always been part of the regional material culture. Originally the string bags were straw-colored. However, sales have increased since the fiber has been colored with "traditional," that is, natural (!) dyes. Equally successful now are the sales of the "traditional / Aboriginal coiled baskets" although they were previously unknown in the area. The artists are eager to explain that during the 1950s and 1960s the inmates of a regional leprosarium learned the techniques from patients coming from Arnhem Land. They brought the knowledge with them when they returned to the Mission.

[40] Duelke, *Same But Different*, 114–40.

[41] Jan Assmann, *Das kulturelle Gedächtnis: Schrift, Erinnerung und politische Identität in frühen Hochkulturen* (Munich: Beck, 1992), 16.

[42] Peter L. Berger and Thomas Luckmann, *The Social Construction of Reality: A Treatise in the Sociology of Knowledge* [1966] (Harmondsworth: Penguin, 1985), 110ff.

[43] Aboriginal Land Commissioner, *Finniss River Land Claim* 22; Basil Sansom, "Aborigines, Anthropologists and Leviathan," in *Indigenous Peoples and the Nation-State: "Fourth World" Politics in Canada, Australia and Norway*, ed. Noel Dyck, Social and Economic Papers 14, Institute of Social and Economic Research (St. John's: Memorial University of Newfoundland, 1985), 81; Maurice Halbwachs, *Das Gedächtnis und seine sozialen Bedingungen*, trans. Lutz Geldsetze (Frankfurt a/M: Suhrkamp, 1985), 368.

[44] Alfred Schütz and Thomas Luckmann, *Strukturen der Lebenswelt* I, 25ff.

[45] Christina Toren, "Making the Present, Revealing the Past: The Mutability and Continuity of Tradition as a Process," *Man*, new series 23 (1988): 712ff.

[46] Jonathan Friedman, "The Past in the Future: History and the Politics of Identity," *American Anthropologist* vol. 94, no. 4 (1992): 837.

[47] Ricoeur, "Memory-Forgetfulness-History," 3.

[48] Ricoeur, ibid.; cf. Koselleck, *Vergangene Zukunft* 33ff., 349–59.

[49] Berger and Luckmann, *The Social Construction of Reality*, 208.

[50] Sansom, "Aborigines, Anthropologists and Leviathan," 91ff.

[51] Elsewhere, in a comparison of the Mission with the Home Community, I have shown that the normative and meaning-generating "functions" of the kinship system equally open the door to the possibility of both order and disorder; in other words, it opens up a multiplicity of orders, each

depending on the context, not all of which are neatly or "stably" articulated—see Duelke, *Same but Different*, 130–40, 248ff.

[52] Aleida Assmann, *Fluchten aus der Geschichte: Die Wiedererfindung von Tradition vom 18. bis zum 20. Jahrhundert*, Report vol. 3, no. 94 der Forschungsgruppe "Historische Sinnbildung. Interdisziplinäre Untersuchungen zur Struktur, Logik und Funktion des Geschichtsbewußtseins im interkulturellen Vergleich" (Bielefeld: Zentrum für interdisziplinäre Forschung der Universität Bielefeld, 1994), 18.

[53] Berger and Luckmann, *The Social Construction of Reality,* 110ff.

[54] Aleida Assmann, *Fluchten aus der Geschichte,* 16ff.

[55] Greg Dening, *Islands and Beaches: Discourse on a Silent Land, Marquesas 1774–1880* (Honolulu: University Press of Hawaii, 1980), 63.

[56] It goes without saying that a "culture of remembrance" is inevitably linked to an equally selective "culture of forgetting," the latter being an universal phenomenon which, in a different context, J. A. Barnes once referred to as "structural [!] amnesia"—Barnes, "The Collection of Genealogies," *Rhodes-Livingstone Journal* 5 (1947): 49, 52. However, Barnes also made clear that it is the ascription of social importance and contextual relevance which is decisive when selections are made, i.e., when it comes to determine which parts of the cultural "repertoire" are given priority to either remain in, to enter or to vanish from "collective memory"—see Barnes 52f. For further North Australian examples, see Basil Sansom, "Aborigines, Anthropologists and Leviathan," 81ff.; for more detailed theoretical considerations see Duelke, *Same but Different*, 232–236; J. Assmann, *Das kulturelle Gedächtnis* 29ff.; Martin Scharfe, "Erinnern und Vergessen," 19–46; see also Maurice Halbwachs, *Das Gedächtnis und seine sozialen Bedingungen* 163–201, 368.

[57] Jan Assmann, *Das kulturelle Gedächtnis*, 29; Halbwachs, *Das Gedächtnis und seine sozialen Bedingungen*, 163–201.

[58] Berger and Luckmann, *The Social Construction of Reality*, 110.

[59] Merlan, "The Limits of Cultural Constructionism," 351.

[60] Nicholas Thomas, "The Inversion of Tradition," 214.

[61] Koselleck, *Vergangene Zukunft,* 125ff, 132.

[62] Duelke, *Same But Different* 232–235, 243–249; see also Babadzan, "Anthropology, Nationalism and 'the Invention of Tradition'," 142ff.

[63] T.S. Eliot, "Tradition and the Individual Talent," in Eliot, *Selected Prose*, John Hayward, ed. (Harmondsworth: Penguin, 1965), 22ff.

[64] Aleida Assmann, *Zeit und Tradition*, 154.

[65] It seems worthwhile to cite the relevant passages in more detail: "Tradition … cannot be inherited, and if you want it you must obtain it by great labour. It involves, in the first place, the historical sense … and the historical sense involves a perception, not only of the pastness of the past, but of its presence… This historical sense … is a sense of the timeless as

well as of the temporal and of the timeless and of the temporal together …

The existing monuments form an ideal order among themselves, which is modified by the introduction of the new (the really new)… The existing order is complete before the new work arrives; for order to persist after the supervention of novelty, the *whole* existing order must be, if ever so slightly, altered; and so the relations, proportions, values of each work … toward the whole are readjusted; and this is [the] conformity between the old and the new."— T. S. Eliot, "Tradition and the Individual Talent," 22ff.

Taking Time

Temporal Representations and Cultural Politics

Richard Terdiman

On the one hand, the exhilarating temporality of revolution. On the other, the baleful "burden of the past." These images identify—at their polar limits—two contrasting projections of time. Of course they are very different. But both are grounded in the materiality of social life. Here I contrast them with a conflicting conception of time that emerges in the projection of Postmodernism. My objective is to suggest what is at stake, theoretically and politically, in contemporary representations of time.

The past's dead hand, and our experience of the inertia with which it weighs us down, clashes with one of the most optimistic fantasies that modernity brings us. Since the Enlightenment we've followed a paradigm of temporality that conceives of time as the medium in which human purposiveness and social action are made possible and become effective. At its extreme this figure of productivity and renewal is the time of revolutions. It is the temporality of Robespierre's resonant appeal to the making of history in his speech on the Revolutionary constitution of 1793: "The time has come to call upon each to realize his own destiny. The progress of human Reason has laid the basis for this great Revolution, and the particular duty of hastening it has fallen to you."[1]

The conflict I'm tracing here takes this form: against hopeful projections of a winged *productivity* of time stands an antagonistic paradigm, the involuntary *reproduction* of history, an arrest of the present in surrender to past contents. In this we recognize the nightmare "tradition of all the dead generations" that in *The Eighteenth Brumaire* Marx evoked to reflect on the constraints on change imposed by the frustrating colonization of "now" by "then." The arrested temporality defined by this blind persistence of the *constituted* reappears in Nietzsche's preoccupation with such disabling historical survivals in "The Use and Abuse of History," and in Freud's theory of neurosis.

It has been one of modernity's tenacious preoccupations. As its emblem, we might think of Baudelaire's image in "Spleen" [II]: "I have more memories than if I'd lived a thousand years"[2]. Such investment of "now" by the proliferating traces of "then" overwhelms our present and tangles our sense of time.

The resonances or reflexes of such representations are pretty clear. Since Flaubert's *Sentimental Education* at least, we've had to deal with the *de*-realization of our most intense personal and political dreams and desires. These are the narratives of modernity. Such plots *plot time* in a characteristically baleful mode. They project a temporality indifferent or positively hostile to human projects—a paradigm of time that elevates to the level of a master representation lurking behind everything Flaubert's perception that irony is the supreme rhetoric of existence. This projection of a temporality that repels politics then becomes the subterranean story of modernity itself. It only deepens in the familiar claims of Postmodernity to dispense with meta-narratives altogether.

Thus the perfusions and distensions of memory converge with representation of modernity's time. In particular, the occupation of the present by the past through our hypertrophic and seemingly unsuppressible recollection complicates and illuminates the model of time on the basis of which we seek to make sense of modernity's paradoxes. This blind persistence of the past is the Enlightenment's bad dream, ceaselessly reproducing the burdensome contents that indenture us "now." The whole dynamic within Modernism that privileges forms that break with tradition and seek to "make it new" collides with the past's own colonization of present experience. This collision threatens to collapse the whole ideology of modernity itself.

Derrida wrote a book about specters. They model this problem. A specter is spectral because it shows up at the wrong time. But this improper temporality of the specter is only a dramatized case of *all* human temporality, which is always underlain and rewritten by other times. Every moment arises not as a liberation from the past's persistence, but as its confirmation, *as the form the past takes now*. Every time thus exhibits the fundamental characteristic of all human time: never single; constitutively multiple and, like the specter, always "inappropriate." Time always has the serried and disorderly character that memory does too; time always contains more than itself. Much more than the present is present in the present. This variegated and multifarious sedimentation, this complex layering of disparate temporalities, is essential to modeling human time in the modern period.

Why can we not simply make temporality fit our desires? It is uncanny that time, though impalpable, so arduously resists us. An old quip has it that "Time is nature's way of preventing everything from happening at once." But why do things *take* time? And what is the time that they take? How do we understand, how do we represent the interval between beginnings and ends, between desire and fulfillment, between intention and result, between action and consequence? What can we say about the refractoriness of time when it is conceived as the form that compresses, deflects, resists, and so often defeats our desire?

Two characteristics are regularly identified with modernity's time. The first is obvious: the *acceleration* that has been a constant element in accounts and critiques of the frenetic quality of modernity since the nineteenth century. The second, on the other hand, is less discussed but probably more significant for my purposes. This is a paradigmatic *metamorphosis* of our image of temporality—specifically, a defensive or apotropaic tendency to transform its representation into the timeless time of synchrony, stasis, or even invisibility. This second characteristic of time's figuration in the contemporary period, the metamorphosis into stasis or spatiality, veils or suppresses what I've been calling the density and distension of time. The obvious way of understanding it is as a variety of reaction-formation, an antidote to the dire finitude of our existence—to the threat that the Greeks figured as Atropos finally snipping our string, and to the very speeding-up of time itself that so defines us in the modern period. Against the dead end of mortality or the anxiety of speeding-up, we take refuge in synchrony, we grind time to a halt.

Imagining time as synchronous is always an available strategy. Many theologies represent time this way, under the aspect of eternity, just as they conceive all of history to be equally present to the perception of divinity. But the *meaning* of such synchrony is not itself transhistorical: it changes in response to the situation of temporality in any moment. Our work is always in dialogue with the ways of seeing and interpreting—with the theories and models—that preoccupy our present. We *construct* our understandings and our meanings—that much is clear. But does this imply that everything is up for grabs, that anyone's meaning carries the same legitimation as anyone else's? Is there a principled basis for resisting or reformulating the relativism that many contemporary social constructivist positions entail? I question such relativism from two related directions.

The theorists of theoretical relativism plausibly maintain that there is no neutral place from which hierarchies of models can be objectively judged. But most arguments about whether your theory is better than mine are badly formed. We need to find the grounds of commensurability that could make a judgment—of range, of effectiveness—possible. This means a much greater attention to specifying the instrumental *means and objectives* of models than is often the case in the theory wars. We need to restore *agency* to the making of our paradigms. If a theory is always *constructed*, that means that models imply *choices*. We cannot treat these critical effects of volition and creation as if they were no more than the incomprehensible result of caprice. A randomizing or arbitrary Saussurian model won't do here at all. For while the attribution of reference to any given phoneme is theoretically unmotivated, this is a disastrous analogy for the way that models of social or cultural understanding arise and function.

To say that no referent can unambiguously determine the theory that accounts for it is *not* the same thing as to say that every theory has equal status in the marketplace or before the law of paradigmatic existence. The reason is that a theory's referent, the object of its modeling work, isn't the only actor in the process. As the product of theoretical labor that works in the way all labor does, a model is constructed within a web of intentions and determinations. The choices that form it are motivated and selective. We need to insist that the acts of choosing that construct the model be brought to light, and that the consequences of this complex of choices be explained. Your right to construct your theory shall not be abridged. But I don't think you have a right to heedless indiscrimination in your choosing. And after your choice is made, we others must ask what might be the range and the strength of the model it determines, and what are its areas of blindness or underdevelopment. Finally, if for social or ethical reasons we believe that seeing the things a given theory blurs is imperative, then we might find a principled and cogent basis for a critique of any model that obscures them.

So let's consider the models of time the tradition makes available to us. A long critique of the Left that describes modernity's obstructed and occluded temporality stretches from Simmel, Lukács and the Frankfurt School to Fredric Jameson today. It projects the experience of contemporary capitalist reality (often figured as Postmodernism) as the determinant of projections of time as dying or already dead. It threads through Jameson's *Postmodernism* book, but he states it nowhere more resonantly than in *The Seeds of Time*:

> We are like people only able to remember their distant pasts, who have lost the whole dimension of the recent and the most familiar. ...This incapacity to imagine change...also stands as the very allegory of the dilemmas we have outlined here: the Identity of a present confronting the immense unthinkable Difference of an impossible future...[3]

This Modernist emptying-out of diachronicity is likewise figured in Freud's uncanny temporalities. The psyche's time folds back on itself, ceaselessly convolutes in neurosis, in *Nachträglichkeit* or "deferred action," in screen memories, and in a number of other labyrinthine structures that defeat the nominal rationality of any straightforward timeline. But at the deepest level of Freud's theory, in the *id* or system *Ucs.*, time simply disappears. As Freud put it bluntly in the *New Introductory Lectures* (1933): "In the id there is nothing corresponding to the idea of time, no recognition of the passage of time and...no alteration of mental processes by the passage of time."[4]

This paradigm ought to seem familiar. Since the Greeks, rationality has been associated with the bracketing of time. Conversely the contingent or the changing have been judged inferior aspects of reality. Parmenides and his student Zeno argued that time and change were illusory, and sought to discredit the ideas of transformation advocated by Heraclitus.[5] Plato and Aristotle agreed in identifying change and movement with inferior ontological states.[6] Augustine described the lamentation of our imprisonment in finitude as essential to Christian consciousness.[7] Boethius defined God's knowledge of the world as a *totum simul*, in which the successive moments of all time are "copresent in a single *perception*."[8] Spinoza wrote that "It is the nature of reason to perceive things under a certain form of eternity."[9] Manifestly we have been flattening the diachronous for a long time. Yet the meaning of our flight from temporality itself needs to be historicized.

This theoretical antipathy to time as protraction and metamorphosis has an intense relation to two other Modernist discourses with which we are familiar: aesthetic formalisms, particularly those from the period of high Modernism; and language-based structuralisms, which continue (whether in convergence or in counter-discursivity) to overdetermine our critical practice today. We pay an enormous cost for the rationalist or the aesthetic satisfactions such paradigms claim to provide.

The temporality of Modernism, of *L'Art pour l'art* and all its formalist progeny, is fundamentally synchronic. A dominant tradition conceives time—with its inevitable dynamics of erosion and decline—as a flaw, as a taint. When poets seek to produce a beautiful object, their project thus entails the task—or at least the fantasy—of withdrawing this object from time itself.

This dynamic becomes radicalized in structuralism. We remain powerfully under its influence. Language still seems the hegemonic paradigm for understanding social entities and meanings. But the projection of linguistic character onto all forms of social practice carries with it a complex of assumptions that bear on the way we see process, change, and the forces that make such change *take time*. What happens when we consider all social processes as acts or facts of language?

On the one hand, language appears to be malleable in unlimited ways. Language performances, everyday reflexes and instantiation of thought's instantaneous lability and scope, seem to arise effortlessly, as if they *were* the real world. In this latter guise, they seem the practical analogue of Idealism's vision: a paradigm that elides time and difficulty.

On the other hand, beginning with Saussure's decision to purge diachrony from his system,[10] language presents itself as a vast structure so intricately and massively architectonic that it appears characterized not so much by stability as by *immobility*. When Saussure insisted that the relationship between signifier and signified was an arbitrary one, he was claiming that no process, hence no *time*, entered its establishment.

This is the temporality of logic. It enforces a crystalline divorce from real-world contingencies like history, desire and irrationality. Crucially, such systems *take no time*. Through their rule-boundedness, logics repel temporality, and structuralist models aggressively repudiate it. So from one angle, language's lability (seen under the aspect of *parole*) appears to make time irrelevant in the direction of instantaneity: by making our practice of utterance seem immediate. From another, time appears to play no role because the overarching system of *langue* appears to exist in a world conceived as eternity.

Western modernity has increasingly seen the world as language. But here is a crucial point. Paradigms based on language have a low aptitude for modeling time in its productivity, and—most crucially—in the perverse inertia with which temporality shows its refractoriness. This inertia is what makes events seem to run on beyond their pertinence or their appropriateness and to create that resistance of the world to change that all social action internalizes as its most obsessive nightmare.

Extraordinary insights have arisen from our languaging of the world. *But language is an unpropitious paradigm with which to model temporality*—at least in its productivity and its density. When we use our seemingly weightless words to figure the world's inertia, a fundamental mismatch occurs. "Language is compliant, things are resistant."[11] We need to struggle against this disequilibrium if we seek to draw the inertia of the world into our discourse, and express the refractoriness of temporality through the gossamer lability of words.

"History is what hurts..." In one of his most memorable formulas, Fredric Jameson expressed a version of our problem about time.[12] To restore time's density to our account of culture, to rectify modernity's partiality to synchronism, we need to be able to model *why* and *how* the world constrains the seemingly uncontainable lability of our discourse about it. For it is the world's *resistance* to us—and hence to our freedom to capture it in language—that we experience in time's intractability. Between our *saying* and our *doing* an irreducible distance opens.

The mismatch between the malleability of language and the obstinacy of the world that language seeks to represent then becomes critical. The traditional way of construing this incommensurability is as a paradigmatic contrast between Idealism and Materialism. The terms of this polarity have become unfashionable in the contemporary period. In particular, Postmodernism seeks to blow the problem away. I want to suggest that one reason this has happened is to obscure the difficulty language has in seizing the *constraint* that we experience in the irreducibility of time.

Since Nietzsche a massive cultural effort has been expended to resist the theses concerning the character of time that were promulgated by Hegel. Such an effort has continued up to our own period. But from the perspective of a model of temporality able to figure time in its rich and irreducible *density*, the contention between Idealism and Materialism becomes relevant again. In terms of this debate, Materialism's most fundamental intuition may turn out to be not about matter at all, but about *time*. Consider the imagery by which the materialist position itself was figured.

Here is a version from just the moment when Marx was trying to clarify his own position in relation to Hegel: "The 'mind,'" Marx wrote in a sentence dripping with irony, "is from the outset afflicted with the curse of being 'burdened' with matter..."[13] The tone here is in keeping with the sarcasm of the polemic in *German Ideology* against the Hegelians and their romance with the de-corporealized Idea. But if we unpack Marx's sentence, the point is fundamental. Our model of the world needs to be able to theorize the world's *resistance*—to our action, as to our efforts to understand it. *Weight*, *density*, *inertia*—these refractory attributes of the world of concrete experience that *are* matter's burden are then invoked, as if automatically, when we seek to build a paradigm that can make sense of this resistance. Without them our model fails to capture that *taking time* which is our most powerful experience of the *non*-sovereignty of mind and thought, of how the world talks back to language. Consequently Idealist systems are obliged to build in complex epicycles to explain why ideas do not instantaneously self-actualize.

Hegel surely knew the world's and history's resistance to the self-realization of Mind. But he made it difficult for us to see the power of this strand of his thinking clearly. Often in his quasi-Platonic mode he seems to suggest that the laborious process of history is merely epiphenomenal. But Hegel also makes it clear that the model of such deduction (or the complementary model of teleology) are not adequate to figure the *production* of knowledge in time as he conceives it: "The movement of carrying forward the form of its self-knowledge is the labour which [*Geist*] accomplishes as actual History."[14] We need to attend better to this side of Hegel's own dialectic.

What preoccupies me about temporality is time's *laboriousness* itself, the difficulty of temporal passage that *requires* the expenditure of human effort and that, in the absence of such expenditure, reciprocally empties out any concept of time. Time without work is abstract and converges nowhere with our lives. That is why human time is irreversible and directional, and why it is characterized by that viscosity that I have been emphasizing. Against this resistance, work—production—changes things in non-reversible ways.[15] For human beings, time *is* the labor that we do within it. And our history is the history of the effort we exert.

This association of time and work might help to resituate the classic enigma of the relation of theory and practice, about which much has been said and written since Marx put the issue into question in the *Theses on Feuerbach*—notoriously, in the eleventh thesis: "The philosophers have *interpreted* the world in various ways; the point is to *change* it."[16] Ho, hum, you might say. But for those of us in the Anglophone world, the interest of Marx's text has recently been renewed from another, unexpectedly philological, direction that bears on my concern. Gayatri Spivak invites us to look at the Eleventh Thesis again by recalling us to a resonance of the image in its second clause: The German for what has been translated as "The point is..." has quite a different flavor. The original text says, "es kommt darauf an...." What Spivak emphasizes is the *progressive, productive* character of this locution, its ability to evoke a *temporal* (rather than a merely logical) connective. Then, we'd have to re-translate it something like "The moment approaches...." or "It is becoming time..." to change the world.[17] This is the moment at which philosophy converges with politics, when logic jumps out of timelessness and invokes human productivity, history, and practice.

The world of practice is always *contingent*, always *embedded*. It cannot be deduced or unfolded out of considerations of form or first principles. This makes it a messy theoretical matter. Such untidy temporality stresses every human moment. Despite hopeful Enlightenment progress myths, for us today human time has much more the character of a laborious Freudian "working-through" (*Durcharbeitung*) than of any kind of unfettered liberation of the Ideal. We are always weighed-down with the past.

Then it begins to seem clear why the vision of language from the side of Idealism, privileging language's *lability*, its apparently untrammelled capacity to express *any* relation to *any* referent in *any* utterance at *any* time, entails a critical and consequential reductionism. In our own period Freud has often been claimed as the inspiration for the most unrestricted hermeneutic libertinage. But as over and against such linguistically labile conceptions of human activity, let me recall a resonant insistence from Freud's own work concerning the *constraints* that must limit our discourses: "What we call things," Freud wrote in an unmistakable reinforcement of his own Materialism, "are externalities that resist thought."[18]

Ideas and language are not sovereign or absolute. This resistance is why language experiences the pressure of elements of reality outside its own control. And indeed, something outside of language has got to norm our discourses, or we are condemned to float without anchor in a subjectivist or solipsist fantasy. This *resistance* of the extra-discursive world—however inaccessible it has been made to seem, for example in the paradigms of poststructuralist textuality—is what a capable theory of culture must be able to model. That's why I'm bearing down on *time*.

Hegel distills the movement of history into a vast process of negativity, an epic of the formative power of contradiction—finally, into a logic of Death. Mortality has no doubt always been the background and the stimulus for any investigation of time. But Hegel's account of it is remarkably abstract. His evocation of death in a celebrated passage from the *Phenomenology* Preface names the ultimate dread that arises in time's refractoriness and defines its power over us. But he seems to depreciate the affect or the lived experience of such a mortal apprehension:

> The activity of dissolution is the power and work of the Understanding.... But that an accident as such... should attain an existence of its own and a separate freedom—this is the tremendous power [*Macht*] of the negative; it is the energy of thought, of the pure 'I'. Death, if that is what we want to call this non-actuality, is of all things the most dreadful, and to hold fast to what is dead requires the greatest strength.... But the life of *Geist*—or Mind, or Spirit—does not shrink from death or keep itself untouched by devastation, but rather endures it and maintains itself in it.... Spirit is this power only by looking the negative in the face, and sticking with it [*bei ihm verweilt*]. This sticking with the negative is the magical power [*Zauberkraft*] that converts it into being.[19]

Hegel thus presents time's refractoriness as what we might term a *high-theoretical epic*, a conceptual allegory, a contention of vast abstract forces that transcend particulars. So while the drama and the difficulty of time's conflicted progression are abstractly expressed, there is something determinately unsatisfying about his narrative, something that seems not to realize its own potentiality for seizing the density of temporality in phenomenological terms.

No one has captured the stakes in such representations, or the differences dividing them, more revealingly than Georg Lukács did in *The Theory of the Novel* (1914–15). There, Lukács frames the opposition between what I have been describing, respectively, as the static time of the concept and the developmental time of experiential process as an antithesis between the temporality of the epic and that of the novel. And he sees their relation not primarily as logical or paradigmatic, but as *historical* and *determined.* In other words, Lukács's theory internalizes and deploys the very consciousness of temporality that he seeks to expound and to understand through its structures. In turn these structures can help us understand both the pertinence of a preoccupation today with *developmental* time, and the meaning of the long depreciation or eclipse of such temporality.[20]

"The greatest discrepancy between idea and reality is time: the process of time as duration,"[21] Lukács wrote in *Theory of the Novel.* He resonantly identifies the aspect of temporality that concerns me here. But his objective in making this seemingly orthodox Hegelian point is to *extend* the scope of Hegel's notion of abstract development through negation in order to motivate his—Lukács's—own engagement with the canonical narratives of nineteenth-century Europe. For these narratives are the most acute and highly developed form in which the consciousness of time as density, has found expression in the Western tradition. In this sense the theory of the novel *is* the theory of developmental time. Both as narrative representation and as conceptual reflection, such a paradigm arises in the wake of the nineteenth century's twin revolutions and their massive material transformation of the structures of social or economic life, and quintessentially of the experience of temporality itself. Conversely, it would not be an exaggeration to suggest that the depreciation of time as duration, which has been a fundamental characteristic of Modernity and Postmodernity, internalizes the process of social restabilization that liberal societies in the West have undergone in the aftermath of the heroic revolutionary period of early capitalism.

From our perspective an entire historical epoch later than these crucial transitional moments just before World War I, it is striking how much the arrested or frozen temporalities of modernity that I characterized earlier converge with Lukács's characterization of the time of the novel's *other*—the time of the epic. Epic time, Lukács writes, "has as little reality, as little real duration, as time has in drama; men and destinies remain untouched by it.... The heroes do not experience time within the work itself.... What they experience and the way they experience it has the blissful time-removed quality of the world of gods."[22] This analysis presciently captures the tendency to ontologize—one might even say theologize—our figuration of stasis in the contemporary period. Lukács then goes on to anticipate many more recent treatments of this temporality by construing its *spatialization.*

Scarcely a decade later, in his celebrated "Reification" essay (1922), Lukács extended and historicized the point with memorable force. His notion of reification stigmatizes the tendency toward temporal stasis, and opens it up to a socio-political critique. "Time sheds its qualitative, variable, flowing nature; it freezes into an exactly delimited, quantifiable continuum filled with quantifiable 'things'...: in short, it becomes space."[23] In this space you can see the crystalline tableaux of Modernism and even the Postmodern tropings of such representations of temporality.

To conclude, I want to urge that our stakes in models, and in temporal paradigms particularly, are consequential on precisely this level. The way we figure time figures us. If we project changing the givens of the world, necessarily this means de-ontologizing temporality. If you think you are living in a time of stasis or at the end of history, you are probably either a victor satisfied to project the continuation of your advantage, or a victim hopeless about ever rectifying your deprivation. Only if you credit time's resistance can you also conceive its productivity—that is, its mediation of circumstances in which a seemingly frozen state of things can be overcome and transformed. We need our theories, but when they blind us to the most consequential possibilities of our lives, we need first of all—*it is becoming time*—to change them.

Notes

[1] "Sur la Constitution, 10 mai 1793"; *Oeuvres complètes* vol. 9: 495. Quoted from Reinhart Koselleck, *Futures Past: On the Semantics of Historical Time*, Keith Tribe, trans. (Cambridge, Mass.: MIT Press, 1985), 7.

[2] "J'ai plus de souvenirs que si j'avais mille ans." Baudelaire, *Oeuvres complètes*, Claude Pichois ed. (Paris: Gallimard-Pléiade, 1975), 73.

[3] Fredric Jameson, *The Seeds of Time* (New York: Columbia University Press, 1994), 70–71.

[4] Sigmund Freud, *New Introductory Lectures on Psycho-Analysis* (1933),

[5] J.T. Fraser, *Of Time, Passion, and Knowledge*, (New York: G. Braziller, 1975), 16.

[6] Ibid., 18–20.

[7] Paul Ricoeur, *Time and Narrative*, trans. Kathleen Blamey and David Pellauer (Chicago: University of Chicago Press, 1984–99), 3: 265.

[8] Ricoeur, *Time and Narrative* 1: 159–60.

[9] Fraser, 93.

[10] See Timpanaro, and Brøndal quoted in Richard Terdiman, *Present Past: Modernity and the Memory Crisis* (Ithaca, New York: Cornell University Press) 53. For an alternative view, cf. Derek Attridge, "Language as History / History as Language: Saussure and the Romance of Etymology," *Poststructuralism and the Question of History*, eds. Derek Attridge, Geoff Bennington, and Robert Young (Cambridge: Cambridge University Press, 1987), 183–211.

[11] Terdiman, *Present Past*, 279.

[12] Fredric Jameson, *The Political Unconscious: Narrative as a Socially Symbolic Act* (Ithaca, New York: Cornell University Press, 1994), 102. The passage runs: "History is therefore the experience of Necessity, and it is this alone which can forestall its thematization or reification as a mere object of representation or as one master code among many others. Necessity is not in that sense a type of content, but rather the inexorable *form* of events.... Conceived in this sense, History is what hurts, it is what refuses desire and sets inexorable limits to individual as well as collective praxis...."

[13] Marx and Engels, *The German Ideology* (Moscow: Progress, 1976), 49. In the alternative C. J. Arthur translation (New York: International Publishers, 1970), the passage appears on 50–51.

[14] Georg Friedrich Wilhelm Hegel, *Phenomenology of Spirit*, A.V. Miller, trans. (New York: Oxford, 1977), 408, § 803.

[15] See Terdiman, *Present Past*, 54–55.

[16] The text in Marx's notebook reads: "Die Philosophen haben die Welt nur verschieden *interpretiert*; es kommt darauf an sie zu *verändern*." Engels later edited these texts. See *The German Ideology* (Moscow: Progress, 1976), 617, 621.

[17] "Limits and Openings of Marx in Derrida," *Outside in the Teaching Machine* (New York: Routledge, 1993), 97; see also 101, 113.

[18] "Was wir Dinge nennen, sind Reste, die sich der Beurteilung entziehen"; "Project for a Scientific Psychology" (1895; *SE*1: 334). See Terdiman, *Present Past*, 257–258. Perhaps surprisingly, but strikingly, Proust made a parallel point in *Sodome et Gomorrhe II*. There he writes as follows: "Toute action de l'esprit est aisée si elle n'est pas soumise au réel." [The activities of our mind are effortless so long as they are not obliged to answer to the real.] *À la recherche du temps perdu*, ed. Jean-Yves Tadié (Paris: Gallimard-Pléiade, 1988) 3:51.

[19] Hegel, *Phenomenology*, 18–19; translation modified.

[20] Lukács is another unfashionable figure today. Like many other Communists he was implicated at various moments in an apologetics for Stalin's totalitarianism—though we tend to forget how powerfully many of these figures were driven to their acceptance of Stalinism by a mortal fear of Germany in general and, as the Thirties wore on, of Hitler in particular. Lukács recalls this fear on the first page of his 1962 Preface to *Theory of the Novel*, trans. Anna Bostock (Cambridge, Mass.: The MIT Press, 1971).

[21] Lukács, *Theory of the Novel*, 120.

[22] Ibid., 121–122.

[23] "Reification and the Consciousness of the Proletariat," *History and Class Consciousness*, trans. Rodney Livingstone (Cambridge, Mass.: MIT Press, 1971), 90.

Image-Times, Image-Histories, Image-Thinking

Catherine M. Soussloff

> "And what determines vision at any given historical moment is not some deep structure, economic base, or world view, but rather the functioning of a collective assemblage of disparate parts on a single social surface."[1]

Derived from the Latin, i.e., *imago* (f. noun) and *imagines* (pl.), the terminology and the concepts adhering to "image" have both fascinated and perplexed scholars, most of them philosophers, psychoanalysts, and historians of religion, art, and film. To explore these etymologies today, however, seems superfluous since the term has an expanded significance in all media studies and practices, including computer technologies, digital and analog photography and film, television, and video. While the conceptual intricacies associated with the image and its cognates in the Western tradition stretch back to ancient Greece and have led to a wide discrepancy of views regarding both the ethical and aesthetical value of images, the technological background of and for the image has a more recent, and a more securely historical foundation, accounting for the current fascination with the image. While no account of the image in today's rapidly expanding image-world can be definitive, no discussion of the image should be considered adequate without an understanding of the conceptual issues invoked around its name in the past. I begin here with image-technology and the image-world in order to return to the conceptual issues with the present firmly in place.

All of the newer image-technologies commonly referred to today are visual, because the bulk of our experience with them begins with their appearance on a computer, television, monitor, movie screen, or with the aid of a variety of camera imaging apparatuses. Image-technologies construct and serve virtually every aspect of society today: arts, entertainment, education, business, science, and medicine. Images produce and construct knowledge; they have been naturalized into a global culture for generating and monitoring information. Any description of the image or of the discourse that surrounds it today must account for this pervasiveness of visuality—a fact which makes the image-world possible.

The technologies presently associated with the factual omnipresence of visuality in our world can be historically situated in the West beginning with the invention of photography c. 1840.[2] Indeed, as Susan Sontag stated so clearly as early as 1974, the analog technologies of the image—photography and film—have been seen as both congruent and synonymous with the modern era.[3] Sontag thought that the image and historical time, or chronology, converged at the very end of the nineteenth century, after the adolescence of the new photographic technologies. According to her, a new concept of time, and therefore of history, adheres like super glue to the image-technologies of modernity. She wrote: "In the real world something *is* happening and no one knows what is *going* to happen. In the image-world, it *has* happened, and it *will* forever happen in that way."[4] Society's and the individual's relationship to the past, present, and future transformed as a result of the photographic image. This new relationship to the image also meant that clear distinctions had to be made between the image in modernity and the older traditions of visual representation that had formerly constituted the history of art: drawing, painting, sculpture, and architecture.

Although Sontag did not make the point explicitly, the birth of both moving picture film and the X-ray in 1895 account for the time–image match; the thing or state of being she termed the "image-world." The photographic visualization of the inside of the human body began with the invention of the X-ray in 1895, which scholars have understood as an image on the threshold; between the inside and outside of subjectivity, or between interiority and exteriority.[5] However, images could not constitute a world, or a worldview, without both the visualized medical body and the body in motion determined by the companion invention to the X-ray: motion picture technology.

Clearly, then, a comprehensive and ultimately useful theory of the image must have room for the altered relation to time and space the image-world inaugurates. Film theorists have long understood the importance of such a conceptualization for their medium. Gilles Deleuze went so far as to establish a chronology for film history based on the earlier era of "the movement image," and the post-World War II era of "the time-image."[6] However, with the exception of film and video, the traditional histories of the early modern through post-modern pictorial art, including photography, privilege space over time through the trope of the invention of perspective at the beginning of the Renaissance in fifteenth-century Florence.[7] This history understands as central the development of increasingly complicated apparatuses for supporting the representational regime of monocular perspective: from Alberti's "veil" of the early 1400s to the *camera obscura* in seventeenth-century Netherlands, to the zooscope and praxiscope, to the movie camera, and so forth. For many writers on the new visual technologies of today, including computer imaging, this traditional and fundamentally art historical account continues as the foundational one.[8]

Sontag, however, had good reasons to ignore linear perspective, commonly considered as "a method of representing three dimensions on a two-dimensional surface," in her discussion of the image-world.[9] First, the image based on perspective calls for a standpoint that is ideal and usually fixed; in this sense timeless. Sontag's persistent interest in the time-based technologies of film required another interpretation.[10] Second, when the French cultural theorist Roland Barthes, Sontag's influential teacher, sought to define the precise character and attraction of the photographic image in his book, *Camera Lucida*, he did not rely on perspective. Rather, Barthes insisted on the uniqueness of the photographic image by using the terms *studium* and *punctum* to refer, respectively, to the "general" impression the image makes and to the detail that draws us into the interpretation and pleasure of the image.[11] Barthes' approach relied exclusively on how the photograph, rather than the photographic apparatus, affected the viewer. His interpretations of the photograph—whether from mass media sources or family albums—moved from the impact of the single image on the individual to the broader ideological implications insinuated in the image.[12]

Sontag's interpretation of the image in modernity recognized that the discourse on linear perspective beginning in the Renaissance both underpins and indicates visual representations understood as "art." This is not to say that perspective did not change the history of art, for it surely did. Yet, if the image is understood as technology and world in an altered relation to time and space, as it correctly was by Sontag, then linear perspective becomes a historical theory pertaining only to art making. Perspective can now be seen as historical artifact, albeit form-shaping, of the historiography of the media of painting, sculpture, and architecture; the very media supplemented by the new image-world.

Again, the historiographical point is significant here: if an account of the image relies on perspective, as the *doxa* has insisted, the image cannot be said to refer to or to encompass time, particularly historical time. Rather, any such account will be doomed to construct only a spatiality—whether virtual or not—for the image; one without or outside of time. As the recent research on the Frenchman Etienne Marey (1830–1904) and the American Eadweard Muybridge (1830–1904) has shown, the earliest experiments in image animation frequently address temporality through movement as their foremost concern.[13] Thus, visual art genres of an earlier period that rely fundamentally on perspective, such as the ideal landscapes of Claude Lorraine, looked to space rather than to time for their generic shape. The protagonist in Sebald's novel, *Austerlitz*, makes this last point as he gazes at the vista of Greenwich Park, described in the passage as if a painted landscape, with a desire to erase the consciousness of time:

> A clock has always struck me as something ridiculous, a thoroughly mendacious object, perhaps because I have always resisted the power of time out of some internal compulsion which I myself have never understood, keeping myself apart from so-called current events in the hope, as I now think, said Austerlitz, that time will not pass away, has not passed away, that I can turn back and go behind it, and there I shall find everything as it once was, or more precisely I shall find that all moments of time have co-existed simultaneously, in which case none of what history tells us would be true, past events have not yet occurred but are waiting to do so at the moment when we think of them, although that, of course, opens up the bleak prospect of everlasting misery and never ending anguish.[14]

The longing for the erasure of time in the genre of painting known as landscape and evoked in this passage by Sebald, required a narrative that placed the image in a seamless line drawn from the congruence of the invention of perspective and pictorial representation to an ideal reality represented by the classical landscape, to a virtual digital image on a computer screen. This desire for timelessness embedded in the historiography of the image allows the screen image—whether wall fresco, oil on canvas, or light pixels on a monitor—to appear transparent to the reality it represents. That reality remains unchanged, as it *had been*, without the effects of time manifested in it.

An outcome of the image-world is nostalgia for timelessness that increases as the image comes into its own in modernity. Indeed, one could argue that the historical amnesias so omnipresent in the Postmodern period, from the 1950s on, respond directly to the overwhelming imposition of time-based images in culture. Douglas Crimp's argument that the "resurrection of painting" in the Postmodern era came about for ideological reasons pertains here.[15] I would add that when the primary historical support for painting's theory, i.e., its approach to perspective, serves as the basis for a critique of the technological image as an imposition on culture, then it will also be for conservative ideological reasons. Painting and image-technologies are fundamentally different. We cannot find the proliferation of "virtuality" puzzling if we do not base image-technology upon an earlier theoretical model used for the construction of "real space," even if that construction was illusion. Just as W. J. T. Mitchell says: "The idea is to release the study of media from a misplaced emphasis on the material support (as when we call paint, or stone, or words, or numbers by the name of media) and move it toward a description of the social practices that constitute it"; so too the technological image must be released from perspectival space.[16]

Sontag's dystopic view of the image-world in her essay built on Walter Benjamin's "The Work of Art in the Age of Its Technological Reproducibility."[17] This essay of the 1930s has been seminal in all later discussions of image-technologies. Although Benjamin's argument never contrasted reality and photography with the severity found in Sontag, he did lament the passing of the aura of the art of an earlier period and its ritual significance, but with a sense of gain too. Reproducibility, a key aspect of the photographic image in Benjamin's essay, led him to conclusions, some of which hold liberating potential, regarding the moving image in film, not only its characteristic repetition, but also something Sontag termed its "proliferation."[18] For Sontag, however, this fact of visual discourse made the modern photograph and the moving picture ideal tools for the culture of late capitalism:

> A capitalist world requires a culture based on images. It needs to furnish vast amounts of entertainment in order to stimulate buying and anesthetize the injuries of class, race, and sex. And it needs to gather unlimited amounts of information, the better to exploit natural resources, increase productivity, keep order, make war, give jobs to bureaucrats.[19]

Thus, overall, Sontag's criticism of the image-world called for wariness towards the power of the image and its perpetrators, a position taken by many in the twentieth century.[20]

In 1990, Jonathan Crary's book, *Techniques of the Observer: On Vision and Modernity in the Nineteenth Century*, made a major contribution to the understanding of the periodization of new visual technologies by rewriting the story of the birth of photography and cinema. Crary's argument divorced the account, and to a certain extent the theory, of the modern image medium, photography—and by extension, film, video, or digital image—from the longer history of pictorial art. Crary's immediate purpose was "to delineate an observing subject who was both a product of and at the same time constitutive of modernity in the nineteenth century."[21] Somewhat contrary to Benjamin, he argued that changes in technology are determined by changes in the subjectivities of vision in the observer-subject, rather than that technological apparatuses *tout court* determined the subjectivity of the observer or the outcome of art. Crary's approach to the production of knowledge through images relied on the earlier work of Michel Foucault and, like that work, sought to offer an alternative history of the image.[22] As did Sontag, Crary disassociated the account of the image-world from perspective, thereby emphasizing time, but he also offered an account reliant on the body of the viewing subject.

In this latter regard, Crary's approach must be associated with the twentieth-century philosophical movement or school known as phenomenology, which understood the visual arts, particularly painting, as determined by the experience of the body. As Merleau-Ponty so eloquently put it:

> The painter 'takes his body with him,' says Valéry. Indeed we cannot imagine how a *mind* could paint. It is by lending his body to the world that the artist changes the world into paintings. To understand these transubstantiations we must go back to the working, actual body—not the body as a chunk of space or a bundle of functions but that body which is an intertwining of vision and movement.[23]

Techniques of the Observer brings phenomenology to image-technology, thereby allowing both temporality and spatiality to be interactive on the social surface of the image-world through human bodies.

In this view the reality in the image resides in the viewer and the maker, rather than in the image's faithfulness to an exterior object. As art historian David Summers has long argued, such a view upsets an influential approach taken towards the history of Western visual representation proposed by E. H. Gombrich in his famous books *The Story of Art* (1972) and *Art and Illusion* (1960), and it has wide-ranging and deep implications for the culture at large.[24] Gombrich's understanding of visual representation required a method of "making and matching" in which the "thing" made succeeds best when it "matches" the exterior object in nature to which it refers or which it represents. From the perspective of the image-world of today where virtuality, rather than naturalism, predominates, this account may be seen to maintain a regime where representation depends fundamentally on what has been seen, rather than on what is or could be envisioned. Some "thing," usually called an "object," had to have been stopped in time in order for it to be noticed and measured against the natural, actual, or "real" world. Furthermore, this view supports the historiography of Western art, as Gombrich had located it. In Gombrich's day, this theory privileged Western culture over others not possessing or adhering to such a model of realist representation for their visual creations.[25]

Crary brilliantly demonstrated that it is precisely because of the image-world's omnipresence that the image requires a theory of the body, or a body-centric theory, rather than a theory located in the representation of the object. W. J. T. Mitchell extended this position with the statement that image-technology is incapable of addressing the viewer.[26] The object does not speak, on the other hand for illusionism—including the role of perspective—the contrary is true. If, as Mitchell asserted, it is we who address the image and are addressed by it, then we can have an image-world with ethics and principles. Thus, when cultural theorist Kaja Silverman called for "an ethics of the field of vision" in 1996, she could do so only after assessing the implications of inserting the body into the massively proliferating image-technologies of the 1990s.[27] Silverman's field of reference, the psychoanalytic theory of Jacques Lacan, allowed her to think of the Postmodern image through both the "mirror-image" of narcissism and "the scopic domain," e.g., through psyche and vision.

Foucault's re-working of the idea of the theoretical progress of visual representation in *The Order of Things* (*Les Mots et les Choses*) sowed the seeds for Crary's focus on the observer as constructed by his society and the technologies of the image in modernity. Basing his argument on a close analysis of Diego Velasquez's monumental oil painting, *Las Meninas* of 1656, Foucault argued that the so-called "Classical age," epitomized by this painting, was a time of parity and cohesion among the domains of representation, language, science, wealth and value.[28] In a thorough re-investigation of Foucault's claims for *Las Meninas*, art historian Victor Stoichita demonstrated the ways in which the painting predicts a fascination with the qualities of representation later identified with "the image," i.e., the references to "framing," the "mirroring" of subjectivities, the "presentness," of images, the auto-circulation of images.[29] These *image elements*, rather than Foucault's meditation on visual representations that give us a "Classical" synthesis, suggest the stakes involved in seeing, as Crary did, a true rupture between representation and image, between the past and the present, between the Renaissance or Early Modern *mentalité* towards the image and modernity's.[30]

In his study Crary concluded that the new visual analog technologies that were invented in the Modern period do not form a continuum with the past. Moreover, he argued, as had Sontag, that they constitute a history and an area of theory separate from that given for the image in earlier periods. According to Crary, the photographic image was organized around a different visual regime. This regime resulted from the historical situation and psychological position of the observer: his (the pronoun is intentional) "subjective vision."[31] Crary saw at the dawn of modernity a subjectivity in the observer different from that found in earlier times; one that could be termed introspective to the observer rather than mimetic of the object. This modern subjectivity focused on the image and the alterations of time and space it proposed or that extended from it, rather than on the representation.

I have already suggested above, with the example of the animation experiments of Marey and Muybridge at the end of the nineteenth century, scientists embraced the possibilities of visualizing a changed consciousness of time and space. For the remainder of this essay I will be stressing the significance of the innovations made by *image-makers* for our current understanding of the image-world. Importantly, Crary's interpretation of the interaction of the new photographic technologies and a modern subjective vision was already at play in avant-garde visual arts in the first decades of the twentieth century. For example, Dada artists, and others working in a completely metropolitan aesthetic, contributed a combinatory method towards various media, appropriating images from printed matter, photography, and art in order to reassemble them into compositions that spoofed the present and delineated its ideological absurdities.[32] The Constructivist photography of Alexander Rodechenko, such as his image of the writer Osip Brik (1924), succeeded particularly well in uniting the image-technology with an overt political or social critique. Other examples of a modern subjective vision, different from the past, may be found in the portrait photography of Heinrich Kühn. His posed, group portraits of his children in active interaction with each other derive from the milieu of Viennese Expressionism, where the inter-subjectivities of sitters and viewers were stressed.[33]

László Moholy-Nagy's experiments in light, sculpture, and film of the late 1920s provide another prime example of the subjective viewer engaged with the image. His film of his *Light Display Machine* used mirrors, light reflections, and superimpositions of image on image to convey subjectivity through abstractions of light and movement. These demonstrated both an anti-representational emphasis and the possibilities that the new technologies held for an inner "vision," one identifiable with the modernist subjectivity. In his influential book, *Painting, Photography, Film* (1924) Moholy-Nagy stated that a major purpose of the new photographic technologies was: "To identify the ambiguities of present-day optical creation."[34] This stress on ambiguity should be contrasted with the precision demanded by, and elicited from, traditional perspective and its theories. Because subjective, these new kinds of images could only always be contingent. In both his artwork and writing Moholy-Nagy sought to exploit qualities of the image peculiar to black and white moving picture technologies, such as light penetration and saturation, chiaroscuro relationships, movement, and tempo, with the dual purpose of explaining and promoting them as art. So too, the Dada artists had already made serious inroads towards a redefinition of the work of art that allowed for the contingency of the new image-technologies.[35]

Using a larger historiographical compass, we can see that Crary's argument for the new subjectivity of modernity and its technologies opened up the possibility that the later, Postmodern visual technologies, such as digital imaging or non-analog technologies, yield other, quite distinctive subjectivities than their predecessors.[36] Just as photography ushered in modernity, so too, the Postmodern era has often been said to have begun with television in the 1950s.[37] But what characterized the new kind of subject tied to television? No one has yet adequately or systematically presented an account of what this Postmodern subjectivity in regard to the image consists, much as we might agree with Margaret Morse's statement: "Television was a prime shifter of value from one ontological state to another, in the socioeconomic and cultural circulation of material objects and bodies, money, and other symbols by means of images."[38]

An interesting critique of the issue of subjectivity in the narrative of visual technology appears in the documentary film, *Videogramme of a Revolution* (1994), directed by Harun Farouki, about the Romanian rebellion against Ceauşescu. Farouki follows several days of demonstrations in Bucharest Central Square by providing filmed sequences of the events shot by a variety of different camera persons, and from cameras controlled by a variety of individual and institutional filmmakers. He shows many uninterrupted takes of the action from various positions, taken at the same time. There is no true chronology of the action and no identity given beyond the titular source. This minimal narrativizing calls attention to the usually maximal effect of editing and sound. In playing with these subjectivities and the viewers' expectations Farouki reveals how much an understanding of the medium's possibilities determines the reception of images. He warns us about expecting an "uncontrolled" or "unmediated" version of history from this tele-visual technology, one without subjectivity or subjects of revolution per se. As a result of his use of the medium, our belief systems about both the technology and history are called into account here. The director shows that the expectation that photographic technologies have certain truth bearing qualities relates directly to the subjectivities of each maker and the viewer. Farouki argues that the power to interpret the events portrayed is directly proportional to our power to interpret the technology used.

Based on the work now underway in all areas of media studies, it seems reasonable to assert that *image-thinking* emerges as the most striking sign of the Postmodern subject. In some sense this understanding of our era's attitude towards the image must be seen as a continuation of Foucault's emphasis on the visuality present in the monumental oil painting in his study of the ways in which society organized knowledge before the modern period. Here, one could say that Foucault betrayed his own Postmodern historical position in regard to the image, when he used *Las Meninas* as a central shifter in the historicization of representation as knowledge, a point made obvious in Stoicita's work cited above.[39]

Moreover, today, image-thinking stresses a relationship between the image and the contemporary subject that occurs as *subjection to the image*, rather than as a *subjectivity engendered by the image*, such as that found in modernity.[40] The discourse that surrounds the theorization of the Postmodern image clearly betrays a point of view *under subjection*. The growth in sales of computerized games and gaming, and the proliferation of theories on them, serve as excellent indicators of today's image-thinker under subjection.[41] Players assume the identities of pre-determined avatars in order to "play" in a world of images that also have been pre-determined. But there is a whole other world of players who re-write these games by fiddling with the software, thereby subverting the intended subjections within the limits of technology.

An interpretation of the *image-as-subjection* requires a concept of "the spectacle." Here, the historical situation of capitalism's effects on society is tied directly to the image-world in modernity through a concept of "visibility." Such a positioning set the stage, so to speak, for the rupture between representation and the image that Sontag, Crary, and others saw between modernity and earlier historical periods. The political action group known as the Situationists first theorized the spectacle in the 1960s, although many of the thinkers discussed above have also contributed to our thinking about it.[42] In 1967, Guy Debord, the group's best-known spokesman, wrote: "The real movement that abolishes reigning conditions governed society from the moment the bourgeoisie triumphed in the economic sphere, and it did so *visibly* once that victory was translated onto the political plane...And everything that had formerly been absolute became historical."[43] Image-technologies could be understood, then, as an outcome or aftermath of the history of modernity. They had helped international capitalism to prevail and to render its ideology all but invisible, obfuscated, one could say, by the very proliferation of images that stimulate buying and anesthetize injury, as Sontag put it, in the Postmodern image-world. The Situationists argued that the image and its technologies presented capital with the ideal means for the coercion of the masses into a culture of commodities, produced by multinational corporations, at the expense of a truly democratic society. At first the Situationists offered admonishments, while at the same time their political philosophy held out the possibility for a utopian society, free from the image-world. However, by 1977, the date of Debord's reconsideration of this earlier position, it seemed clear to him that the negative aspects of the society of the spectacle had only magnified, becoming what he termed an "integrated spectacle."[44] This kind of spectacle, so the Situationist would say, subjects all consumers of images to a global economy. As with much critical commentary on the image-world, a dark shadow stretches to the far horizon.

This critique of the *image-as-spectacle* emerged at a key moment in the Postmodern period: when political activists could look back from a distance on the earlier technologies and theories of modern image-making. The Situationists undertook their critique of mass media in the golden age of both television and Madison Avenue-style advertising. No critic was more aware of the implications for the image proposed by the Situationists than John Berger, whose *Ways of Seeing*, originally envisioned as a television program, attacked the entire history of art and mass media, from Renaissance painting to Postmodern advertising as the gradual emergence of a society of European capital, which, according to Berger, can only be described as a system of appropriation.[45]

This highly skeptical or dystopic position towards the image-world continues today in the era of digitization and the Internet, usually directed towards the same corporate sponsorship in the newer forms of the image.[46] The architect and critic Paul Virilio brands this way of understanding new media "technological fundamentalism."[47] To experience the reality of being human in a culture that has already made the turn to the visual means, according to Virilio, we need to question the story of the ways in which we receive images, not only how they are transmitted. The circulation history of the digital images of the atrocities at the Abu Ghraib prison in Iraq illustrates the importance of the investigation of digital image reception.[48] These photographs of torture circulated as shared files before their release as photographic representations to the public, but if they were utilized as part of sexualized rituals among the perpetrators, as alleged in court, this cannot be known from the images themselves. In this initial form of circulation, the images clearly served the immediate purposes of the individual perpetrators and their investments in "the practice of violence." Another kind of circulation occurred when these images were printed and shown on television. They were known by the public first through their release to the news media via television on CBS *60 Minutes* on April 28, 2004, then in an article by Seymour Hersh released on April 30, 2004 for the May 10 issue of *The New Yorker*, and finally in the *Washington Post* of May 6 and 21, 2004. As shown on TV they appeared as typical, enframed shots, rather than as televised images. Television "quoted" them as photographs, rather than as digitized images. In a comparative analysis of all of the published photographs of the Abu Ghraib tortures, André Gunthert has convincingly demonstrated that the images also circulated to the print media as traditional photographic prints, and not as digital files.[49] The initial release to the public by CBS was delayed and then carefully monitored by the Department of Defense. This official censorship, which included their enframing, must be considered a key sign of the restricted range of such imagery to circulate publicly, even when deemed newsworthy.[50]

At the same time, in the twenty-first century the spectacle—at once analog and digital, projected and projecting—presents for some the possibility of transformation through the very volume of interactivity taking place over the Internet. In an otherwise scathing critique of contemporary global society based on capital, Michael Hardt's and Antonio Negri's book *Empire* recently proposed some salutary prospects for political transformation—through the images produced in and by a community of "users" formed by the new digital technologies:

> The society of control (Modern-Postmodern) might thus be characterized by an intensification and generalization of the normalizing apparatuses of disciplinarity that internally animate our common and daily practices, but in contrast to discipline, this control extends well outside structured sites of social institutions through flexible and fluctuating networks.[51]

Networks or communities emerge from the interactivity presently characterizing image-technologies and suggest ways subjection may be limited or contested.

Earlier popularizing accounts of the new digital technologies—most of them coming from within this high-tech industry itself—made the same kinds of claims regarding interactivity and freedom, proposed as a utopian and communal possibility by Hardt, Negri and others; the main difference being that such transformations were seen as liberating for the individual rather than for society as a whole. In *Transforming Higher Education: A Vision for Learning in the Twenty-first Century* (1995) by Michael G. Dolence and Donald M. Norris, the authors argued that the university and the teaching infrastructure must adapt to the technological age in order to survive. The authors put it this way: "Those who realign their practices most effectively to Information Age standards will reap substantial benefits. Those who do not will be replaced or diminished by more nimble competitors."[52]

Notwithstanding this euphoric approach to the image-world, the potentially transformative image-world had already been the subject of experimental video art and installation from the early 1970s, just as the avant-garde artists of the early part of the century had demonstrated so clearly the possibilities for photography and film. The influence of tele-visual images appropriated by Pop artists, particularly by Andy Warhol, beginning in the 1960s, provides a point of departure for a deeper interpretation of the pervasiveness of image-thinking in the Postmodern period. A recent exhibition of early single-channel videos demonstrated the powerful impression that the "instantaneous feedback of a moving image" had on the imagination of the avant-garde art world of the 1960s and early 1970s.[53] The availability of relatively affordable, portable video equipment beginning in the late 1960s allowed artists to experiment with moving images in ways that the earlier film technologies had not permitted, giving video "a key role in both facilitating and critiquing the circulation of media images and ideas between the worlds of art and popular entertainment."[54] The video artist Peter Campus remembered:

> 1969–79... it was a very particular time. The art world was somehow very briefly looking for anti-marketplace ways of being.
>
> ...I was interested in what the medium could do and what was different about the medium... what the contrast was between video and film. Somewhat perversely at that time I wasn't interested in editing. Editing video was weird stuff

> then... It wasn't precise whereas film editing was down to the single frame, the way video is now. I always had a good sense of timing, I thought. And I wanted to put the timing into performance rather than into shot editing.[55]

Much of this early experimental video critiqued the industry of television, while using its apparatus to forge independent visions.

By the early 1980s, however, and precisely at the time that the personal computer made its appearance, MTV had taken up many of the precedent artistic innovations formed by video art in order to forge the music industry's own televised transmissions, thereby bringing into visuality the music that for a globalized culture remains its lingua franca still today. Not surprisingly, then, music required the image to achieve its current status in the image-world, one that was further enhanced in the 1990s by digital recording and players, such as Walkmans and iPod's.

A major aspect of the Postmodern image may be located in the history of video until 1980: the oscillation between art and commerce. This became a transparent interface characterized by the rapid transferability of commercial technologies to the artistic and private cultural realms and their re-appropriation, equally transparent and rapid in character, back to the industrial side of media production. Between 1968 and 1978 the critical video documentaries of the artistic collective, Ant Farm, revealed much about the interactions between images, culture, and the mass media.[56] *The Eternal Frame* (1975), Ant Farm's video on the image of President John Fitzgerald Kennedy and his wife Jacqueline Kennedy, during the event of his assassination, makes the point about filmic and tele-visual images in the artistic crossovers of image-technology. The artists rehearsed and re-enacted, on site in Dallas, the assassination of the President, including interviews conducted at the time of taping.[57] In addition, they inserted the infamous amateur Zapruder film of the original event, first seen by millions of Americans on television, into their taped performance. The Ant Farm video predicts the interactivity later associated with digital media; transcending the notion of the individual "user" while permitting conventions of the image-world to circulate for whole-sale or partial appropriation, whether by other individuals or corporate entities.

The celebrity culture that may be understood as characteristic of the image-world also played a role in the critique of *The Eternal*

Frame. The "look" or image of Jacqueline Kennedy, for example, had been generated through the mass media throughout the JFK presidency. Warhol made this point in his Pop portraits, based on images generated by technologies, of celebrities, including film stars, artists, past cultural "icons," and the rich and famous. Film theorist Stephen Heath defined image as precisely that which gives film its cinematic or projecting qualities, those elements that put the images into our consciousness: "image—The person, the body, in its conversion into the luminous sense of the its film presence, its cinema."[58]

These few examples from experimental video and the art world that demonstrate the crossover and circulation of images in the Post-modern period indicate their metonymic function for the kind of cognition that visual images achieve—what I have called *image-thinking*. Ed Ruscha, another artist who has made this situation of image-technology visible to us, works in the traditional medium of paint and canvas to make his point in the age of digital technology. Using images drawn from the landscape of the Postmodern urban environment, Ruscha demonstrates that crossover visuality typifies the American environment. His series called *Course of Empire*, undertaken for the United States Pavilion of the 51st Venice Biennale, juxtaposed his earlier views of buildings in the industrial landscape with their present-day global mutations.[59] For example, *Blue Collar Trade School* of 1992 has transformed into *The Old Trade School Building* of 2005, a prison-like edifice behind barbed wire and perpetually lit windows. These comparisons rely on the art historical method, developed in the early twentieth century for the purposes of university instruction, of simultaneously projecting two slide images side-by-side in order to provoke interpretation of the images shown.[60] Ruscha's series of paintings in the compare-and-contrast mode demonstrated the chronology of the Post-modern image-world in regard to the associative operation of the image itself, another example of image-thinking.

To conclude this investigation of image-technologies and image-world with image-thinking is to return to the ethical and aesthetical value of images in its pre-modern history. The conceptual issues that the term image invoked in the past cannot be said to have completely disappeared during the period investigated by this essay, when image became subjective and humans subjected to it. In Latin, *imago* carried the meaning of likeness, resemblance, or copy. Thus, in the Latin literature the word *imago* referred to visual representation in general, but specifically in its reference to resemblance it also meant a portrait, whether portrait bust, portrait statue, or any figure of a known individual.[61] Before modernity, the reference to resemblance made *imago* a placeholder for those artifacts of visual culture that referred to the real, more specifically to what I would call the realest of the real, i.e., the self. In a self-portrait drawing of piercing intensity and fluid draftsmanship, at National Gallery, Convent of St. George in Prague, the Italian Renaissance artist Arcimboldo writes his name in the upper corner "Joseffi Arcimboldi imago," as if to say my name, my face, and my gaze reveal myself in as complete an image of myself as I can give.[62]

It is in this anachronistic and revelatory sense that we find the term, *imago*, used in 1912 by Sigmund Freud's Viennese associates, Otto Rank and Hanns Sachs.[63] They entitled their new professional journal *Imago* and dedicated it to the "psychic powers" of dream images as model for the revelation of the motivations of mankind in the fields of "aesthetics, literary history and art history, mythology, philology, education, folklore, criminology, theory of morals, and theology."[64] In the late 1930's French psychoanalyst Jacques Lacan used the ancient concept of *imago* in his essay on the formation of an individual's identity through identification with an image, or reflection, of the self. Lacan wrote: "We have only to understand the mirror stage *as an identification*, in the full sense that analysis gives to the term: namely, the transformation that takes place in the subject when he assumes an image—whose predestination to this phase-effect is sufficiently indicated by the use, in analytic theory, of the ancient term *imago*."[65]

This persistence of the earlier concepts of the image took place in the field of psychoanalysis where a concept of the subject, the action of identification, and the image are all entailed in the interpretation of the modern individual. If we deem image-thinking symptomatic of the image-world, then this use of the archaized word *imago* by psychoanalysis may also be understood historically, according to the account given here for the Modern and Postmodern image.

In conclusion, aspects of pre-modern theories of the image, except for those of psychoanalysis, have been extensively explored in recent scholarship.[66] However, while the emergence of an extensive fascination with the image and its significance may well be part of the image-world that came into being with modernity, as this essay has attempted to show, the historical consciousness of the image remains relatively unexamined. A concept of the power of the image—whether to subject, to transcend the text, to cause the destruction of itself via iconoclasms, to bring into visibility that which is invisible, to sell, to transform, to memorialize, or to anesthetize—has played a predominate role in the ways in which modernity and post-modernity have understood the images of their own times and those of the past.[67] That idea of the power of the image has also served to occlude a historiography that explores the variety of ideological ends that the image has been made to serve. Psychoanalysis resurrected the image of the image, the *imago*, to serve the exploration first of culture and then, of the unconscious. Political theory has taught us to be suspicious of the image-world and its technologies. The *doxa* continues to understand images as illustrations and art history as the means of their elucidation. The image-world has engulfed the previous structures by which humans gave representation to their being. The ramifications of this epistemic turn towards image-thinking challenge and question earlier systems of belief. Clearly, then, images call upon us to think about how the ground beneath us has shifted.

Notes

[1] Jonathan Crary, *Techniques of the Observer: On Vision and Modernity in the Nineteenth Century* (Cambridge, Mass.: MIT Press, 1990), 6.

[2] See Graham Clarke, *The Photograph* (Oxford: Oxford University Press, 1997). Of course there are numerous books on the history of photography and its centrality for the modern world, but I find Clarke to be the most precise in regard to the convergence of image, technology, and subjectivity undertaken in this essay, see in particular his "Timeline," 234–237.

[3] Susan Sontag, "The Image-World," in *On Photography* (New York: Farrar, Straus, Giroux, 1977), 153: "...a widely agreed-on diagnosis: that a society becomes 'modern' when one of its chief activities is producing and consuming images, when images that have extraordinary powers to determine our demands upon reality and are themselves covet substitutes for firsthand experience become indispensable to the health of the economy, the stability of the polity, and the pursuit of private happiness." See also, Jacques Aumont, *The Image*, trans. Claire Pajackowska (London: British Film Institute, [1990] 1997), 241: "So, the real image revolution (if there was one) happened a long time ago, when images lost the transcendent power they used to have and were reduced to mere records, however expressive, of appearances." Aumont's useful compilation of the history of images is singularly disciplinary in its focus on film and art history and Western oriented, thereby not addressing the ritualistic significance of images in local contexts, nor accepting the results of a global culture that has produced an "image-world."

[4] Sontag, "The Image-World," 168.

[5] On the X-ray and visuality in modernity, see Akira Mizuta Lippit, *Atomic Light (Shadow Optics)* (Minneapolis: University of Minnesota Press, 2005), 35–59, particularly his discussion of Lisa Cartwright, "Screening the Body: Tracing Medicine's Visual Culture" (Minneapolis: University of Minnesota Press, 1995).

[6] Gilles Deleuze, *Cinema 1: The Movement-Image*, trans. Hugh Tomlinson and Barbara Habberjam (Minneapolis: University of Minnesota Press, 1986) and idem, *Cinema 2: The Time-Image*, Hugh Tomlinson and Robert Galeta, trans. (Minneapolis: University of Minnesota Press, 1989). The first of these books appeared in France in 1983.

[7] For the best treatment of perspective as developmentally related to the later photographic technologies, see Martin Kemp, *The Science of Art. Optical Themes in Art from Brunelleschi to Seurat* (New Haven: Yale University Press), 1990. For a critique of the dominance of perspective as the founding trope of the history of art, see Hubert Damisch, *The Origin of Perspective*, trans. John Goodman (Cambridge: MIT Press, 1994). For the important arguments made on behalf of this spatial history see Anne

Friedberg, *The Virtual Window: From Alberti to Microsoft* (Cambridge, Mass.: MIT Press, 2006).

[8] Lev Manovich, "The Automation of Sight: From Photography to Computer Vision," in *Electronic Culture: Technology and Visual Representation*, ed. Timothy Druckrey (New York: Aperture, 1996), 229–295. Idem, *The Language of New Media* (Cambridge, Mass.: The MIT Press, 2001). Manovitch's later book presents a more subtle approach that both acknowledges and rejects earlier models of visualization for a history of new media technologies, see pages 6–10.

[9] Erika Langmuir and Norbert Lynton, *The Yale Dictionary of Art and Artists* (New Haven: Yale University Press, 2000), 525. The literature on perspective in art and art theory is enormous and will not be cited here.

[10] On this standpoint see Whitney Davis, "Seminar Paper on Standpoint and Viewpoint," Seminar in Visual and Performance Studies, Directed by Catherine M. Soussloff, University of California, Santa Cruz, Cowell College, April 2006.

[11] Roland Barthes, *Camera Lucida: Reflections on Photography*, trans. Richard Howard (New York: Farrar, Straus and Giroux, 1981).

[12] In this regard, his study of the advertising images in his book, *Mythologies*, first published in France in 1957, is highly significant for the historiography I outline here, see Roland Barthes, *Mythologies*, Annette Lavers, trans. (St. Albans: Paladin, 1973; reprint 1976).

[13] See especially the excellent collections of essays in *Actes du Colloque Marey/Muybridge: Pionniers du Cinéma, Rencontre Beaune/Stanford* (Beaune: Conseil Regional de Bourgogne, 1996) and *La Passion du Mouvement au XIXe siècle: Hommage á E.J. Marey* (Beaune: Musé Marey, 1991).

[14] W. G. Sebald, *Austerlitz*, Anthea Bell, trans. (New York: Random House, 2001), 102–103. W. G. Sebald is, in this regard, most illustrative as a writer and a maker of images. He uses photographs without captions in his novels. These photographs do not explicitly identify their referent, thus making the referent mutable and subjective to the reader/viewer. Many of the photographs were taken from his own family archive, or in familiar locations, or from found documents that relate to his itinerant writing practices. These photographs are not explicitly referential to the text but they augment it. They exist on the visual level as the narrative proceeds on another level. Montage in film is transformative according to Sergei Eisenstein, so too photography collaged into text is transformative for the reader. I would speculate here that Sebald or any artist who opens herself up to this difficult visual probing does so in order to enhance the experience of empathy required for the understanding of others, whether the author, the characters in the novel, the loved ones, or fellow humans who suffer. It is worthwhile noting that Sontag was a great admirer of Sebald.

[15] Douglas Crimp, "The End of Painting," *On the Museum's Ruins* (Cambridge, Mass.: MIT Press, 1993), 84–106.

[16] W. J. T. Mitchell, *What Do Pictures Want?: The Lives and Loves of Images* (Chicago: University of Chicago Press, 2005), 204. In this section of the book Mitchell makes an extended plea for a theory of media, and by extension of the image, that locates in the "vernacular media themselves." (210)

[17] Walter Benjamin, "The Work of Art in the Age of Its Technological Reproducibility," *Selected Writings Volume 4 1938–1940*, trans. Edmund Jephcott et al., eds. Howard Eiland and Michael W. Jennings (Cambridge, MA.: Harvard University Press, 2003), 251–283.

[18] Sontag, "The Image-World,"178.

[19] Sontag, "The Image-World," 178. Sontag's view here should also be related to Benjamin's understanding of photography as the perfect technology for the rising class of the bourgeoisie, see Benjamin, "Little History of Photography," *Selected Writings Volume 2 1927–34*, trans. Rodney Livingstone et al., eds. Michael W. Jennings, Howard Eiland, and Gary Smith (Cambridge, MA.: Harvard University Press, 1999), 507–530.

[20] For the seminal study of the history of negative views towards the image, see Martin Jay, *Downcast Eyes: The Denigration of Vision in Twentieth-Century Thought* (Berkeley: University of California Press, 1993). For a good example of an extreme and negative attitude towards the image, one whose radicalism causes it to verge on irony, see Jean Baudrillard, *The Evil Demon of Images* (Sydney: The Power Institute of Fine Arts, University of Sydney, 1987).

[21] Crary, *Techniques of the Observer*, 9.

[22] See especially, Michel Foucault, *Discipline and Punish: The Birth of the Prison*, trans. Alan Sheridan (New York: Pantheon, 1979) and idem, *The Order of Things* (New York: Pantheon, 1980). For further thoughts on Crary's book and Foucault, see Catherine M. Soussloff, "Review Article: The Turn to Visual Culture," *Visual Anthropology Review* vol. 12 (Spring 1996): 77–83. In a recent essay on the image, Crary has moved closer to Foucault and further from phenomenology, e.g., "Many of those origins have much less to do with reproductive techniques (photography, for example) than with the emergence of new institutional requirements and social imperatives through which many kinds of images merged with dominant economic networks and the industrialization of cultural production" idem, "Image," in *New Keywords: A Revised Vocabulary of Culture and Society*, eds. Tony Bennett, Lawrence Grossberg, Meaghan Morris (Oxford: Blackwell, 2005), 178.

[23] Maurice Merleau-Ponty, "Eye and Mind," *The Merleau-Ponty Aesthetics Reader*, ed. Galen A. Johnson, trans. Michael B. Smith (Evanston: Northwestern University Press, 1993), 123–124.

[24] David Summers, "Real Metaphor: Towards a Redefinition of the 'Conceptual' Image," *Visual Theory: Painting and Interpretation*, eds. Norman

Bryson, Michael Ann Holly, Keith Moxey (New York: HarperCollins, 1991), 231–259 and idem., *Real Spaces: World Art History and the Rise of Western Modernism* (London: Phaidon, 2003).

25 Summers, *Real Spaces*, 654–655.

26 Mitchell, *What Do Pictures Want?*, 217–221.

27 Kaja Silverman, *The Threshold of the Visible World* (New York: Routledge, 1996), 2.

28 Foucault, *Order of Things*, xxiii.

29 Victor Stoicita, *The Self-Aware Image: Insight into Early Modern Meta-Painting*, trans. Anne-Marie Glasheen (Cambridge: Cambridge University Press, 1997).

30 "The Trouble with Painting: Image (less) Text," *Journal of Visual Culture*, Special issue on the state of visual culture studies, ed. Martin Jay, 4 (Summer 2005): 203–236.

31 Crary, *Techniques of the Observer*, 7.

32 The recent comprehensive Dada exhibition that originated at the Beaubourg in Paris makes this claim consistently valid, see Leah Dickerman, *Dada: Zurich, Berlin, Hannover, Cologne, New York, Paris* (Washington, D.C.: National Gallery of Art and New York: D.A.P., 2005).

33 Catherine M. Soussloff, *The Subject in Art: Portraiture and the Birth of the Modern* (Durham, N.C.: Duke University Press, 2006).

34 László Moholy-Nagy, *Painting, Photography, Film* (Cambridge, Mass.: MIT Press, 1987).

35 For the elaboration of this point and others related to it, see Thierry de Duve, *Kant After Duchamp* (Cambridge, Mass.: MIT Press, 1996).

36 Much as this statement is a valid one, it is not the opinion now held by Crary, see above note 14.

37 For the earliest expression of this argument see, Raymond Williams, *Television: Technology and Cultural Form* (New York: Shocken, 1975).

38 Margaret Morse, *Virtualities* (Bloomington: Indiana University Press, 2000), 4.

39 Joel Snyder has made a similar point about Foucault's anachronism, see "Las Meninas and the Mirror of the Prince," *Critical Inquiry* 11 (June 1985): 539–72.

40 For the genealogy of the concept of subjection in modernity and post-modernity, see Judith Butler, *The Psychic Life of Power: Theories in Subjection* (Stanford: Stanford University Press, 1997).

41 See "UCSC Creates New Major in Computer Game Design" at *http://www.ucsc.edu/news_events/press_releases/text.asp?pid=900*: "The University of California, Santa Cruz, has approved a new major in computer game design, the first of its kind in the UC system. The new major, leading to a B.S. degree, provides students with a rigorous background in the technical, artistic, and narrative elements of creating interactive computer games. 'We are pleased to be able to offer this new degree program, which provides a unique combination of technical and artistic training,'

said Ira Pohl, professor and chair of computer science in UCSC's Baskin School of Engineering. The Department of Computer Science will administer the new interdisciplinary program, which will also involve faculty in the Department of Film and Digital Media in UCSC's Arts Division. Students can enroll in the new major beginning this fall. 'Millions now play massively multiplayer online games, which constitute a new cultural force—a new medium. Digital media courses will provide students with the tools they need to understand this cultural transformation in conjunction with its technological and artistic possibilities,' said Warren Sack, assistant professor of film and digital media."

42 On the Situationist critique of the image and capitalism, see Guy Debord, *Society of the Spectacle*, trans. Donald Nicholson-Smith (New York: Zone Books, 1995).

43 Debord, *Society of the Spectacle*, 73, 48.

44 Guy Debord, *Comments on the Society of the Spectacle*, Malcolm Imrie, trans. (London: Verso, 1998).

45 John Berger, *Ways of Seeing: Based on the BBC Television Series* (Middlesex: BBC and Penguin, 1972). As Berger states, his book is very much indebted to Walter Benjamin's essay cited in note 17 above.

46 Morse, *Virtualities*, 20: "As the time and space between the act of enunciating and receiving images closes in, it becomes more proper to speak of an image-world with which we will interact more or less continuously unless we make the effort to disengage ourselves from it."

47 Paul Virilio, *The Art of the Motor*, trans. Julie Rose (Minneapolis: University of Minnesota Press, 1995).

48 For a thorough examination of the images of torture from Abu Ghraib prison, see my "Who is Being Tortured? Rituals of Viewing at Abu Ghraib," *Ritual and Event: Interdisciplinary Perspectives*, ed. Mark Franko (London: Routledge Press, 2006).

49 André Gunthert, "L'image numérique s'en va-t'en guerre: Les photographies d'Abou Ghraib," *Études Photographiques* vol. 15 (November 2004): 124–134. I am grateful to the author for bringing this essay to my attention and for his subtle analysis regarding the meaning of the circulation of the images of torture at Abu Ghraib.

50 Marianne Hirsch, "Editor's Column: Collateral Damage," *PMLA* vol. 119 (October 2004): 1209–1210: "General Richard B. Myers, chairman of the Joint Chiefs of Staff, even went so far as to ask CBS's *Sixty Minutes II* to delay showing the photographs of prison abuse to the public lest their dissemination do harm to the troops abroad, and throughout May and June of this year pictures from Abu Ghraib were printed and distributed slowly and with an unusual degree of cautionary framing. Congress members first saw hundreds of images in closed, tightly protected settings. They *described* them to the press but were prohibited from showing them to the public. The ethics of revelation was debated among com-

mentators and politician weighing freedom of information versus the victim's right to privacy."

51 Michael Hardt and Antonio Negri, *Empire* (Cambridge, MA.: Harvard University Press, 2000), 23. This view is echoed, but also critiqued by the collective Retort: "The spectacle is rampant, but at its epicenter less and less persuasive. It accelerates, like capitalism in general, as a result of the falling rate of illusion. The disenchantment of the image-world may follow the disenchantment of the world in general. The spectacle is the key form of social control in present circumstances, but also a source of ongoing instability. The sheer proliferation of image-technology means that too much of the texture of everyday life is captured and circulated, and subject to inflection outside the normal circuits of power," *Afflicted Powers: Capital and Spectacle in a New Age of War* (London: Verso, 2005), 188. For further thinking about Retort's argument on image technologies, see Soussloff, "Who is Being Tortured? Rituals of Viewing at Abu Ghraib."

52 Michael G. Dolence and Donald M. Norris, *Transforming Higher Education: A Vision for Learning in the Twenty-first Century* (Ann Arbor: Society for College and University Planning, 1995), 23: The authors write: "An essential characteristic of learning in the information Age is that it be highly individualized. Learning needs to occur at the time, place and pace of the individual's choosing. The individualized character of learning will need to be established early in the learner's career and to be continually recalibrated to meet changes in the learner and the environment."

53 See Constance De Jong, "Introducing the Early Show," *The Early Show: Video from 1969–1979* (New York: The Bertha and Karl Leubsdorf Art Gallery, Hunter College, 2006), n.p.

54 John G. Handhardt and Maria Christina Villaseñor, "Video/Media Culture of the Late Twentieth Century," *Art Journal* Vol. 54 (Winter 1995): 21. I continue to be struck by the importance of this essay in the historiography of new media because the authors situate the contributions of video art squarely at the center of the emergent digital media and argue for the significance of the political critique embodied in the early history of the artistic use of video technology. The video image shows the potential for the tele-visual image to disrupt and interrupt the "commercial flow of messages and images" found in television. (21) In this sense, the video image responded to Sontag's negative criticism of the "image world."

55 Interview with Peter Campus in *The Early Show*, 25, 26.

56 A recent history of Ant Farm may be found in Constance M. Lewallen and Steve Seid, *Ant Farm* (Berkeley: University of California Press, 2004).

57 For a good summary of the complex issues involved in *The Eternal Frame*, only some of which can be mentioned here, see Art Simon, *Dangerous Knowledge* (Philadelphia: Temple University Press, 1996), 145–157.

[58] Stephen Heath, *Questions of Cinema* (Bloomington: Indiana University Press, 1981), 181–82. Heath continues: "The high point of the image is the *star*; it is as such that Gabor is used in *Touch of Evil*: an appearance of a few seconds, highlighted on the stairs that sweep down into the cabaret, picked out in a theatrical space with men gazing up at her—and with a chandelier in the form of a star hanging down beside her as a kind of literalized metaphor, a token of her image-status, the perfection of its achievement. The image can determine the narrative construction of a film: Hitchcock is obliged to end *Suspicion* in a way commensurate with the image of Cary Grant; films are developed as 'vehicles', 'showcases' for this or that star. Equally, the image can be taken up in a film in a play from character to image, drawing the narrative into the spectacle of its impersonation, resolving there: think, in this respect of the *Towering Inferno* with its clash of architect and firemen and simultaneously of Paul Newman and Steve McQueen, who come together in the end—watched by the crooked and the ordinary, the cowardly and the powerless—as a god letting flow the waters from on high, a pact of the stars much more than of characters, and effective, providing a resolution for the narrative, as such."

[59] See the catalogue for this exhibition, Linda Norden and Donna De Salvo, *Course of Empire: Paintings by Ed Ruscha* (New York: D.A.P., 2005).

[60] For a discussion of the role of slide photography and projection in the discipline and instruction of art history, see Robert Nelson, "The Slide Lecture, or the Work of Art in the Age of Mechanical Reproduction," *Critical Inquiry* vol. 26 (Spring 2000): 411–434.

[61] Harriet Flower, *Ancestor Masks and Aristocratic Power in Roman Culture* (Oxford: Clarendon Press, 1996). I thank Karen Bassi for this reference.

[62] Here we might agree with the French film theorist André Bazin, "The Ontology of the Photographic Image," in *What is Cinema?*, trans. Hugh Gray (Berkeley: University of California Press, 1967), 10: "No one believes any longer in the ontological identity of model and image, but all are agreed that the image helps us to remember the subject and preserve him from a second spiritual death."

[63] Jean-Paul Sartre, *The Psychology of Imagination* (New York: Philosophical Library, 1948) has made my understanding of the significance of the anachronistic use of the image in psychoanalysis possible. According to Sartre, dominant views of the image from Hume to 1945 contradicted common sense. Importantly, Sartre gives portraiture as the exemplary genre, or site, with which to illustrate this radical incongruity between what we "know" about the image of someone and what we know: "When I say that "I have an image" of Peter, it is believed that I now have a certain picture of Peter in my consciousness. The object of my actual consciousness is just this picture, while Peter, the man of flesh and bone, is reached but very indirectly, in an "extrinsic" manner, because of the fact that it is he whom the picture represents. Likewise, in an exhibition, I

can look at a portrait for its own sake for a long time without noticing the inscription at the bottom of the picture "Portrait of Peter Z..." (6)

[64] Otto Rank and Hanns Sachs, "Prospekt," *Imago* vol. 1 (March 1912): n.p. I express my gratitude to my colleague Mark Franko for alerting me to the significance of the title of this journal. I thank Dr. Theo Honnef for his work on translation.

[65] Jacques Lacan, "The mirror stage as formative of the function of the **I** as revealed in psychoanalytic experience," *Écrits: A Selection*, Alan Sheridan, trans. (New York: Norton, 1977) 1–7. The first version of this essay was delivered in Marienbad in 1936, see Elisabeth Roudinesco, *Jacques Lacan: Outline of a Life, History of a System of Thought*, Barbara Bray, trans. (New York: Columbia University Press, 1997), 513. Given the significance of terms here, it will be useful to provide the French version of this citation: "Il y suffit de comprendre le stade du miroir *comme une identification* au sens plein que l'analyse donne a ce quand il assume une image, dont la predestination a cet effet de phase est suffisamment indiquée par l'usage dans la théorie, du terme antique d'*imago*." (In *De l'Art et de la Psychanalyse Freud et Lacan*, Claude This, ed. (Paris: École Nationale Superieure des Beaux-Arts, 1999), 171.) For further thinking about the linguistic alliances between Lacan and the Vienna School of Art History, see Soussloff, "Review of *The Vienna School Reader: Politics and Art Historical Method in the 1930s* by Christopher Wood," *Centropa* (Fall 2002): 230–32 and idem, *The Subject in Art*, Chapter 1.

[66] See, for example, Hans Belting, *Likeness and Presence: A History of the Image Before the Era of Art*, Edmund Jephcott, trans. (Chicago: The University of Chicago Press, 1994), first published in German in 1990; and Alain Besançon, *The Forbidden Image: An Intellectual History of Iconoclasm*, trans. Jane Marie Todd (Chicago: University of Chicago Press, 2000), first published in French in 1994.

[67] See, in particular, Louis Marin, *Des Pouvoirs de l'Image: Glosses* (Paris: Éditions du Seuil: 1993) who explores the power of the image in relation to Medieval and Early Modern texts; and David Freedberg, *The Power of Images: Studies in the History and Theory of Response* (Chicago: University of Chicago Press, 1989) who examines the cultural history of images in regard to prevalent psychological and behavioral responses towards them. In my opinion Freedberg's book deserves re-assessment in light of both the historiography of art and the history of digital technologies since its publication.

Documentary Re-enactments

A Paradoxical Temporality That Is Not One

Bill Nichols

"Could you do the kiss again?"
Lonely Boy (Wolf Koenig, Roman Kroitor, 1962)

Re-enactments, the more or less authentic recreation of prior events, provided a staple element of documentary representation until they were slain by the "verité boys" of the 1960s (Ricky Leacock, D.A. Pennebaker, David and Albert Maysles, Fred Wiseman and others) who proclaimed everything, except what took place in front of the camera without script, rehearsal or direction, to be a fabrication—inauthentic. Observational or direct cinema generated an honest record of what would have happened had the camera not been there, or what did happen as a result of recording what happens when people are filmed. Observational temporality possessed the flat, one-dimensionality of real time divorced from its past, but it gloried in the complex dynamics of interactions whose future was unknown at the moment of filming.

Times have changed. Re-enactments once again play a vital role in documentary, be it of a Solidarity movement that cannot be filmed in *Far from Poland* (Jill Godmilow, 1984), a murder for which radically disparate accounts exist in *The Thin Blue Line* (Errol Morris, 1988), the schematic simulation of a harrowing escape from captivity in *Little Dieter Needs to Fly* (Werner Herzog, 1998), or events during the final days of Salvador Allende's socialist government in *Chile, Obstinate Memory* (Patrizio Guzmán, 1999). Apart from the occasional controversy, such as that surrounding the use of re-enactments that cannot be distinguished from actual footage of the historical event—a turn that shifts the debate from strategies of representation to questions of deceit—re-enactments are once again taken for granted. They pose, however, a number of fascinating questions about the experience of temporality and the presence of fantasy in documentary. These are the issues this essay explores.

Re-enactments occupy a strange status; it is crucial that they be recognized as a representation of a prior event, while also signaling that they are not the representation of a contemporaneous event. They stand for something, but are not identical to what they stand for. This is akin to the status of play compared to fighting, at least in the formulation provided by Gregory Bateson in his essay on "A Theory of Play and Fantasy": "These actions in which we now engage do not denote what would be denoted by those actions which these actions denote."[1] The controversy surrounding the 2004 Academy Award short documentary winner, *Mighty Times: The Children's March* (Robert Houston, 2004), involved charges that re-enactments blended imperceptibly with authentic footage of civil rights activity in the South in the 1960s, as did archival footage of other events, such as the Watts Riot in Los Angeles. Viewers must recognize a re-enactment as such, if issues of deception are to be avoided, and if the re-enactment is to function effectively, even if this recognition also dooms the re-enactment to its status as a repetition of something that already occurred, elsewhere, at another time and place. Unlike the contemporaneous representation of an event—the classic documentary image, where an indexical link between image and historical occurrence can be claimed—the re-enactment forfeits its indexical bond to the original event and yet, paradoxically, it draws its phantasmatic power from this very fact. The shift in level prompts awareness of an impossible task: to retrieve a lost object in its original form even as the very act of retrieval generates a new object and a new pleasure. The viewer experiences the uncanny sense of a repetition of what remains historically unique. A specter haunts the text.

What constitutes a lost object is as various as all the objects toward which desire may flow. The working through of loss need not entail mourning: it can also, via what we might call the phantasmatic project, evidence gratification, or pleasure, of a highly distinct kind. Documentary film generally entails a performative effort to register what is or will immediately become past in a *mise-en-scène* produced by a desire to retrieve, re-experience, master or enjoy. Re-enactments double this performative dimension by hinging themselves on the viewer's awareness that what is represented, represents what has already occurred.

As a collective activity, fantasy—from the *folie à deux* to documentary film—functions as ideology.[2] It provides the psychic foundation for imaginary relations and actual social relations through which we invest ourselves in the world around us.[3] Fantasy thus functions as a structuring agent, something that reshapes what it takes up, and it is this action that becomes of key importance. As Jean Laplanche and J. B. Pontalis note, "...it is the subject's life as a whole which is seen to be shaped and ordered by what might be called, in order to stress this structuring action, a 'phantasmatic' (*une phantasmatique*)."[4] In this sense, fantasy is where the subject represents and becomes itself. And yet, the subject finds itself dispersed throughout the fantasy, in the *mise-en-scène*, or in the syntax: "...all distinction between subject and object has been lost."[5] It is a subject that is not one, but one that is now subject to and the subject of desire. The dispersed subject, like the re-enactment, draws near to the gap between then and now, seeking to go through the motions that will allow it pass across, but finding instead the new pleasure that constitutes the *phantasmatique* itself. Fantasy always entails a temporal trajectory, a version of narrative structure and action, in the service of desire. The *mise-en-scène* constructed is one that straddles the narrow gap between need and desire. In such a case, shared fantasy or ideology is no longer a subject for idle curiosity.

As classically understood by Laplanche and Pontalis, fantasy operates within another gap as well, in the space between before and after: the before of what has already occurred, namely, the pacification of a need, and the after of a new gratification, occasioned by the fulfillment of a desire. This marks the phantasmatic as, in fact, "the moment of separation between before and after."[6] It inaugurates the lived experience of temporality in the form of desire. It is where we fill time with the duration of a longing. It is the moment, or, better, the duration of a moment between "the object which satisfies and the sign which describes both the object and its absence...".[7] The latter does not provide the same type of satisfaction as the former, but, as the "between" extends across the *mise-en-scène* of a phantasmatic space, it comes to be entirely real, psychically.

A stunning example of this process unfolds in *Capturing the Friedmans* (Andrew Jarecki, 2003). The film explores the complex web of family relations and submerged desires that lie behind the formal charges of pedophilia against the father, Arnold, and his son, Jesse. They are alleged to have fondled, seduced, abused, raped, and sodomized dozens of young boys who took computer classes in the family's home. Andrew Jarecki draws on home movies, shot over the course of the family's lifetime, video diaries, shot mainly during the period of tumult precipitated by Arnold and Jesse's arrest, television news reports and Jarecki's own interviews with most of the involved parties. If the trial of Arnold and Jesse sought to achieve the either/or clarity of guilt or innocence, Jarecki is far more concerned to capture the ambiguity and confusion that this very process produces within the family.

Une phantasmatique, as the duration of a moment between the object of an original satisfaction and the sign that describes both this object and its absence, occurs within the *mise-en-scène* of some of the family's more recent video diaries. These are scenes shot by sons Jesse or David as they attempt, with their father, to re-enact the form of spontaneous family togetherness that has become a lost object. Home movie footage, shot in Super 8, has shown the boys and their father in moments of carefree bliss, dancing, singing, and generally cavorting together with a casual acceptance of the camera as an embodied participant (it is held by a family member). The recent video or digital footage—distinguishable from the home movies by its absence of film grain, lack of color fading, and greater contrast—however, demonstrates the impossibility of stepping into a temporal river for the first time twice. The boys and their father are markedly older, their dancing and clowning slightly forced, the father visibly burdened by the weight of his arrest, and their mother more emphatically excluded than simply absent. The video footage is the sons' attempt to re-enact their own past. They are clearly aware that their attempt is a re-enactment rather than a genuine return to a lost object and irretrievable moment: the video footage stands as sign that describes both the lost object (the unqualified pleasure of physical cavorting that once was theirs) and its absence (the effort that must now be made to re-enact what was once spontaneous exuberance). This is nowhere more evident than in the refusal of Arnold's sons to see his haunted and mortified expression, an expression, which, if acknowledged, would thwart their own desire to go through the motions that will generate the compensatory pleasures they desire.

These extraordinary moments, in which the participants attempt to will themselves back to the past and yet know very well that the effort must fail, border on the act of mourning which cinema or video makes possible. They compound that semi-acknowledged work with the production of a phantasmatic pleasure, for the sons at least, that lessens the sting of that which is lost and cannot be retrieved. They go through the motions that locate them within a *mise-en-scène* of desire, a phantasmatique their mother can no longer share. They do, once again, now, what they once did, then, and derive from this act not the original satisfaction of a need, but the gratification of a desire that stems from the sequence of images or signifiers they fabricate for themselves.

Laplanche and Pontalis stress the importance of a temporal convolution in which the past and present are inextricably woven together. Fantasy is not the mere retrieval of something past, not the recovery of a real object, nor, as in the example they adopt, the milk a baby may have ingested, but "the breast as a *signifier*..." (italics mine).[8] What was once an external object transforms into an internal image or signifier. The signifier bears an emotional weight. What fantasy restores in this example is not the act of actually obtaining the mother's milk: "not the act of sucking, but the enjoyment of going through the motions of sucking."[9] Such motions, necessarily corporeal, produce, when successful, their own pleasure.

This enjoyment, or pleasure resides in the realm of psychic reality. Like the re-enactment, it involves a pleasure that comes from "going through the motions" associated with a past event, but doing so in a distinctly different, phantasmatic domain. Pleasure flows from an act of imaginary engagement in which the subject knows that this act stands for a prior act, or event, with which it is not one. A separation that entails a shift from physical needs and their pacification to psychical desires and their gratification, between before and after, then and now, is as integral to the phantasmatic experience as it is to the efficacy of ideology.

A telling moment of this sort occurs in *Chile, Obstinate Memory*, when four of the personal bodyguards for President Allende re-enact their role in a Presidential motorcade prior to the military coup d'etat that toppled his government on Sept 11, 1973. Guzmán cuts between the footage of the men re-enacting what they used to do, and shots of them actually guarding Allende some thirty-five years ago. Then, Allende and others sit inside a large, topless black limousine, crowds line the way, and the four men trot alongside, eyes scanning the surrounding scene, as each keeps a hand in contact with one of the four fenders of the car. Now, the men walk alongside an economy size, red, hard top sedan, on a deserted country road, with no crowd in sight, but each with a hand in contact with the car and their eyes once again scanning the surroundings.

At one point Guzmán freezes the image of the motorcade "then," and the guards identify themselves and compatriots from the still image. The authentic image becomes remote, an instigation for memory and identification, whereas the re-enacted image allows the men to "go through the motions" of guarding the President one more time. It clearly does not fulfill an official state need this time; instead it gratifies a personal desire, it makes possible "the enjoyment of going through the motions of guarding," as it were, when guarding itself remains squarely lodged in the past. Nothing captures this temporal knotting of past and present better than a close-up image of the hand of one of the guards slowly fluttering up and down on one of the half-open car windows: the rhythm follows from the cadence of his gait beside the car but the camera's close-up view of his delicate grip, the rise and fall of his fingers, and the overt absence of an engulfing crowd attest to the psychically real but phantasmatic linkage of now and then.

Despite the gulf between now and then, and as a precondition for the gratification of re-enactment, like fantasy in general can provide, the subject becomes "caught up in the sequence of images" that populate the *mise-en-scène* of desire. This holds for the bodyguards in this striking scene from *Chile, Obstinate Memory*, just as does for the Friedman sons and father in their video re-enactments, but it is also true of the viewer, immersed in an experience in which he knows very well that the re-enactment is not that which it represents and yet, all the same, allows it to function as if it were. Above all, however, the filmmaker is the one caught up in the sequence of images; it is his or her fantasy that these images embody. The filmmaker need not be physically present in the image, as he or she is in many participatory or interactive documentaries. "… [T]he subject, although always present in the fantasy, may be so in a desubjectivized form, that is to say, in the very syntax of the sequence in question."[10]

This de-subjectivization is acutely true of the video recordings by David and Jesse Friedman in *Capturing the Friedmans*. As filmmakers, rather than as participants before the camera, they take on a de-subjectivized form in the syntactical parallelisms they construct with the home movies of a decade or more before. The camera functions not as an omniscient observer or third person narrator, but reiterates the function of the home movie camera generally as participant and instigator in scenes, here, of camaraderie and high-jinks. These same images subsequently double-up to become part of the phantasmatique structured by Andrew Jarecki. In his case, psychic pleasure seems to stem from the construction of ambiguity about what happened in the past, what these social actors have said and done, and how they understand the actions of others as well as their own. Jarecki complicates the far more literal, linear and binary logic of the judicial system that sets out to determine "what really happened." He re-inscribes the ambiguity of perspective, and voice that separates such judicial determinations from the plane of fantasy.

Guzmán, too, in his re-enactment of guarding the Presidential car inhabits the syntax of a sequence that he causes to flutter between past and present, that restores specificity (names, relationships) to the past and brings phantasmatic gratification to the present as he goes through the motions of re-enacting the past to new ends. To me, this makes the subject's presence, in re-enactments, and documentaries more generally, a function of what I have described as the documentary voice of the filmmaker rather than his or her corporeal appearance before the camera. The documentary voice speaks through the body of the film: through editing, through subtle and strange juxtapositions, through music, lighting, composition, and *mise-en-scène*, through dialogue overheard and commentary delivered, through silence as well as speech, and through images as well as words.

The voice of an orator, or documentarian, enlists and reveals desires, lacks and longings. It charts a path through the stuff of the world that gives body to dreams and substance to principles. Speaking—giving voice to a view of the world—makes possible the necessary conditions of visibility to see things anew, to see, as if for the first time, what had, until now, escaped notice. This is not objective sight, but seeing as a way of being "hit and struck," caught in that precarious, fleeting moment of insight when a *Gestalt* clicks into place. Meanings arise from what had seemed to lack meaning, or are already filled to capacity with all the meaning it could bear.[11] Such an insight does not occur however, until given external shape—the shape provided by the film's voice as it addresses others.[12]

The documentary voice is the embodied speech of the historical person—the filmmaker, caught up in the syntax of re-enacted images through which the past rejoins the present. Voice, given in re-enactments partially as an awareness of the gap between that which was and that which now returns to what has been, also affirms the presence of a gap between an objectivity/subjectivity binary and a phantasmatique. Subjectivity suggests it is added to something and could also be subtracted. Voice is the means and "grain" with which we speak and can never be separated from what is said.

Objectivity desires a fixed relation to a determinate past, the type of relation that permits of "guilty/not guilty" verdicts or other definitive answers to the question of "what really happened." Voice, in the form of re-enactments that embody the "I know very well but all the same" formulation at the heart of psychic reality, imposes recognition of the relentless march of a temporality that makes the dream of a pure repetition and an omniscient perspective impossible. The very syntax of re-enactments affirms the having-been-thereness of what can never, quite, be here again. Facts remain facts, their verification possible, but the iterative effort of going through the motions of re-enacting them imbues such facts with the lived stuff of immediate and situated experience.

Re-enactments also foil the desire to preserve the past in the amber of an omniscient wholeness, the comprehensive view we like to think we have that accounts for what came to pass, without attaching this account to a distinct perspective. This can be the source of a sense of dissatisfaction: our view is too partial, too incomplete or too cluttered (it may contain a body or bodies too many, as contemporary figures fill in for their historical counter-parts). Re-enactments are clearly *a* view rather than *the* view from which to apprehend the past. They insist on the cumbersome limitations of an embodied perspective. Re-enactments produce an iterability to that which belongs to the singularity of historical occurrence. They reconcile this apparent contradiction by acknowledging the adoption of a distinct perspective or point of view. Such perspectives can proliferate indefinitely, but each of them can also intensify an awareness of the separation between the lost object and its re-enactment. Re-enactments belong to a situated phantasmatique that nullifies the status of that other phantasmatique of objectivity, omniscience and finality that haunts the documentary film.

Irene Lustig's film, *Reconstruction* (2002), demonstrates the fragile veneer of objectivity that serves the interests of the state in her remarkable discovery of a re-enactment of an extremely rare event: a bank robbery in Communist Romania. One of the robbers, it turns out, was Lustig's grandmother, the only person not executed for this crime against the people, or the people's capital. Lustig incorporates portions of the Romanian government's re-enactment of the crime, also called *Reconstruction*, which apparently sought to reaffirm the power, of what was at that time a Communist state, to write and control the past. In this case the phantasmatic quality of the re-enactment pursues what is more clearly than usual an ideological issue: the, at least temporarily, lost object of state power. It seeks the gratification of staging a *mise-en-scène* within which that power can reconstitute itself. The robbers, once caught, were compelled to re-enact their planning, the robbery, their confessions, trial and sentencing. They must once again go through the motions of their defiance of the state but, this time, without any anticipation of success.

The state's film had apparently been intended to demonstrate the folly of breaking the law, but it was never shown publicly for reasons that remain unclear. In it the suspects exhibit a decidedly despondent manner, a sign that they know the pleasure of this re-enactment will not be theirs. Their eyes are vacant, their gaze unfocused, their words slow in coming and stilted in tone. In one scene, a prosecutor attempts to pry a confession from one of the men (Lustig's grandmother was the only woman involved). The suspect resists; he denies knowledge of the crime. Then the prosecutor produces hard evidence: the actual weapons used in the crime. The game is up and the suspect, with downcast eyes and the same hopeless tone, quickly admits his guilt: "You have those too? I see we've been discovered. Until now I've been hiding the truth." With that, his confession begins.

This state sponsored re-enactment introduces a sense of the ritualistic quality to re-enactments of significant but past events. The robbery must be represented as a remarkable exception, and the power of the state brought into play as another iteration in the eternal ritual of justice fulfilled. The capacity of this machinery to remain invulnerable requires the criminal participants in its re-enactment to be shorn of defiance, calculation, and volition. Their own bodies serve as the surface for a textual rewriting in which agency reverts entirely to the state. This triumph of judicial invulnerability, however, unwittingly betrays the very condition of its being in the barely animated bodies of the criminals, who must go through the motions of a past event in a context where need and pleasure, desire and gratification accrue only to the state. A "curse" continues to haunt the text but in the form of a disavowal that Lustig, by re-contextualizing the original re-enactment from a distinct perspective of her own, exposes. Whereas re-enactments may allow for an owning or owning up to the past, in *Reconstruction* the owning of the past takes the more literal form of the state coming into physical control, or ownership, of the bodies and minds of those who defied it.

Re-enactment takes another fascinating turn in Werner Herzog's *Little Dieter Needs to Fly.* Shot down on a bombing run over Laos and captured, Dieter Dengler, after a series of extraordinary adventures, escapes his captors and returns to the United States. This is the story he tells to Herzog, but in the course of doing so he decides to re-enact what he began to recount. Dengler and Herzog return to Laos where local villagers play his captors and Dieter plays his former self.

Unlike the bank robbery re-enactment in *Reconstruction*, the walrus hunt in *Nanook of the North* (Robert Flaherty, 1922), the re-enactments of detention at Guantanamo in *The Road to Guantanamo* (Winterbottom and Whitecross, 2006), or the "pre-enactments" of what might happen in the event of nuclear attack in Peter Watkins' *The War Game* (1965)—all of which adopt the performative qualities of suspenseful, dramatic intensity—the re-enacted scenes in *Little Dieter* exhibit a Brechtian sense of distanciation. In one scene, for example, the recruited villagers stand listlessly around Dengler, "going through the motions" of guarding him as their prisoner by wearing uniforms or displaying weapons. Their half-hearted, good-natured performance clearly conjures what Dengler went through, without compelling prisoner or guards to re-enter the psychic and emotional space of the original event. Neither Dieter nor Herzog seek to render suspense dramatically or verisimilitude perfectly. The necessary awareness of a gap between past event and present re-enactment remains altogether vivid, as it gradually does in *The Thin Blue Line* where the series of re-enactments of the original murder of a policeman construct an Escher-like impossible space of conflicting narratives.

Dieter transports himself back to that which now functions as a lost object through the social gests he puts into motion.[13] It allows him to own his past in a corporeal but phantasmatic form that does not require the presumably therapeutic dramaturgy that Charcot inaugurated in his treatment of hysterics and that so many re-enactments imitate. The sense of mastery that arises from this iteration in which the outcome is now known allows him to go through the motions of a triumphant passage already complete; it is this passage that the film within a film in *Reconstruction* denies to those whose bank robbery attempt failed. Dieter Dengler, the one who survived what once put survival in question, now occupies a phantasmatic *mise-en-scène* that affirms his very survival. "Going through the motions" takes on a formal, ritualistic quality that nonetheless spans the moment between before—when need prevailed, and after—when these social gests function as signifiers of what was but is now, at the moment of signification, past. The gests or signifiers both embody the lost object of a former experience and gratify the force of desire.

A similar form of re-enactment occurs in the 1979–80 film, *2 Laws* (Strachan and Cavadini, 1982), about aboriginal land rights in Australia.[14] Divided into four parts, the first part introduces the main participants and then proceeds by a mixture of recounting, re-enacting and reflecting on past events. The focal point is the arrest of several Aboriginal people for the theft of a cow in 1933. One woman is beaten to death by a white constable during a forced march to the nearest jail. Although some events are recounted as the participants in the film reconstruct and relay the story to the younger members of the group, some are re-enacted such as this beating. As with Dieter Dengler's re-enactment of his captivity in Laos, no effort is made to render the event dramatically. Instead the constable swings a long club in slow motion toward the woman, who plays the original victim, without coming close to touching her. The entire action is filmed with a long shot that refuses the conventional *découpage* of cutting from a master shot to tighter, more highly charged shots to capture the violence in graphic form.

The re-enactment has the distanciating quality of a Brechtian gest. No one is compelled to act out the original pain and trauma. Nor does the re-enactment facilitate the work of mourning the past as much as the process of reclaiming it. The lost object remains lost, but the social gest that replaces it restores it to a condition of visibility through a phantasmatic signifier that emerges from and embeds itself within the bodies of the participants. The duration of the moment between before and after hinges on this signifier, that clearly stands for that to which it refers historically without any attempt to replicate that for which it stands, now, in the present. That there is a body too many is a crucial condition of the process in this case. The other—Australian law—clashed with aboriginal law and, in 1933, won. Now, the re-enactment of this event serves to give corporeal embodiment to what happened, as those who remember it go through the motions that populate the *mise-en-scène* of their desire. A specter and a curse may haunt this space, but it also serves as the site from which a desire for justice might achieve, in the future, the gratification that this event from the past set in motion.[15]

These various re-enactments begin to suggest some ways in which re-enactments tend to cluster into different types. Some are highly affective, some far less so. Some make their status as re-enactment obvious, some do not. These differences do not establish hard and fast divisions but do suggest different nodal points within a diffuse and overlapping universe of possibilities. Some particularly common variations include the following.

1. Realist Dramatization. The most contentious, because it is the least distinguishable from both that which it re-enacts and the conventional representation of past events in fiction, be it in the form of a historical drama, docudrama, "true story," or flashback. It is the suspenseful, dramatic re-enactment in a realist style. *Nanook of the North* may be the first recognized instance of this—with the heightened tension of its will-he-or-won't-he-succeed structure—turning the hunt for salmon, walrus, and seals into high drama. These scenes re-enact how hunting occurred in pre-contact Inuit culture, although this point is never made explicit. The status of the scenes as re-enactments, as well as drama, remains shrouded, the better to intensify their affective dimension. That such a strategy can continue to prompt controversy is evident in the case of *Mighty Times II* mentioned earlier.

An important model for many recent re-enactments of realist dramatization occurs in the powerful documentary, *Las Madres de la Playa de Mayo* (Muñoz and Portillo, 1985). As one of the mothers who gather every day at the Argentine White House (La Casa Rosada), recounts how armed men abducted her son in the dead of night, the film cuts to a re-enactment of this event. The re-enactment possesses the surreal tones of a nightmare with its grainy, high contrast and slow motion imagery in which individual figures are unrecognizable. The distortions work to impede realist transparency. These formal devices shift the re-enactment toward the fourth category—

the stylized re-enactment—but Portillo's and Muñoz's expressive rendering of what happened stresses the emotional impact on the mother as something that was not part of the event itself, but part of its affective reverberation ever since. Dramatization comes to a focus within the subjectivity of individual memory.

2. Typifications. Problematic for similar reasons to realist dramatization is the re-enactment of typifications. In this case there is no specific event to which the re-enactment refers, and the sense of separation between event and re-enactment fades as a sense of typifying past patterns, rituals and routines increases. Such re-enactments characterized many early documentaries, including *Nanook of the North*, where the suspenseful dramatization of events—presented as if they were present-day—re-enacted the typical processes of the Inuit's pre-contact past. The fur trapping and igloo building, for example, did not re-enact specific historical occurrences as much as characteristic ones. To the extent that the viewer recognizes that the re-enactment's claim to authenticity resides not in their depiction of present-day activity—carried out despite the presence of the camera—but in the re-enactment of pre-contact activity, staged for the sake of the camera, this very claim of authenticity undergoes erosion. The indexical quality of the image anchors it in the *mise en scène* of the filmmaker's desire, as it does in fiction. John Grierson, the commonly acknowledged father of the British documentary of the 1930s, adopted this technique wholesale. Re-enactments, as typifications, proliferate. *Coalface* (Cavalcanti, 1935) has several sequences of coal miners mining, or taking their lunch break, that possess an aura of present-day reality simply observed when they are, in fact, staged. *Night Mail* (Watt and Wright, 1936) is the most famous example, postal workers sort mail, on a sound stage dressed to simulate the Postal Express as it makes its overnight journey from London to Glasgow. This scene re-enacts, cites or reiterates the typical, and quotidian, quality of their labor. Like other films from this period and like their companion films from the Soviet Union such as *Salt for Svanetia* (Mikhail Kalatazov, 1930) or *Enthusiasm: Symphony for the Don Basin* (Dziga Vertov, 1931) the films exhibit a vivid desire to idealize the common working man as a vital part of a larger social whole.

Such scenes in *Coal Face*, *Night Mail*, *Listen to Britain* (Jennings and McAllister, 1942), *Fires Were Started* (Humphrey Jennings, 1943) and other films function as "typical particulars" in precisely the way Vivian Sobchack applies this term to film. The specific actions and objects viewed in a fiction may be highly concrete as relayed by indexical images, but they are not usually understood to have a concrete historical referent "...unless something happens to *specifically* particularize these existential entities as in some way singular.... [Instead] they will be engaged as what philosophers call *typical particulars*—a form of generalization in which a single entity is taken as exemplary of an entire class."[16] This displacement from the singular to the exemplary, if recognized as such, forfeits some of the distinctive peculiarities of the documentary re-enactment, most specifically the heightened sense of viewer responsiveness, and responsibility, that attends to the historical world.[17] It reintroduces characteristics of narrative fiction into the documentary domain in a furtive manner. This re-introduction calls for a study of its own.

3. Brechtian Distanciation. The re-enactment of social gests (such as those in the pioneering *Far from Poland*, but also abundantly evident in *Little Dieter Needs to Fly* and *2 Laws*, greatly increases the separation of the re-enactment from that specific historical moment which it re-enacts, giving greater likelihood that the phantasmatic effect will come into play. The sense of a here and now couples with an awareness of a there and then, so that the being-there-ness of the moving image and the having-been-there-ness of historical occurrence converge in the social gest. The gests of Dieter Dengler going through the motions of his captivity in a way that clearly re-animates its affective charge for him, but without benefit of a full-blown realist dramatization, allows the viewer to recognize the tangled temporality that has surrounded Dengler's ever since. It is as if the very act of distanciation brings the spectral quality of the past into greater relief.

4. Stylization. Highly stylized re-enactments such as those in *The Thin Blue Line* of Randall Adam's interrogation or of the Dallas police officer's murder, in which, most memorably, a perfectly lit container holding a malted milkshake tumbles through the night air in slow motion as if to blatantly over-dramatize one subject's account, also achieve a sense of separation. This need not be in the ironic key so prevalent in Morris's work. For example, *His Mother's Voice* (Denis Tupicoff, 1997), an animated documentary from Australia, couples the radio interview of a bereaved mother as she is

asked how she learned of her son's shooting, with an animated version of her journey to the house where her son had just been shot. These animated images sever any indexical linkage to the actual event, but give voice to·the acutely selective and pained perspective from which she experienced this tragic event. What is all the more remarkable in this mesmerizing film is that the sound track—her radio interview—repeats itself. As her account begins again, the animation shifts in tone and quality, and instead of taking us to the scene of the shooting it lingers within the family home, especially within the room of the now deceased son, as if to retrieve the memories that haunt this space. This highly stylized rendering of "what really happened" functions in a manner not unlike the abduction or disappearance re-enacted in *Las Madres de la Playa de Mayo*. It carries the element of stylization even further. The viewer remains vividly suspended in that moment between before and after, embodied in signifiers that possess an iconic rather than indexical relation to what has already happened.

5. Parody and Irony. Some re-enactments adopt a parodic tone that may call the convention of the re-enactment itself into question or treat a past occurrence in a comic light. Mockumentaries such as *Forgotten Silver* (Botes and Jackson, 1997), about an early but fictitious pioneer filmmaker in 1900s New Zealand, often adopt this tack. Errol Morris skirts on the edges of this characteristic, but his ironic perspective takes aim more at the subjectivity of his interview subjects than at the capacity of the re-enactment to capture the authenticity of past events. Caveh Zahedi, on the other hand, adopts the parodic re-enactment whole-heartedly in *I Am a Sex Addict* (Caveh Zahedi, 2005), a semi-serious account of his struggles with sex addiction, and the confusion it wrecks on his longer-term relationships and attempts at marriage. At one point, speaking to the camera, he tells the viewer that he lacked money to go back to Paris to re-enact his first encounters with prostitutes and so "this street" in San Francisco (the street on which he stands) will have to stand in for a Parisian street. The film cuts to another view of an evenly spaced line of about eight young prostitutes in front of a red brick wall as Caveh walks past, asking each of them the same questions about what they will do and how much they will charge, before hesitating, appearing ready to take them up on what they offer, but then deciding against it and going on to the next woman.

The Karen Carpenter Story (Todd Haynes, 1987), an underground cult favorite that cannot circulate legally because director Todd Haynes failed to secure permission for the sound track of songs by the Carpenters, tells the story of her eating disorders, dysfunctional family dynamics, addictions and ultimate death. Haynes does so by re-enacting key scenes from her short life with Ken and Barbie dolls. For the most part these scenes have the quality of typical particulars, exemplifying pivotal moments without reference to historically singular events. The posed shots of dolls, however, add a powerfully ironic edge to the representations: as with *His Mother's Voice*, this choice forfeits the impression of indexical authenticity in the image. At the same time it compels the viewer to assess this tragedy as something both beyond the reach of any re-enactment and as something typically reduced to a cautionary tale about the perils of anorexia and bulimia. The parodic edge puts the mass media's penchant for the realist dramatization of tragedy on display as a potentially exploitative trope. The doll figures, by maintaining a clear separation between re-enactment and prior event, may actually mobilize a more complex form of understanding of what this tragedy actually entails than more straightforward representations that confuse the boundary between the two.

Similar points might be made about *The Eternal Frame* (Uthco and the Ant Farm collective, 1975), the Ant Farm collective's parodic re-enactment of the Kennedy assassination. This video documents the re-enactment and the behind the scenes preparations that went into it far more than it purports to be a documentary about the assassination itself. Unlike *JFK* (Oliver Stone, 1991), which attempts to fit the black hole of what really happened with the round peg of a definitive explanation, *The Eternal Frame* calls the very act of re-enactment into question. By exaggerating the separation between then and now, before and after, this work functions to bare the device of re-enactment itself, rather than rely upon this peculiar form to achieve any final answer to the question of what really happened or generate a *mise-en-scène* in which the desire for a lost object might find a gratification.

Re-enactments within these clusters of overlapping and fuzzy categories do not do what archival footage and other images of illustration do.[18] They do not provide evidentiary images of situations and events in the historical world. If they allow viewers to think that they do, they lay the groundwork for feelings of deception. The indexical bond, which can guarantee evidentiary status—but not the meaning or interpretation of images taken as evidence—no longer joins the re-enactment to that which it stands for. This bond joins the image to the production of the re-enactment: it is evidence of an iterative gesture but not evidence of that for which the re-enactment stands. It is, in fact, an interpretation itself, always offered from a distinct perspective and always carrying, embedded within it, evidence of the voice of the filmmaker.

Re-enactments do not contribute inartistic proofs—in Aristotle's description of rhetoric—but artistic ones. Not evidence but invention. Although it is possible, especially with realist dramatizations and typifications, to think that they do contribute evidence, what they actually contribute is persuasiveness. They fulfill an affective function. For documentaries belonging to the rhetorical tradition, re-enactments intensify the degree to which a given argument or perspective appears compelling, contributing to the work's emotional appeal, or contributing to its rational appeal by means of real or apparent proof.

As pathos or logos, re-enactments enhance or amplify affective engagement. Re-enactments contribute to a vivification of that for which they stand. They make what it feels like to occupy a certain situation, to perform a certain action, to adopt a particular perspective visible and more vivid. Vivification is neither evidence nor explanation. It is, though, a form of interpretation, an inflection that resurrects the past to re-animate it with the force of a desire. They provide a key element of the *mise-en-scène* within which a documentary phantasmatic takes shape.

Re-enactments effect a fold in time. They vivify the sense of temporal duration that extends from here to there, now to then. Lived experience thickens and twines about itself. They take past time and make it present. They take present time and fold it over onto the past. They resurrect a sense of a previous moment that is now seen through a fold that incorporates the embodied perspective of the filmmaker and the emotional investment of the viewer. In this way re-enactments effect a temporal vivification in which past and present co-exist in the impossible space of *une phantasmatique*. Re-enactments stand for something different from that for which they would stand. This difference manifests itself in the resurrected ghosts that both haunt and endow the present with psychic intensity. Re-enactments, like other poetic and rhetorical tropes, bring desire itself into being. They activate the phantasmatic domain wherein the temporal duration of lived experience and the efficacy of ideology find embodiment.

Filmography

Capturing the Friedmans (Andrew Jarecki, 2003)
Chile, Obstinate Memory (Patrizio Guzmán, 1999)
Coalface (Alberto Cavalcanti, 1935)
Enthusiasm: Symphony for the Don Basin (Dziga Vertov, 1931)
The Eternal Frame (T. R. Uthco and Ant Farm Collective: Doug Hall, Chip Lord, Doug Michels, and Jody Proctor, 1975)
Far from Poland (Jill Godmilow, 1984)
Fires Were Started (Humphrey Jennings, 1943)
Grizzly Man (Werner Herzog, 2005)
His Mother's Voice (Denis Tupicoff, 1997)
I Am a Sex Addict (Caveh Zahedi, 2005)
JFK (Oliver Stone, 1991)
Las Madres de la Playa de Mayo (Susana Blaustein Muñoz and Lourdes Portillo, 1985)
Listen to Britain (Humphrey Jennings and Stewart McAllister, 1942)
Little Dieter Needs to Fly (Werner Herzog, 1998)
Lonely Boy (Wolf Koenig, Roman Kroitor, 1962)
Mighty Times: The Children's March (Robert Houston, 2004)
Nanook of the North (Robert Flaherty, 1922)
Night Mail (Harry Watt and Basil Wright, 1936)
Reconstruction (Irene Lusztig, 2002)
Salt for Svanetia (Mikhail Kalatazov, 1930)
The Road to Guantanamo (Michael Winterbottom and Mat White-cross, 2006)
The Thin Blue Line (Errol Morris, 1988)
2 Laws (Carolyn Strachan and Alessandro Cavadini, 1982)
The War Game (Peter Watkins, 1965)

Notes

[1] This formulation derives from Gregory Bateson's insightful essay on the difference between play and fighting among animals where a nip no longer means exactly what a bite, to which it refers, would mean: "The playful nip denotes the bite, but it does not denote what would be denoted by the bite." "A Theory of Play and Fantasy," *Steps to an Ecology of Mind* (Chicago: University of Chicago Press, 2000). The distinction is akin to Gilbert Ryle's discussion of the difference between an unintended blink and a fully intended wink in "The Thinking of Thoughts: What is 'Le Penseur' Doing?" *University Lectures*, No. 18 (University of Saskatchewan, 1968).

[2] I am deeply indebted to Terry Castle's essay "Contagious Folly: An Adventure and Its Skeptics," in a special section of *Critical Inquiry*, vol. 17, No. 4 (Summer 1991): 741–772, devoted to "Questions of Evidence" in which she makes clear the imbricated relationship between a particular *folie à deux* and the work of ideology.

[3] The formulation derives from Louis Althusser's essay, "Ideology and Ideological State Apparatuses," in *Lenin and Philosophy* (New York: Monthly Review Press, 1971). The emphasis here is on the subject's psychical investment in a given state of affairs, rather than on the idea of ideology-in-general, as that which constitutes the subject. Though related, the latter point leads to numerous questions about the degree to which such a subject is essentially determined by and subject to the work of ideological apparatuses. However that question is answered, the centrality of a psychical investment is what provides a linkage between ideology and a fantasmatic.

[4] Jean Laplanche and J.-B. Pontalis, *The Language of Pscyho-Analysis* (New York: W.W. Norton, 1973), 317.

[5] Jean Laplanche and J.-B. Pontalis, "Fantasy and the Origins of Sexuality," *The International Journal of Pscyho-Analysis*, vol. 49, part 1 (1968): 16. The essay also appears in a booklet that reprints the original essay first published in *Les Temps modernes* in 1964: *Fantasme originaire, Fantasme des origins, Origines du fantasme* (Paris: Hachette, 1985).

[6] Laplanche and Pontalis, "Fantasy and the Origins of Sexuality," 15.

[7] Ibid.

[8] Ibid., 15, fn 36.

[9] Ibid., 16.

[10] Ibid., 17.

[11] "In great moments of cinema you are hit and struck by some sort of enlightenment, by something that illuminates you, that's a deep form of truth, and I call it ecstatic truth, the ecstasy of truth, and that's what I'm after in documentaries and feature films." Werner Herzog, "Fresh Air," NPR radio, July 28, 2005.

[12] Some speak of subjectivity in documentary. This, to me, represents something of a slippery slope, slipperier than the use of subjectivity in narrative fiction which is usually related to the perspective of characters and the voice of the narrator. Both are different concepts from documentary subjectivity that sets out to admit that documentaries represent situated, emotionally and politically informed views of the world. Though true, they become inevitably contrasted with an alternative idea, a different way of representing or engaging with the world, objectivity. The issue of objectivity enters, like a Trojan Horse, in ways it does not do in fiction and it causes endless trouble.

Objectivity in relation to a fictional world might seem a peculiar notion since the fiction is a subjectively endowed creation by definition, or it may seem like a way to identify a scrupulously neutral, detached mode of representing it, in the spirit of *écriture blanche*, a zero degree of style. In documentary, objectivity implies a lack or subtraction of subjectivity as if subjectivity could be put on, taken off or stepped beyond, as if it were a bias. Unlike a fiction, the actual world, it is argued, can be viewed objectively, unless the decision is made to "add" subjectivity. In some instances, like scientific investigation, subjectivity can be subtracted to a great extent but these instances are not the instances in which an I stands before a You; they are instances of I's embedded within institutional procedures and discourses that objectify or analyze, that have instrumental effects—for good or ill—but that cloak the I or You in ways the voices of these films refuse to do. Voice affirms the presence of an embodied subject who is necessarily and inescapably in possession of a subjectivity. Objectivity catapults us into another realm entirely.

[13] Brecht regarded social gests as physical actions that revealed social relations, or as Roland Barthes put it, "What then is a social gest? It is a gesture or set of gestures (but never a gesticulation) in which a whole social situation can be read. Not every gest is social: there is nothing social in the movements a man makes in order to brush off a fly; but if this same man, poorly dressed, is struggling against guard-dogs, the gest becomes social." "Diderot, Brecht, Eisenstein," in *Image, Music, Text* (New York: Hill and Wang, 1977), 73–74.

[14] Originally shot in 16 mm film in 1978, the filmmakers resisted transferring it to video or laserdisc because of concern for the resulting quality of the image. Facets Video is now in the process of releasing a freshly prepared DVD version of the film, with additional commentary and new scenes bringing the original struggle to recover tribal lands up to date.

[15] In 2006 a court victory allowed the Aboriginal people to reclaim several islands off the shore of north-central Australia as tribal lands.

[16] Vivian Sobchack, *Carnal Thoughts: Embodiment and Moving Image Culture* (Berkeley, University of California Press, 2004), 281.

[17] Sobchack develops this point on page 283 in relation to fiction film and moments when the image ceases to function as a typical particular and takes on the full force of a singular moment, such as the image of a real rabbit shot during the fictional hunting scene in *The Rules of the Game*. This attitude seems a default value for documentary film in general.

[18] Images of illustration comprise those images utilized to support a typically verbal argument or perspective. They offer particular instanciations of points that may imply broader application or offer what appears to be evidence in support of a specific assertion.

Shapes of Modernity

Time and Progress—Time as Progress

An Enlightened Sermon by William Robertson

László Kontler

In the introductory studies of his seminal *Futures Past*, Reinhart Koselleck offers an engaging and succinct illustration of the course of what he calls the "temporalization of history" in European thought during the early-modern period. Koselleck conceives the process as a whole in terms of the changes in the perception of the "compression" (or "acceleration") of time that, supposedly, precedes the onset of the future in the thought of these past generations: "For Luther, the compression of time is a visible sign that, according to God's will, the Final Judgment is imminent, that the world is about to end. For Robespierre, the acceleration of time is a task of men leading to an epoch of happiness, the golden future."[1] In the intervening period, experience showed that religious and civil wars did not herald the Final Judgment, at least not in the direct manner previously envisaged. The absolutist state suppressed prophecy, while humanists and skeptics revealed its psychology, undermining oracles and associated superstitions; as a "counter-concept" of prophecy, rational prognosis marked out new horizons for the future by both remaining within the dimensions of the (political) situation and attempting to change it or "relat[ing] to events whose novelty it releases." Finally, in the eighteenth century, the appearance of the philosophy of the historical process, exploiting the notion of progress in order to combine rational prediction with salvational expectation, "inaugurated our modernity with a new future." "Acceleration," Koselleck writes, "initially perceived in terms of an apocalyptic expectation of temporal abbreviation heralding the Last Judgment, transformed itself—also from the mid-eighteenth century—into a concept of historical hope."[2]

This paper, as might be described in Koselleck's terms, is an attempt to grasp the very moment of this transformation through looking at a single text by a single eighteenth-century author, and also to point to certain limitations in this transformation that existed within the discourse of the Enlightenment itself. However, my aims are not merely illustrative but analytic and critical as well: the test case chosen also allows the author to express some reservation, and to offer correctives to the approach that associates the Enlightenment with "secularization" and a "critical spirit" which, despite the profoundly nuanced character of Koselleck's presentation, is a *Leitmotif* in his work.[3] Despite its sophistication, the Koselleckian typology of conceptualizations of time is somewhat teleological in the sense that in its terms any approach which marries Christian stories—endeavors and expectations—of salvation with those of the improvement of the temporal condition of man, is likely to be divested of its intellectual distinctiveness and to be discussed as a transitional position, at some distance both from pre-modern "origins" and modern "culminations." My uneasiness with this perspective, is conceived within a tradition of Enlightenment studies perhaps most vigorously represented by John Pocock. Pocock and others have called attention to the fact that in the Enlightenment there was a significant element of, and a strong desire for, continuity and conservation as well as (in some cases, even more than, and anyway mutually dependent on) change; and also that "among the phenomena which 'Enlightenment' connotes for us there are some whose effect may be termed 'conservative' in the sense that it was to strengthen elites—some of them clerical—in their capacity for civil control."[4]

A considerable portion of Pocock's *oeuvre* is also devoted to the plurality *versus* the uniformity of the European Enlightenment, and in a sense the present study addresses some aspects of this debate. This problem is approached on the basis of the evidence that the Enlightenment as an intellectual movement embraced the *whole spectrum* of efforts to confront the challenges of commercial modernity, of the erosion of the Christian and republican ethical foundations of western societies, from the "crisis of the European mind"[5] in the late seventeenth century to the era of the French Revolution—at least those segments of the spectrum that were not confined to a mere repudiation or negation of these challenges. The questions were the same, or very similar to one another for those who contributed to taking them up. How is it possible to eliminate the religious and political conflict inspired by the extremes of "superstition" and "enthusiasm" that had marred the social and political atmosphere of the sixteenth and seventeenth centuries? How is it possible to accommodate commerce, which had become inevitable and indispensable for modern societies, but equally inevitably reinforced the self-regarding impulses inherent in human nature, with the moral imperatives of cooperation, sympathy and public spirit in human collectives? How is it possible to enshrine the dignity of man in constitutions that also allow for strong government and stability? These, and a great many other questions defined themes, fields, and pursuits in eighteenth-century intellectual inquiry that were central to what we know as Enlightenment: toleration and the "natural history" of religion, political economy and conjectural history, natural law, and so forth.

An influential stream in recent Enlightenment studies has been inspired by the idea that the plurality mentioned above can be most adequately conceived in national or regional terms;[6] others try to give substance to this approach by stressing that "enlightenments" were fundamentally distinguished by their ideological commitments.[7] It is also possible to adopt a somewhat different perspective while accepting the fundamental premise shared by these positions, namely that the Enlightenment can no longer be understood as the "influence" radiating from the *salons* of Paris with versions of their "free-thinking" more or less diluted according to time and place. In this case one might suggest, as it indeed has been suggested by introducing further sophistication in the comparative study of the Enlightenment, that while the questions asked were fundamentally similar, the answers could differ quite considerably.[8] Besides regional and local variation, this also implies that from intellectually sophisticated and (in the sense outlined) "enlightened" confrontations with the above-mentioned issues it was equally possible to arrive at conclusions *protective* and *subversive* of the political, social and intellectual status quo of the European old regime. Without necessarily using the Pocockian label of "conservative Enlightenment," this is the thrust of a sizeable body of scholarship devoted to the Scottish Enlightenment. For instance, it has been demonstrated that it was perfectly possible for individuals embedded in cultural and ecclesiastical elites and eschewing religious and political heterodoxy, such as the "moderate literati" of Edinburgh, to pursue ends usually acknowledged as being typically enlightened.[9] My test case in this paper will be the first published work and only printed sermon, on *The Situation of the World at the Time of Christ's Appearance, and its Connection with the Success of His Religion* (1755), of one of the members of this group: William Robertson (1721–1793), ecclesiastical leader, long-serving Principal of the University of Edinburgh, historiographer royal and the author of some of the historical bestsellers of the eighteenth century, who was ranked with Hume, Voltaire and Gibbon by his contemporaries.

In more ways than one, Robertson was a synthetic figure. As a historian he stood at the intersection of several traditions of historical studies in the eighteenth century: national and cosmopolitan, narrative and "philosophical," stadial or "conjectural."[10] In his *History of Scotland* (1759), Robertson endeavored to write a kind of national history which transcended the traditional confines of partisanship and vainglory by establishing a European context for it through a comparative analysis of social structures. The slow dismantling of feudalism in Scotland, relative to the pace at which this took place elsewhere, seemed to him the chief cause of Scotland's sixteenth- and seventeenth-century calamities; while as a progressive Whig he thought that the Union of 1707 had sufficiently remedied this predicament.[11] The European horizon, already adopted in his first book, was later expanded into a study of the emergence of an European state system, and the beginnings of balance of power against a background of religious strife and the struggle for hegemony between high-taxing monarchies maintaining large standing armies in the *History of Charles V* (1769). The volume-length introduction of this work, *A View of the Progress of Society in Europe from the Subversion of the Roman Empire to the Beginning of the Sixteenth Century*, which traced Europe's emergence from feudal barbarism and the growth of refined manners and the institutions of the rule of law through commerce and chivalry, was considered a model for the grafting of stadial history on philosophical history, and perhaps Robertson's most famous text. His subsequent books employed the same approach and aimed at exploring the ties forged through commercial and cultural exchange, as well as imperialism between the unit thus created (i.e., Europe) and the rest of the globe (*History of America*, 1777;[12] *An Historical Disquisition of the Knowledge which the Ancients had of India*, 1791). This made Robertson one of the quintessential authorities of eighteenth-century "cosmopolitan history."[13]

From yet another point of view—and most importantly for the purposes of this paper—Robertson must be regarded as a Christian historian who was at the same time one of the outstanding masters of enriching the "enlightened narrative" with the perspective of "stadial history," most commonly associated with Adam Smith and the French physiocrats.[14] He understood the history of the western world as the unfolding of the great plan of Providence, a gradually increasing accessibility of the divine revelation, a process which in his view crucially depended on, but also furthered, the improvement of the means of subsistence, and the consequent refinement of manners and enlightenment of the human mind. There is reason to believe that Robertson's almost unparalleled contemporary popularity as an author of historical works was to a considerable extent due to his power in representing this synthesis—for which, however, taking account of the problematic relationship between Christian and secular understandings of time, there was an important theoretical condition. This, I want to suggest, is one of the tasks performed in Robertson's early sermon, published at a time when he was also busy working on the historical narrative which first established his literary fame.

Before turning to the sermon, it will be useful to address two questions. The first concerns the making of Robertson as a historian, in particular the nature of the challenges and dilemmas that the adoption of a stadialist-relativist position implied for a scholar desirous of retaining a Christian framework of interpretation. The second, not unrelated to this issue, is Robertson's theology—or rather the little that can be known about the theology of an influential minister whose public statements about the church concerned its social role rather than its doctrine, who left no autobiography, whose commonplace-books disappeared, and whose surviving correspondence is predominantly businesslike and silent on matters of personal sentiments, convictions and faith.

In the late 1730s Robertson attended, and took a great deal of inspiration from, the classes of Charles Mackie, the first professor of universal history at the University of Edinburgh.[15] Mackie was a practitioner of philosophical history, with a scholarly and skeptical approach, in the style of Guicciardini, Machiavelli and other humanist—and ancient—masters: a version of history where the task was to provide a selective narrative of events with a view to revealing men's moral and political character, and to highlight to readers or listeners, "by example," the principles that uphold or, on the contrary, undermine public good.[16] Adopting this scheme of explanation, however, depended on the historian's willingness to regard past historical agents as his contemporaries whose actions could be judged by timeless standards of morality. Even the kind of relativism introduced by Machiavelli could be accommodated in this scheme: the circumstances that warranted an otherwise questionable morality were themselves entirely contingent in the sense that they might occur with equal probability in any age or society.

It was not until the late 1740s that Robertson, already having embarked on the research towards his first great work, was confronted with the kind of historical approach that he would embrace and apply consistently in his own narratives. Montesquieu's preoccupation with the effects of the physical environment and the historical principles which animated the laws and customs of an age and which determined the spirit of the people made a deep impression on him; as did Hume's alternative to Montesquieu, suggesting that the moral determinants of national character had to do with political, legal, religious and cultural institutions, rather than geography or climate. Above all, Robertson became inspired with the system of historical jurisprudence which the young Adam Smith outlined in a series of lectures in Edinburgh in 1748–50. Smith endeavored to show that the principles of justice and politics depended on sentiments, manners and customs which, in turn, were themselves functions of the means of subsistence and the distribution of property. He combined these observations with a theory of the stadial progress of civilization from rudeness to refinement, or hunting-gathering through shepherding and agriculture to commerce.[17]

We can now grasp the challenge which this approach constituted for the philosophical historian, especially if that historian was as good a Christian as it may be expected from a devoted (and ambitious) minister of the established *Kirk* of Scotland. In terms of stadial or "conjectural" history, people who lived in civilizations different from our own were separated from us by a cultural chasm to the extent that they not only possessed different manners and opinions, but even different minds and selves. Any attempt to assess their moral, political or other virtues by standards other than their own was wrong, not only in the sense of being unfair towards the objects of the investigation, but also methodologically incorrect and inevitably doomed to failure. Most disturbingly, then, it became unclear what lessons the modern reader could learn by studying the past. Especially perplexing was the question what the knowledge of the progress of civilization could reveal about the eternal and unchanging mind of God.

Perhaps the difficulty is not so great as it might at first seem. After all, even Augustine stressed the significance of context: he displayed an acute awareness that man could only act in his own age, that humans before and after Christ could not be expected to be the same, and that good and evil ought to be judged in terms of the conditions necessary to the individual at the particular time and place.[18] The point, however, is that, this is still possible to explain in terms of a conscience that places the highest priority on personal spiritual progress occurring within a narrative of creation, fall, incarnation and redemption. While these truly cataclysmic events may certainly be identified with points in time, the succession of particular events between them is not rendered intelligible, nor is any special importance ascribed to time itself as the dimension of that succession. The time-bound experience of individuals is contrasted to a timeless and eternal God, occupying a *nunc-stans*, a standpoint from which he can see every moment in time as simultaneously present. To man, whose intelligence is imprisoned in one moment, the knowledge of another is neither quite possible, nor quite relevant. Insofar as it is still both possible and relevant, it has to do with Providence. It is our awareness of divine foreknowledge that persuades us about the meaning of each apparently insignificant episode in the flow of history from one cataclysm to the other.[19]

Centrally important for the topic of this paper, Robertson's own views of Providence were heavily influenced by his early acquaintance, through his father's library, with the work of late seventeenth- and early-eighteenth-century Arminian authors such as the Huguenot refugee Jean Le Clerc, John Locke's Dutch friend Philippe van Limborch, or the Swiss Samuel Werenfels.[20] Theologically, Arminianism was defined by its opposition to the absolute predestination that Calvin had argued, and by a greater emphasis on man's free will. Philosophically, it was based on a constructive and mitigated skepticism that established a permanent suspension of judgement (rather than doubt) as a means of arriving at truth. For Robertson the minister and church politician, the import of Arminianism was its being instrumental in combating the Calvinist orthodoxy prevailing in the Presbyterian *Kirk*, and to reshape it as a moderate and tolerant establishment. For Robertson the historian, Arminianism was a way to accommodate human agency with God's sovereignty, the central tenet of Calvinist theology. Even for Limborch, it had been possible to acknowledge God's power in ordering the universe, while finding in that ordered universe a scope for independent human action: actions by human individuals making free choices, but ones which invariably contribute to the plan of God.[21] God does not coerce or decree absolutely, but orders the interaction of the parts of the universe in accordance with His grand yet varied design, which admits some flexibility regarding how His ends will be accomplished.

From this it is possible to develop a synergetic view of historical agency, according to which human actions may be seen as expressions of divine providence, while at the same time God's providence may be conceived as offering so many opportunities for the exercise of human will. This is what Robertson was doing in the sermon. In fact, he sought to accommodate dramatically differing views within a larger whole, in order to give an account of the sequentiality of events and of the rhythm of historical changes that precede and prepare the cataclysmic events of Christian history and fill the time gaps between them. To make the whole scheme even more paradoxical, he also relied on the incipient, essentially materialist, interpretations referred to above, which portray human beings essentially as creatures of need. Hume and Smith argued that our needs and our understanding of needs are historically determined, and that our minds will only develop insofar as we need them to develop in order to go about the business of seeking the satisfaction of our needs. Robertson's move that aimed to marry these views with his providentialism proposed to shift the argument from the mind itself to the circumstances in which the mindful human being finds him or herself. From this line of reasoning, our understanding will only develop in proportion to the development of the faculties we possess to improve the world in which we find ourselves. With improvement comes an increased understanding of the material and the spiritual world, and only then can God be expected to display more of His being and nature to us. To orthodox Presbyterians, with whom Robertson was trying to build bridges, the theological consistency and rigor of this position may have looked shaky. But this did not upset Robertson who, in fact, took pains to evade the immensely difficult metaphysical and theological issues at stake, and strove, instead, to provide a pragmatic scheme in which the emphasis was on social progress and on the intended impact of civic harmony—objectives in whose attainment he did not fare too badly.

The sermon was preached before the Society for the Propagation of Christian Knowledge on 6 January 1755. From the outset, Robertson leaves no doubt that his preoccupation is with the problem of design in human history, and shortly thereafter it is also made clear that he intends to confront the problem in terms of "before" and "after"—i.e., before and after one of the epoch-making events of sacred history, the advent of Christ and the preaching of the Gospel.

> There is no employment more delightful to a devout mind than the contemplation of the divine wisdom in the government of the world. The civil history of mankind opens a wide field for this pious exercise. Careful observers may often, by the light of reason, form probable conjectures with regard to the plan of God's providence, and can discover a skilful hand directing the revolutions of human affairs, and compassing the best ends by the most effectual and surprising means: But sacred history, by drawing aside that veil which covers the counsels of the Almighty, lays open his designs to the view of his creatures; and we can there trace the steps which he taketh towards accomplishing them with more certainty, and greater pleasure. ... The publication and establishment of Christianity in the world is a remarkable event of this kind.[22]

What Robertson sets out to address is the objection by "[Christ's] adversaries ... and modern infidels" that if the Gospel is indeed the truth, why was it "so long concealed from the world"?[23] Robertson's problem, then, becomes a problem of time: Why so late—and not earlier? He seeks to answer the question by reference to the "divine oeconomy" and the "particular juncture to render the discovery of the Christian religion more necessary, or the propagation of it more successful." He is concerned with the urgency of the revelation in a specific historical moment. His particular explanations befit a conjectural historian who was at the same time a Presbyterian minister with a strong Arminian inspiration.

To begin with, Robertson lays down two general principles. First, it is one of the general laws whereby "the Supreme Being conducts all his operations" that "no perfection of any kind can be attained of a sudden. The motion by which his works advance towards their final and complete state is gradual and progressive." He also expresses the same principle in the metaphor of time: "The obscurity of dawn went before the brightness of noon-day." As a consequence, it was "in proportion as the situation of the world made it necessary, [that] the Almighty was pleased farther to open and unfold his scheme."[24] Second, Robertson stresses that although there is a strong and manifest design in human history, direct interventions by God are infrequent, and even then they are organically embedded in a context of processes predominantly triggered by merely human agency. "The Almighty seldom effects, by supernatural means, any thing which could have been accomplished by such as are natural."

The advent of Christ is of course one of these rare supernatural interventions, but the thrust of Robertson's analysis is to demonstrate how it was catalyzed by the confluence of a colorful variety of natural causes that, as it were, increased the density of history or accelerated the flow of time, after a long period of stagnation. Providence and human agency are thus assigned a mutually supplementary role in bringing about the design in human history, the latter, while "ordained in reality by the wisdom of God," still possessing a sufficient degree of independence to create conditions propitious for the working of the former, should that prove "necessary." In the particular case discussed in the sermon, the advent of Christ is at once a supernatural event and an event in the secular world (domains between which Robertson is moving constantly), an event that has been thoroughly prepared for by previous history.

Time, then, itself becomes a dimension not only defined by the rhythm of the "cataclysms," but one also marked by a periodicity emerging from the contemplation of human activity exerted in the intervals between those cataclysms and deriving momentum from the period preceding them—and as a result, contributing to the crucial definitions of "before" and "after." It would be tempting to explore the extent, if any, to which Robertson may have relied on relatively recent philosophical approaches to time, each of which could be easily demonstrated to have been relevant for these perceptions. These include, first, Newton's ideas of "absolute" and "relative" time, the one being an equable flow, in irreversible succession, of a mathematical straight line, independent of matter and motion, while the other is the relation between time and sensible objects, depending very much on motion whose rate varies.[25] Second, Locke's attempt to provide Newton's approach with an empiricist epistemological grounding by explaining time in terms of duration as traced to its source in sensation and reflection.[26] Third, Leibniz's retort to Newton that, were time merely absolute, there would be no reason for things (including the Creation!) to exist at one time rather than at another (and, therefore, all time can only be "relational").[27] However, while these sources were easily available for Robertson, there is no evidence known to this author that he did avail himself of them. What he did employ, with a great deal of ingenuity, was the organizing principle of stadial history: the idea that, because of a number of natural propensities of the human animal, societies have undergone certain stages of progress that can be defined as of the dominant mode of subsistence, and the degree of refinement expressed in their standards of conduct, as well as their ability to comprehend sophisticated and abstract notions of morality, religion, etc., depended on the stage reached in that process.

To be sure, the argument that the Word had not, and could not have been, revealed to the world until it was ready to receive it, is at least as old as Augustine.[28] However, the dynamics that Robertson added to this view were of a peculiarly eighteenth-century character in the suggestion that even the world of primitive Christianity had been unrefined and pre-commercial, inhabited by people who *therefore* could not have possibly understood the laws whereby God exercised His governance of the natural and moral worlds; and, consequently, that God could have only revealed as much of His Word as the primitive Christians were able to understand. It was also necessary to assume that the rest would be revealed gradually as progress made it appropriate. It must be added—and it does not contradict the argument of this paper—that the interdependence of revelation and progress is fully reciprocal for Robertson. He indeed also believed "revelation to be critical for the true refinement of manners and for moral improvement, and that without revelation, human intellectual and cultural development will be limited and inevitably lead to error, delusion, and moral corruption."[29]

Robertson's conjectural history of the propagation of the Gospel starts with the observation that "[t]he world, in the most early ages, was divided into small independent states... Commerce had not hitherto united mankind, and opened the communication of one nation with another. The world may now be considered one vast society... But, in those more simple ages, the intercourse between nations was extremely inconsiderable."[30] Naturally enough, such conditions, in which mankind had languished too long, had by themselves constituted an insurmountable obstacle to the propagation of the Gospel across the whole of the western hemisphere. The catalytic role of removing this obstacle was played by "Roman ambition and bravery" that "paved the way, and prepared the world for the reception of the Christian doctrine": union and tranquility, as well as civilization, all corollaries of conquest and enslavement by the Romans, brought about with them as an unintended consequence in the best Smithian fashion, the moment auspicious for the global spread of Christianity.[31]

Besides the civilizing effects of Roman expansion, there were moral causes too, related to the former in a rather paradoxical way. The Roman empire imposed itself on the small independent states of earlier times in which public liberty rested on the foundation of the private virtues—in regard of which, however, "the conduct of every citizen was subjected to the eye of the magistrate." The Romans themselves were no exception from this rule; "[but], by subduing the world, [they] lost their own liberty … The alliance between morals and government was now broken … Together with despotic power, entered all those odious vices, which are usually found in its train." The corruption characteristic of empires that succeeds upon the republican purity of manners, however, supplied the occasion for God to "manifest the Christian revelation to the world, not to re-establish virtue upon the same insecure foundation of civil government [mere human agency], but to erect it upon the eternal and immoveable basis of religion." In Robertson's account Christianity appeared in order to mitigate the pernicious effects of "despotic and unlimited empire"—as well as the luxury which inevitably proceeds from safe commerce over a vast territory—and to perpetuate virtue among men by divine causes at a time when human causes were no longer sufficient to effect this.[32]

Robertson then considers the state of the world with respect to religion, domestic affairs and what might be called social justice and finds that in these terms, too, it was insufficiently critical "crisis"—in this case denoting a sort of pregnancy of changes—to invite a thorough-going "reformation." Religion languished between extreme forms of corruption as represented by the superstition and hypocrisy of the Pharisees and the libertinism of the Sadducees. The theme is developed by Robertson in terms vaguely resembling the version of the Enlightenment discourse on religion as presented in one of the most famous essays of Hume, first published in 1741.[33] This was a discourse which employed the dichotomy of superstition and enthusiasm, as the two archetypical forms of false religion, to account for the social and political turmoil of the preceding two centuries all over Europe, and offered itself as an antidote. For some, like Hume, this could be skepticism, but for many others it was "moderation," or the virtuous middle: a sober and reasoned commitment to religious truth without subscribing to either the fanatic conceitedness of those sectarians who claimed immediate divine inspiration, or an uncritical submission to authority. Robertson also conceived of two extreme attitudes, between which the force of true religion evaporated. To him as well, the one was superstition; the other, for the time being, he styled as "scandalous libertinism."

It was only shortly thereafter, that he, as an ecclesiastical leader, recognized a militant interpretation of Calvinism, as professed by a considerable party within the *Kirk*, to be an even more dangerous disposition.[34] A mere year after the sermon was preached, the famous *Edinburgh Review*, which boasted Robertson among its founders, came under attack by Calvinist enthusiasts who protested against its criticisms of their theological works. At approximately the same time Robertson and his moderate associates in the church had great difficulty in averting the excommunication of Hume and his cousin Lord Kames as pernicious skeptics. Such struggles occupied Robertson throughout his career as a church politician, until his retirement in 1780—when riots presented a threat to his personal safety. These riots were evoked by the lifting in 1778 in England, and initiated in Scotland too, of some of the centuries-old sanctions against Catholics, and because of which the Scottish Relief Bill was shelved. Shocked, Robertson said in one of his last speeches: "I love to see my countrymen discover that jealous concern for the preservation of their rights which characterizes the spirit of liberty: but I am sorry to behold them wasting their zeal without a cause." He called on the church to denounce "the principle for conscience sake, as repugnant to the spirit of the gospel, and contrary to the genius of the Protestant faith."[35]

As regards the "regular system of superstition" introduced among the ancient Jews by the Pharisees, this type of "false religion" already stands in full armor before the reader of the sermon: the proliferation of traditions, ceremonial prescriptions and rites caused the decline of principles. "Superstition never prevailed among any people, but at the expence of morals. The heathen superstition, far from giving any aid to virtue, seems not to have had the least connection with it." As elsewhere, political degradation is also consequent upon the spread of superstition and the moral decay it occasions: "Tyranny and superstition, like those other destroyers of mankind, famine and pestilence, are nearly allied. Superstition breaks the spirit, and prepares it for servitude. Tyranny, for this reason, encourages superstition, and employs it as a useful auxiliary to illegal power."[36]

Further on, Robertson presents the domestic scene during the times immediately preceding the appearance of Christ, in dark tones, as having been marked by polygamy in the East and by the practice of divorce carried to extremes among both the Jews and the heathens of the West, the one conducive to domestic slavery, and the other bringing the idea of the natural bond between man and woman into disrepute. Finally, in view of "the wants of human society ... far the greater part of mankind is condemned to constant toil and labour, in order to supply them," and the primitive means of subsistence in ancient times, the majority of people were reduced to slavery—a state that became really intolerable under the despotic government of the Roman empire. In other words, the religious attitudes of virtually all, and "the lives of those who are at the head of domestic society, needed reformation," while "the sufferings of those who were subject to them merited relief."[37]

Time in secular terms was then on all fronts—social and domestic, political and moral—ripe in a peculiar sense for the most important event of sacred time between the Creation and Redemption to happen. And indeed, in the time "after" (i.e., after the incarnation) the benevolent potential inherent in Christianity on all of these fronts asserted its corrective effect on the very phenomena in the secular domain whose "unintended consequence" its appearance was. Particularly noteworthy is Robertson's unhesitating ascription of the abolition of slavery to the mild and liberal spirit of Christianity: indeed the Book of Isaiah is cited by him in order to draw a parallel between the spiritual salvation prophesied there and the temporal deliverance from personal servitude.[38] The mildness and humanity of modern manners is summarily represented as having been inspired, or what is more, awakened, by the Christian religion.[39]

Here, however, there seems to be some confusion about cause and effect, at least if the whole of Robertson's historical thought is taken into consideration. Of the entire "great generation" of the Scottish Enlightenment, he was perhaps the most straightforward "progress-and-refinement" thinker. He came closest to believing that progress was irreversible, that the values and virtues of modernity were ultimately superior, and that man's capacity to absorb and comprehend sophisticated truths and develop refined perceptions of his moral and physical environment depended on the advance of civilization in more broadly conceived terms. All of his thinking revolved around the recognition that commerce had a transformative effect on civilization. Market relations and commercial exchange on the one hand functioned as a generic metaphor to describe so many other forms of human intercourse, while on the other hand it was also a very direct form of communication, which, by virtue of its peculiar rules, was especially well suited to grasping the needs and interests of the one party in terms of and as depending on those of the other. The extent that commerce comes to prevail in supplying men's needs enhances the opportunities for intercourse to lead to the growth of sympathy, politeness and sociability, as well as affluence and knowledge, even among otherwise self-regarding individuals. Emulation, inspired by self-regard, which had once been violent, assumes milder forms,[40] until even laws, issued by the civil magistrate to tame such passions and suppress eruptions of violence, cease to be regarded as cumbersome limitations of liberty, but rather come to be valued by polished citizen-subjects as the instruments of the rule of law.

This conspicuously materialist logic could be abundantly documented from the works of Robertson. The following passage, which concerns the awakening of medieval urban communities from their long slumbering, is taken from the classic blend between narrative and stadial history:

> The spirit of industry revived: commerce became an object of attention and began to flourish: the population increased: independence was established: and wealth flowed into cities which had long been the seat of poverty and oppression. Wealth was accompanied by its usual attendants, ostentation and luxury; and though the former was formal and cumbersome, and the latter inelegant, they led gradually to greater refinement in manners and in the habits of life. Together with this improvement in manners, a more regular species of government and

> police was introduced. As cities grew to be more populous, and the occasions for intercourse among men increased, statutes and regulations multiplied of course, and all became sensible that their common safety depended on observing them with exactness, and on punishing such as violated them with promptitude and rigour. Laws and subordination, as well as polished manners, taking their rise in the cities, diffused themselves insensibly through the rest of the society.[41]

From Robertson's views on the formative effect of material progress on manners and the mind it would not necessarily follow that the truth of the gospel could at once triumph among the prevailing conditions of civilization, represented by him as rather primitive. And indeed, while on the one hand he thought that cultural progress itself was of doubtful value, with even dangerous consequences, in the absence of revelation, on the other hand he also believed that it was in his own century that religion, which at the time of the Reformation and the Counter-Reformation was still rooted in a necessarily imperfect understanding of the Word of God, and penetrated by superstition and enthusiasm, could at last be understood as it was intended by the Almighty and his messenger. Developments in secular human history, then, having again prepared the world, if not for a further revelation, then at least to a fuller and more self-conscious grasp of the one already available.

To underpin this, Robertson's conclusion to the sermon opens with reflections upon Europe's special place in the history of Christianity. It cannot be by way of sheer coincidence that Europe, where Christianity first spread, surpasses other regions of the earth in science and improvements. "Of this superiority the Europeans have availed themselves to the utmost, in every project for extending their empire and commerce … Now, the same attainments in science or policy, might be employed to good purpose, on the side of religion."[42] Europe, or at least a part of it, has been privileged by running the full cycle of stadial progress at a quicker pace, and reaching the pinnacles of the commercial and civilized stage earlier than the more and less remote corners of the globe that were opening themselves to the gaze of Europeans in Robertson's lifetime. In his experience and interpretation, the progress of commerce also coincided with the growth of politeness and knowledge, and thus advanced the cause of a more moderate and tolerant version of Christianity than the one which had held souls in subjection throughout the Middle Ages, and subsequently inspired the ravages of a whole continent in the age of religious wars. While fulfilling their civilizing mission in bringing commerce and refinement to the barbarous nations of distant regions, in other words, accelerating secular time for them, Europeans should also pay more attention to rendering their souls to the service of propagating the Gospel in a more systematic manner, thereby also accelerating sacred time—the progress towards their receiving of the revelation, and ultimately for all concerned, of redemption.

Perhaps I might conclude here by recapitulating that Robertson employs the paradigm of Enlightenment stadial history to present a highly dynamic picture of the intersections of secular and sacred time, and of the mutually supplementary roles of human and divine agency in this dynamics. But there is yet another, also very characteristically eighteenth-century dimension to his variations on the theme of time and progress. Underlying the sermon, as indeed virtually all of his works, is the idea that travel in space might easily assume the character of travel in time. In the wilderness of North America one can obtain a fair idea of the life of Tacitus' barbarians, while the Pacific islands are home to various modifications of Adamite man:

> [T]he characters of nations depend on the state of society in which they live, and on the political institutions established among them; and ... the human mind, whenever it is placed in the same situation, will, in ages the most distant, and in countries the most remote, assume the same form, and be distinguished by the same manners. ... Many of the German tribes were more civilized than the Americans. ... The resemblance, however, between their conditions, is greater, perhaps, than any that history affords an opportunity of observing between any two races of uncivilized people, and this has produced a surprising similarity of manners.[43]

Or, in even more simple terms, on account of the theory of the population of America from the old continent: "The character and occupations of the hunter in America must be little different from those of an Asiatic, who depends for subsistence on the chase. A tribe of savages on the banks of the Danube must nearly resemble one upon the plains washed by the Mississippi."[44]

The observation of "primitive" peoples in remote continents and the vast work of collecting data about them contributed immensely to the development of early ethnology,[45] while in the eyes of contemporaries fulfilling the mission referred to above in relation to such peoples passed for a heroic feat indeed. But does one truly need to cross the oceans in order to collect the same kind of anthropological knowledge, and perform the same kind of civilizing service? Far from it, according to the approach displayed in Robertson's texts, but also represented by many others in the eighteenth century. Distance in space and distance in time can be brought to a common denominator, but occasionally the relationship is inverse: crossing just a few hills would sometimes suffice for traversing many centuries. Indeed, Robertson concludes,

> the conversion of distant nations is not the chief care of the Society for the propagating [sic] Christian knowledge: An object nearer at hand demands its more immediate attention. The Highlands and Islands of *Scotland* present to us a scene, which we would little expect in a nation where true religion and polished manners have long flourished. There society still appears in a rude and imperfect form: Strangers to industry, averse from labour, inured to rapine; the fierce inhabitants scorned all the arts of peace, and stood ready for every bold and desperate action.[46]

This primitive state of society among them was indeed exploited by "the enemies of our religion and liberty" to sow among them "the seeds of the worst superstition, and the most pernicious principles of government" (a reference to the support among Highlanders for the Jacobite rising of 1745). Human agency is then again enlisted for the advancement of divine purposes: Robertson urges the legislature to enhance its already existing policy of enacting laws "with the most humane spirit, in order to retrieve that part of the kingdom from ignorance and barbarism"—a course of action from which "the members of the Society expect great assistance in the prosecution of their design."[47]

In view of the textual environment, this is fairly revealing. World and time are both a "given" for Robertson, in the strictest Christian sense of the word. There is design and ordination in the arrangement of both, but in such a way that motion in the one has inevitable consequences for motion in the other; and the character of that motion, Robertson reminds us, depends, as much on what is "given," as on the disposition of the receiving agent who uses the world once given, in the time given, turning the one to the well-being of his body and finding in the other the salvation of his soul. Accordingly, Robertson's ideas on time and the event, especially events of particularly great importance from the point of view of the divine plan, represent a very interesting shade within the thought of the Enlightenment about these questions. As a parallel case, we might invoke that of the transformation of the meaning of "revolution," precisely simultaneously with his own career. At the time when Robertson was born, *revolutio* was still, as Copernicus had described it in the case of the movement of celestial objects, regarded as a circular movement concluding in re-occupying an initial position (such as, in the political world, the Glorious Revolution in England in 1688); or, the sudden and shocking interference of an unpredictable force, beyond man's control, in human affairs (usually in affairs of government). Around the time Robertson's sermon was published, a different version of the same perspective started to take shape, in whose terms such calamities may provide an enlightened people with an opportunity to take their fates into their own hands—without implying that the cataclysmic event is prepared by the people itself, but that using the event as a springboard, they might thereafter become sovereign agents.[48] Robertson's logic, in a certain sense, is the very reverse: men engaged in commerce, refining their manners and discovering the natural, social and moral world around them, further the course of Christian history by these very activities, because doing so they facilitate and abbreviate their own path to the clear understanding of the Gospel while they do not have any influence on the ultimate outcome of that history.

Both approaches are capable of an explanation for the emergence of the modern terminology and conceptualization of historical change, and of the role of human agency in that change. However, it is equally useful and intellectually perhaps more rewarding to regard these conceptual cousins, not, as yet, "imperfect" anticipations of a later, more "developed" idea, but as mature theoretical experiments representing specific shades of opinion within the Enlightenment, claiming our attention in their own right. Paraphrasing Koselleck, one might suggest that Robertson's early sermon catches, for us, the very moment in which the notion of the acceleration of history was not yet quite divorced from the apocalyptic hope attached to the ever shortening periods preceding the Last Judgment, but was already being transformed into a notion of historical hope. But a formulation that would drop the words "yet" and "already," and replace "moment" with "perspective," might in fact far more accurately describe the situation.

Notes

For comments and suggestions, especially for having shared their views on Robertson's specific blend of providentialism, the author is grateful to Dr Thomas Ahnert and Professor Nicholas Phillipson.

1 Reinhart Koselleck, *Futures Past: On the Semantics of Historical Time* (Cambridge, Mass.: MIT Press, 1985), 7.

2 Ibid., 14, 16, 36.

3 See also his *Critique and Crisis: Enlightenment and the Pathogenesis of Modern Society* (Oxford: Berg, 1988).

4 J. G. A. Pocock, "Conservative Enlightenment and Democratic Revolutions: The American and French Cases in British Perspective," *Government and Opposition* vol. 24 (1989): 82. See also ibid., "Clergy and commerce: the conservative Enlightenment in England," in R. Ajello et al., *L'etá dei lumi: studi storici in onore di Franco Venturi* (Naples: Iovene Editore, 1985).

5 To borrow the title of a famous book: Paul Hazard, *La crise de la conscience européenne* (Paris: Fayard, 1935).

6 *The Enlightenment in National Context,* Roy Porter and Mikulás Teich, eds. (Cambridge: Cambridge University Press, 1981).

7 J. G. A. Pocock, "Clergy and commerce"; and his monumental work on Edward Gibbon (now in its fourth volume), especially *Barbarism and Religion. vol. I: The Enlightenments of Edward Gibbon, 1737–1764* (Cambridge: Cambridge University Press, 1999), "Introduction" and "Epilogue."

8 John Robertson, "The Enlightenment Above National Context," *The Historical Journal*, vol. 40 (1997): 667–697. At the same time, in his recent book Robertson uses this perspective to reassert a case for the ultimate unity of the Enlightenment. Idem, *The Case for the Enlightenment: Scotland and Naples 1680–1760* (Cambridge: Cambridge University Press, 2005).

9 Richard B. Sher, *Church and University in the Scottish Enlightenment: The Moderate Literati of Edinburgh* (Princeton: Princeton University Press, 1985).

10 "Philosophical" history, rather than chronicling events and throwing light on the logical relations or immediate causality that linked them, focused on specific political or moral themes and represented them in an episodic manner. Enlightenment stadial history conceived of the history of humanity as a sequence of stages defined in terms of their dominant mode of subsistence, and keenly employed analogies (often supplementing the lack of reliable sources and empirical data through "conjectures" based on the known elements of the environment) between societies at great distance in time and space, but deemed as suitable units of comparison by reference to certain fundamental features.

[11] This made Robertson a pivotal figure in "subverting" traditional Scottish discourse on the national past, or "rescuing" it from remaining locked in nostalgia (at the expense of making it Anglo-British). Colin Kidd, *Subverting Scotland's Past: Scottish whig historians and the creation of an Anglo-British identity, 1689–c.1830* (Cambridge: Cambridge University Press, 1993), and Karen O'Brien, *Narratives of Enlightenment: Cosmopolitan History from Voltaire to Gibbon* (Cambridge: Cambridge University Press, 1997), Ch. 4.

[12] This book, intended as a direct sequel the *History of Charles V*, offered a compelling analysis of native American cultures and traced the history of Spanish colonization in the New World.

[13] O'Brien, *Narratives of Enlightenment*, Ch. 5.

[14] Smith and Turgot have also received special emphasis in the literature devoted to the idea of progress, and by implication to the concepts of time and history. See, for instance, J. B. Bury, *The Idea of Progress: An Inquiry into Its Growth and Origin* (New York: Dover, 1955); Leslie Sklair, *The Sociology of Progress* (London: Routledge, 1998), Ch. 1; Robert Nisbet, *The History of the Idea of Progress* (New York: Basic Books, 1980), Ch. 4; David Spadafora, *The Idea of Progress in Eighteenth-Century Britain* (New Haven and London: Yale University Press, 1990).

[15] On the relationship between Robertson and Mackie, see Nicholas Phillipson, "Providence and progress: and introduction to the historical thought of William Robertson," in *William Robertson and the Expansion of Empire*, ed. Stuart J. Brown (Cambridge: Cambridge University Press, 1997), 57–8.

[16] On Machiavelli and Guicciardini as representatives of this trend in humanist historiography, the best synthesis is Felix Gilbert, *Machiavelli and Guicciardini: Politics and History in Sixteenth-Century Florence* (Princeton: Princeton University Press, 1965).

[17] Montesquieu's *The Spirit of the Laws* was published in 1748, while Hume's *Essays Moral and Political* from 1741, in several, continually rewritten editions (the third, "improved and expanded" edition also exactly in 1748). "Of National Characters" is one of these essays. As for Smith, he never published a text on jurisprudence in his lifetime: his lectures held on this subject ("on the third and fourth part of moral philosophy," i.e., on justice and expediency) after his appointment as professor of logics and moral philosophy at the University of Glasgow in 1751, have been reconstructed on the basis of lecture notes of his students through modern scholarship. See *Lectures on Jurisprudence*, R. L. Meek, D. D. Raphael and P. G. Stein, eds. (Oxford: Clarendon Press, 1978). Previously, however, between 1748 and 1750 the young Smith also held a few public lectures in Edinburgh, allowing Robertson to borrow some of his views—perhaps literally so: Smith is said to have accused Robertson of having plagiarised the backbone of *A View of the Progress of Society in*

Europe. Ian Simpson Ross, *The Life of Adam Smith* (Oxford: Oxford University Press, 1995), 105–6.

18 Cf. Robert F. Nisbet, *Social Change and History: Aspects of the Western Theory of Development* (Oxford: Oxford University Press, 1965), 85 ff. For Augustine on the paradoxes of time, see *Confessions*, R. S. Pine-Coffin, trans. (Chicago: Encyclopaedia Britannica, 1990), Book XI especially sections 10–27, 116 ff.

19 Cf. J. G. A. Pocock, *The Machiavellian Moment: Florentine Political Thought and the Atlantic Republican Tradition* (Princeton: Princeton University Press, 1977), Ch. 1. It must be added that, from patristic times on, the idea of progress asserted itself in various ways in the Christian apocalyptic–cataclysmic conception of history, the more so as both of these approaches also regard time as not merely an exact chronological standard, but the framework of change, which becomes intelligible through understanding the relationship between separate events. A more thorough investigation of this problem is beyond the confines of this study, but for the intersections of apocalyptic and progress-based approaches to time see the pioneering and still relevant work of Ernest Lee Tuveson, *Millennium and Utopia: A Study in the Background of the Idea of Progress* (New York: Harper & Row, 1946). See also Chester G. Starr, "Historical and Philosophical Time," *History and Theory* 6 vol. 6 (1966): 24–35; Elizabeth Eisenstein, "Clio and Chronos. An Essay on the Making and Breaking of History-Book Time," *History and Theory* 6 vol. 6 (1966): 36–64.

20 Jeffrey Smitten, "The Shaping of Moderatism: William Robertson and Arminianism," in eds. Patricia Craddock, Carla H. Hay *Studies in Eighteenth-Century Culture* vol. 22 (1992): 281–300.

21 Philippe van Limborch, *A Compleat System or Body of Divinity* (1713), quoted in Smitten, "The Shaping of Moderatism," 287.

22 William Robertson, *The Situation of the World at the Time of Christ's Appearance, and its Connection with the Success of His Religion*, in *The Works of William Robertson* (London: Thoemmes Press, 1996), vol. XI, 6–7.

23 Ibid., 8–9.

24 Ibid., 9–10.

25 Isaac Newton, *Mathematical Principles of Natural Philosophy* (Chicago: University of Chicago Press, 1952), 8.

26 John Locke, *An Essay Concerning Human Understanding*, ed. Peter H. Nidditch (Oxford: Clarendon Press, 1975), Book II, Chapter 14, "Of Duration, and its simple Modes," 181–196. Cf. W. Von Leyden, "History and the Concept of Relative Time," *History and Theory* vol. 2 (1963): 263–85.

27 The increasingly bitter dispute that started between Newton and Leibniz over the "copyright" of differential calculus, later also concerned other scientific as well as philosophical and theological questions, and

involved Newton's disciples, above all Samuel Clarke, the translator of the *Optics*. The problem of time and space is abundantly discussed in the Leibniz–Clarke correspondence, published in 1717. On the idea mentioned above, see *The Leibniz–Clarke Correspondence*, H. G. Alexander, ed. (Manchester: Manchester University Press, 1956), 27–8. For a comparison of the relevant views of Newton, Locke and Leibniz, see Philip Turetzky, *Time* (London: Routledge, 1998), 71–83; and J. T. Fraser, *Of Time, Passion, and Knowledge: Reflections on the Strategy of Existence* (Princeton: Princeton University Press, 1990), 33–4.

28 Nisbet, *Social Change and History*, 85–91.

29 On this basis, it is further argued by Thomas Ahnert that "Moderatism was not characterized by 'reasonable religion', but by a (theologically inspired) epistemological scepticism, which emphasizes the limitations of human, natural reason in theological questions"—a case for a "pious Enlightenment," not characterized by religious indifference, but integrating the cultivation of "polite" manners with emphasis on religious reform. "Religion and the Moderates' Enlightenment: the Historiography of William Robertson," manuscript (paper read at the conference "Empire, Philosophy and Religion: Scotland and Central/Eastern Europe in the Eighteenth Century," Central European University, Budapest, 23–26 June 2005).

30 Robertson, *Situation*, 14–5.

31 Ibid., 15–9.

32 Ibid., 20–4.

33 David Hume, "Of Superstition and Enthusiasm," in *Essays Moral, Political and Literary*, ed. Eugene F. Miller (Indianapolis: Liberty Press, 1985), 74–79.

34 See Sher, *Church and University*, 67–70, 277–97; Stuart J. Brown, "William Robertson (1721–1793) and the Scottish Enlightenment," in Brown ed. *Robertson and the Expansion of Empire*, 16–31.

35 John Erskine ed., *A Narrative of the Debate in the General Assembly of the Church of Scotland, May 25, 1779* (Edinburgh: William Gray, 1780), 49, 61.

36 Robertson, *Situation*, 25–31.

37 Ibid., 39.

38 Ibid., 44. The obvious objection that slavery did not prove to be incompatible with Christianity in his own times is dismissed by Robertson by claiming that "[t]he genius and tendency of any religion are known by the operation of its vigorous, not of its declining age."

39 Ibid., 47.

40 The Mandevillian formula of "private vices—private benefits" and his extreme statement of the paradigm of unsocial sociability might have been universally rejected as a "system" by eighteenth-century theorists, but its implications were hard to escape.

[41] William Robertson, "A View of the Progress of Society in Europe from the Subversion of the Roman Empire to the Beginning of the Sixteenth Century," in The *History of the Reign of the Emperor Charles V* (London: Routledge/Thoemmes Press, 1996), vol. I., 42–43.

[42] Robertson, *Situation*, 51.

[43] Robertson, *View of the Progress*, 372.

[44] William Robertson, "The History of America," in *The Works of William Robertson* (London, 1835), 806.

[45] See Hans Erich Bödeker, "Aufklärerische ethnologische Praxis: Johann Reinhold Forster und Georg Forster," in eds. Hans-Erich Bödeker, Peter Hanns Reill, Jürgen Schlumbohm *Wissenschaft als kulturelle Praxis, 1750–1900* (Göttingen: Vandenhoeck & Ruprecht, 1999), and the literature quoted there.

[46] Robertson, *Situation*, 53.

[47] Ibid., 54–5.

[48] Cf. Keith Michael Baker, "Inventing the French Revolution," in idem, *Inventing the French Revolution. Essays on French Political Culture in the Eighteenth Century* (Cambridge: Cambridge University Press, 1990), 203–223. See also Neithard Bulst, Jörg Fisch, Reinhart Koselleck, "Revolution," in *Geschichtliche Grundbegriffe. Historisches Lexikon zur politisch-sozialen Sprache in Deutschland* Otto Brunner, Werner Conze, Reinhart Koselleck eds. (Stuttgart: Klett-Cotta, 1984), Bd. 5, esp. 714–24.

Religious Revivals

The Binds of Religion and Modernity in Friedrich Nietzsche's *The Anti-Christ* and Richard Wright's *The Outsider*

Andrew Wegley

One linchpin that holds together Western modernity, the secularization thesis, is crucial in narrating the historical progress of society away from religious control to individual determination. The thesis can be seen as comprising six components: 1) the creation of different institutional apparati that separate religion from politics, 2) the depoliticization and privatization of religion, 3) the decline in religious belief, 4) the development of cultural identity, 5) the rise of the state as governmental form, and 6) the production of capitalist markets.[1] Together these components describe the new phenomenon that constitutes modernity as different from religious and traditional cultures. Moving away from religion, the secularization thesis makes the radical claim that society is free to cultivate spaces of critique and uses of rationality that promote or contest institutions.[2] However, this thesis has undergone scrutiny itself and has been seen as not accurately describing the shifts and phenomena associated with modernity. Here, I will make another attempt at describing how the secularization thesis is problematic and misconstrues the nature of "modernity." But, my attempt will not be concerned primarily with historical data, but rather with literature. I will read two works that already contain complex views on the relationship between "modernity" and "religion," not simply to prove false the secularization thesis, but rather to open up the discursive modalities in which religion is still internally present in the modern.

Friedrich Nietzsche's *The Anti-Christ* and Richard Wright's *The Outsider* are examined with an eye to their related concepts of religion, in order to gauge how they themselves already offer a complicated understanding of the religion/modernity split that underlines the secularization thesis. The readings and interpretations I offer, which contest the secularization thesis before its postmodern critics, call into question the efficacy and name of modernity.[3] The theory of rupture between religion and modernity, I argue, posits a false sense of freedom from religion. Instead, I offer the alternate view that "religion" and "modernity" are both products of modernity.[4] This "doubledness" arises not out of chronological developments but rather results from discursive productions: modernity and religion co-produce and co-construct each other. As Nietzsche and Wright show, the attempt at separation itself is a disavowal: the separation is impossible; instead that which is separated inflects and shapes the proper category, namely, modernity.

Before turning to the textual analysis, I provide a description of these works. Friedrich Nietzsche (1844–1900) wrote *The Anti-Christ* in 1888; however, it was first published in 1895, making this work one of his last publications. It was conceived as the first in his four-volume magnum opus, *The Will to Power*, which was not completed before his mental collapse in 1889. As the title suggests, *The Anti-Christ* is a critique of Christianity, but also of any truth system that squelches the human drive for mastery. But it is equally an incitement to breed a new race of humans whose servile dependency on morals has been exhausted. Drawing inspiration from Nietzsche in order to theorize his historical situation, African-American novelist and former Communist Party adherent Richard Wright (1908–1960) wrote *The Outsider* in 1953.[5] This epic and philosophical novel set out to portray American racism and its victims in unrelenting and unapologetic ways. *The Outsider*, like other great American works, is a story whose hero, Cross Damon, is on a quest for freedom and identity. But Cross is a black man in a white man's world denied the means necessary to realize his goals. Instead, he drinks and becomes a man of ideas, constantly reflecting and thinking about his condition, believing that something meaningful and more important exists in this world apart from the injustices of an externally imposed social system. His break comes, when, after surviving a freak train accident and being presumed dead, Cross assumes a new identity and starts over. Even as Cross seems to find respite in the form of a white girlfriend, Eva, and community in the form of the Communist Party, Cross' search for freedom and meaning eludes him: he still remains enslaved. Moreover, his situation only worsens. After committing a series of murders in retaliation to the corruption of the Communist Party, Cross is hunted down and convicted by district attorney, Ely Houston. Even in punishment, Cross finds no meaning or redemption. The book ends on a note of hopelessness when Cross is murdered; his entire life is appraised as meaningless.

On the face of things, *The Anti-Christ* and *The Outsider* are indictments of religion and other cultural ideologies that prevent and obscure human freedom. Both works also share a blending of literary forms (the philosophic novel and aphoristic, quasi-poetic philosophy) as well as a certain difficulty in reading. Both works follow their respective protagonists, Nietzsche's Anti-Christ and Wright's Cross Damon, through their parallel quests of ideological unmaskings. Even in their names a resemblance exists; Cross is an incarnation of Nietzsche's Anti-Christ.[6] That is, both "supermen" contain within themselves a negation of the mythical Jesus Christ, and, by implication, a metonymic figure for all religions and ideologies. Fleshing this out a little, to speak metaphorically, I stress that both Nietzsche and Wright as they contemplate religion and its effects do so through the lens of sovereignty.[7] Just as in the secularization thesis, religion is viewed in Nietzsche and Wright as an obstacle to achieving sovereignty. But this place freed from religion will also need to be addressed. Their examination of the limits of sovereignty challenges the characterization of religion as unfree and modernity as free. Both texts detect traces of religion in modernity, ultimately questioning the possibility of sovereignty, the goal of modernity. By extensively focusing on Nietzsche's *The Anti-Christ*, I aim to establish thoroughly the terms that Wright further employs in his text. As I argue, *The Outsider* is best understood as furthering the applicability and ramifications of Nietzsche's text.

Nietzsche's work *The Anti-Christ* is a polemic against religion, in general, and aimed at Christianity, in particular. While most of his vitriolic prose is targeted at Christianity, Nietzsche also disparages Buddhism and Judaism. When reading, it is important to remember that its claims are purposely aggravating, extreme, and hyperbolic. To make sense of its extravagant claims along with its scant historical evidence, understanding the function of polemic is crucial; the form contains and reveals Nietzsche's understanding of religion. In his study of Nietzsche's use of metaphors in *The Anti-Christ*, Tim Murphy concurs and suggests: "...Nietzsche's view of religion *is* reductive, but in so being he is claiming that *all* views of religion are reductive, and that religion itself is reductive."[8] Murphy highlights that Nietzsche's understanding of religious discourses is complex.

The complexity arises because the ground of interrogation consists of discourse not only *of* religion (Christianity's claims) but also *about* religion (Nietzsche's claims).

From the very beginning the aggravating quality of Nietzsche's text is actually productive because it establishes a positive correlation between his text and religion. The link between them is that they are reductions. However, what does this reduction entail in Nietzsche's estimate? Nietzsche models his reduction of religion on religion's *reduction of life*:

> Let me repeat: this depressing and contagious instinct stands against all those instincts which work for the preservation and enhancement of life: in the role of protector of the miserable, it is a prime agent in the promotion of decadence—pity persuades to extinction... Of course, one doesn't say "extinction": one says "the other world," or "God," or "the true life," or Nirvana, salvation, blessedness... This innocent rhetoric, from the realm of religious-ethical balderdash, appears a good deal less innocent when one reflects upon the tendency that it conceals beneath sublime words: the tendency to destroy life.[9]

Religion works against human life through the creation of concepts such as pity, salvation, and God. Religion veils and prohibits what is "natural" to humans, the life instincts. Nietzsche investigates Old Judaism to prove his point. However, paradoxically, Nietzsche's Old Judaism is and is not yet religion because the concept does not exist. Old Judaism is a religion where no differentiation or separation exists between the community, nature, and religion. To be sure, reduction occurs in the production of such concepts. Nietzsche returns to this point and summarizes two means whereby religion reproduces its authoritative and effective masking of life:

> [O]n the one hand, *revelation*, which is the assumption that the reasons lying behind the laws are *not* of human origin, that they were *not* sought out and found by a slow process and after many errors, but that they are of divine ancestry, and came into being complete, perfect, without a history, as a free gift, a miracle ...; and on the other hand, *tradition*, which is the assumption that the law has stood unchanged from time immemorial, and that it is impious and a crime against one's forefathers to bring it into question.[10]

For an institution to continue to reduce human nature so effectively, concepts such as religion, tradition and revelation are used to bolster the idea of God. Nietzsche sees religion as securing its reductionist aim in a privileged way. Yet the manner in which Nietzsche presents religion underlines not only religion's reductive role, but also the enormity of the situation: while the goal of religion is easily phrased, its violent tools and effects are far from simple. Religion distorts and disfigures nature and its practitioners. In this very manner, Nietzsche disfigures religion. Reduction only makes religion's absurdity apparent since Nietzsche's imitation of religion works to dismantle the very tools it uses to sustain itself. Nietzsche shows that this first and foremost entails the destruction of life and the creation of ideas. Yet the language of religion and Nietzsche are brought closer together in this respect—they are polemics directed against the reduction of an opposing force.

Nietzsche's text subsequently is concerned with how religion and its discursive productions work. He extracts from Old Judaism the role of the priest, who helps in implementing and wielding these ideas:

> To the sort of men who reach out for power under Judaism and Christianity—that is to say, to the *priestly* class—decadence is no more than a means to an end. Men of this sort have a vital interest in making mankind sick and in confusing the values of "good" and "bad," "true" and "false" in a manner that is not only dangerous to life, but also slanders it.[11]

The priest introduces a certain idea of God as each new crisis emerges in the community. Each crisis produces a gap in the former harmony between god, nature, and the community. It is this distance which religion proposes to close.[12] Accordingly, we need to examine how this gap works. God's interpreted split now casts a shadow on the world. The priest attributes the good to god and the bad to nature.

Nietzsche elaborates on religion's expanding use of reductionist ideas. A new note resounds, as Nietzsche focuses on how these ideas are embodied—particularly how the priest enforces these ideas. He emphasizes the priest's regulative force in terms of punishment:

> The public notion of this god now becomes merely a weapon in the hands of clerical agitators, who interpret all happiness as a

> reward and all unhappiness as a punishment for obedience or disobedience to him, for "sin": that most fraudulent of all imaginable interpretations, whereby a "moral order of the world" is set up, and the fundamental concepts, "cause" and "effect," are stood on their heads.[13]

The introduction of moral order, as Nietzsche recognizes, is violent. Where there was once a harmonious whole, there is now a division between good and bad. These pairs, as Nietzsche indicates, now are used to demarcate different spheres; these spheres have not only moral connotations, but they also map order and truth onto the world. Clearly, these acts are not innocent or without consequences. Nietzsche recognizes this inherent inequality in the terms and the fraught relationship that binds them:

> For this should be noted: that every natural habit, every natural institution (the state, the administration of justice, marriage, the care of the sick and of the poor), everything demanded by the life-instinct, in short, everything that has any value in itself, is reduced to absolute worthlessness and even made the *reverse* of valuable by the parasitism of priests (or, if you choose it, by the "moral order of the world").

As expected of any critical semiotician, Nietzsche notes how the natural is now labeled artificial, while the artificial is now natural. Whereas before this division, there was no artificiality, now the natural and the artificial are confused and subject to future confusion—consequentially becoming bound by invoking the other half. Nietzsche emphasizes this relational quality by focusing on the "reversal." Pressing this last citation further, we can see that the "natural" and "unnatural" become connected by their mutual traffic. Even Nietzsche's own invocation of what is prior—and hence what is "natural"—is bound up in this potentially unending series of inversions and reversions. Religion in saying what is the natural and the unnatural is actually creating both what it is and is not. Religion is tied to the nature that it names bad. As he demonstrates through the priest's punishment of the criminal, this separation from nature is a violent pretense to undo a "natural" connection.

This movement back and forth from the natural and the unnatural is potentially infinite; however, the priest, like Nietzsche, needs to halt this momentum. Or more aptly, the priest needs to employ a tactic to encourage his related ordering of the world in terms of morals, truth, reality—where the priest's map remains essentially unchanged, a topography of "good" and "bad, " "yes" and "no," and "X" and "not X." Nietzsche elaborates in a compact, but powerful manner: "The fact requires a sanction—a power to *grant values* becomes necessary, and the only way it can create such values is by ***denying nature***... The priest depreciates and desecrates nature: it is only at this price that he can exist at all."[14]

The first act that makes this overlapping terrain possible—or what makes these familiar dichotomizations of the world possible—is cast in a new light. The first act of reduction—the priest's interpretation—does not simply reduce nature in favor of looking to god, the *super*natural. Seeing the past in light of the present (a similar position to Nietzsche's when he wrote), we can appreciate how religion's origin story occurs at the same time as nature's story; both religion and nature are bound up in each other's defining moments. Consequentially, any origin tale from Nietzsche's perspective is contaminated with negation. Religion's origin is not just singular but at least double; religion and nature are twins.[15] Nature is produced alongside of religion as well as destroyed at the same time (although never completely). Nature's split, including the bad, the false, and the rest "all the way down" is violent precisely because its co-created twin negates nature.[16] To summarize how this tactic of negation affects religion and nature, I stress that the very act of creating religion—or naming God—is at the same time the breaking and separating of the nature that is being co-constructed. Further, it is religion's original reduction that Nietzsche parodies in his own act of writing and naming religion: that is, the complexity in the act of creation is reduced and hidden. Religion denies the very object it creates. Both "religion" and "nature" as they are co-constructed share the same origins; they both bear the marks of being negated under their simultaneous production.[17]

Each part of the binary produces and prevents the other from being a whole, autonomous entity. Since religion and nature are gestated together, the consequence of such an act is that each bears the mark of negation. Nietzsche acknowledges from the outset as a central part of his project that he wants to uncover the denied links: "I unearth this theological instinct in all directions: it is the most widespread and the most *subterranean* form of falsehood found on earth."[18] Nietzsche lays the groundwork for an archaeological recovery; his search excavates the remains of what religion negated. By exploring this "underworld," Nietzsche transgresses into the realm where what has been construed as bad has been sealed off—yet, this Herculean task has been configured by the priest who initially constructed barriers in the first place. Nietzsche can only cross over and commit his theological crime because the initial traffic was perpetrated by the bearers of religious law.

Nietzsche's archaeological quest to uncover the remains of religion's reduction arrives at psychology. Nietzsche construes psychology as the "natural" product that religion creates to negate. Using psychology as his trope, Nietzsche is not being anachronistic; rather, he is demonstrating the presentism of his work. Here, Nietzsche reveals the motivating concern behind his polemic: that, as moderns, we too might have buried the effects of religion in our psyches. The past becomes a myth, irrecoverable in its totality, barred from its consciousness. Similar to his treatment of religion using nature for its own end, Nietzsche uses psychology to produce the past—in other words he is diagnosing the present by reading symptoms as effects of past deeds. Nietzsche himself makes the connection between his interests in excavation and psychology explicit: "*To that end the thing must be made unconscious:* that is the aim of every holy lie."[19] The use of the word 'unconscious' here does double duty; it is the link between the concerns of archaeology and psychology. It signals that Nietzsche's interest lies in undoing the effects of the 'holy lie' in psychic registers. Nietzsche's psychology translates his work from simply a critique to a therapeutics—but a therapeutics that needs to revive the entombed past.

In chapter 29, Nietzsche first introduces "psychology" as a trope; it is significant to note that "psychology" only emerges now in the work. This concept is introduced after his survey of Judaism's evolution and before his examination of Christianity's advent. Psychological interiorization only fully develops under the rubric of Christianity. While Nietzsche would not deny that Judaism produced psychological effects, he argues that Christianity alone fully develops this process of interiorization. Nietzsche stresses that, whereas Judaism is a religion of the community, Christianity is the religion of the individual. Nietzsche's discourse in this section focuses on an intimate, individual level—again highlighting how his writing performs and parallels what he claims about religion. In Christianity, the denial of nature is translated into the denial of *human* nature. As a result, the life instinct found in human nature is repressed and bound in the unconscious. In effect, the burial of the life instinct is the genesis of individual psychology. As Nietzsche has made clear, the process of negation is anything but simple. A lot of work is required to make this reduction possible. The focus, interest, and production of an individual psychology lead to the fulfillment of religion's promise of negation. He writes:

> Brought to its logical conclusion, such a psychological *habitus* becomes an instinctive hatred of all reality, a flight into the "intangible," into the "incomprehensible"; a distaste for all formulae, for all conceptions of time and space, for everything established—customs, institutions, the church; a feeling of being at home in a world in which no sort of reality survives, a merely "inner" world, a "true" world, an "eternal" world... "The Kingdom of God is within *you*"... [20]

What is typified in this movement in Christianity is Nietzsche's keystone against religion: religion is the ultimate reduction in that it produces nihilism. Whereas in Judaism nihilism destroyed nature, now Christianity reaches the logical conclusion of nihilism: extending the negation of nature now to the negation of individual psychologies produced as a result of God's alleged break from the world—and hence the negation of the self.[21] The final product—a human as a denial of life—then is made in the image of an idea, a greater good, a god-head brought upon by submission to the priest. Nietzsche captures the spirit of this insidious technique: "Prime axiom: 'God forgiveth him that repenteth'—in plain English, *him that submitteth to the priest.*"[22] The invention of psychology enables a space to emerge where new techniques can cultivate a sense of servitude. The priest relies upon manipulating human nature so that submission is possible—one which suits his will-to-power.

Continuing the discussion of psychology, I connect it to the concept of sovereignty. At the beginning of his work, Nietzsche acknowledges that liberty and servitude are at stake in his survey of religion:

> The problem that I set here is not what end shall replace mankind in the order of living creatures (—man is an end—): but what type of man must be *bred*, must be *willed*, as being the most valuable, the most worthy of life, the most secure

> guarantee of the future. This more valuable type has appeared often enough in the past: but always as a happy accident, as an exception, never as deliberately *willed*. Very often it has been precisely the most feared; hitherto it has been almost *the* terror of terrors; —and out of that terror the contrary type has been willed, cultivated, and *attained*: the domestic animal, the herd animal, the sick-brute-man—the Christian.[23]

Nietzsche recognizes two important things in this passage. First, religion is in the category, which molds humanity in the name of something contrary to its nature. The issue is that humanity has been subjected whereas it should be the subject. Through submission to the priest, techniques of confession, and beliefs in sin, this prohibited nature is "repressed" in the interest of "God." Second, Nietzsche wonders about his own role in posing such questions and articulating such concepts: that is, he admits the possibility that by exposing religion's reductions he will reproduce religion's claims—reducing humanity to another use. Nietzsche realizes that not only does religious discourse pretend to be sovereign, but his discourse also poses as sovereign in its ability to make claims about religion. Religion pretends to be sovereign, as Nietzsche writes, in that it destroys life. "In point of fact, the end for which one lies makes a great difference: whether one preserves thereby or destroys."[24] What we see is that destruction is valued over life. Christianity is Nietzsche's exemplar because it reduces *flourishing life* to *life stagnant*.[25] As Nietzsche suggests if we are to recover we need to alter these lies "over which no arbitrary fact, no 'modern idea,' can exert any influence."[26]

The price of religious freedom, as shown in chapter 25–26, is the denial of life, nihilism. Nietzsche is perfectly aware of the paradox of religion's claim to sovereignty. He spells this out accordingly: "...the inequality of rights is essential to the existence of any rights at all."[27] It is precisely this ideology of freedom and sovereignty that Nietzsche condemns at the close of the work.[28] However, the recognition of religion's dialectic of negation coupled with Nietzsche's own negation of religion becomes another instance where Nietzsche implicates his work. Hence, is he repeating the same negative dynamic that makes religion possible? As I have shown, Nietzsche is aware of religion's reductive role. This reduction is best read in light of sovereignty—by attempting to reduce the world to a function and value of God, it postures as sovereign. Accordingly, does Nietzsche's theory move beyond the inequity of religion? Or does it suggest a different practice of sovereignty in modernity?

This awareness suggests that we should scrutinize Nietzsche's own condemnation more closely to see what it says. He himself suggests this close reading:

> When a man feels that he has a divine mission, say to lift up, to save or to liberate mankind—when a man feels the divine spark in his heart and believes that he is the mouthpiece of supernatural imperatives—when such a mission inflames him, it is only natural that he should stand beyond all merely reasonable standards of judgment. He feels that he is *himself* sanctified by this mission, that he is himself a type of higher order!... He has determined the meaning of "true" and "not true"![29]

After having gone through Christianity's "missionizing" history—specifically the conditions in which produces, distinguishes, and ranks the true and the false religions—Nietzsche's own attempts to breed a modern race of Hyperboreans thus lies rather suspect alongside his condemnation of religion since he himself has problematized the simple notion of religion's freedom from nature's servitude. How does one read his statements? How do we get free from religion's mastery? The answer lies in the problematization, which Nietzsche performed on religious discourse. Religion is sustained by posturing as sovereign, and, as Nietzsche demonstrated, this discourse by being reductive has a limited claim on the world. Similar to how Murphy demonstrated that religion loses its primacy in Nietzsche's text because it is mapped onto other metaphoric domains, I find that Nietzsche's polemic gets mapped on to religion's reduction. The very absurdity of the writing a condemnation matches religion's condemnation of the world; they both seem to speak on behalf of reality. Nietzsche knows all too well that through his aphoristic, exaggerated statements there is no outside of the registers of religious discourse without performing and reiterating religious reductions. Whereas religion masks its work, Nietzsche all along unearths, showing the very ways "holy lies" and truths get entangled. Nietzsche's tactic produces religion not simply to negate it—just as religion sought to produce and negate nature. Rather he lets his claims become entwined and bound up with religion's claims. His work becomes a religious manifesto, the *Anti-Christ* (which is not simply *not*-Christ).

By implicating both religion's discourse and his own, Nietzsche intimates that both acts are essentially productive—the question then is how they are employed in modernity. This awareness of their productive and reductive nature is encapsulated in his inauguration of a "transvaluation of all values."[30] Nietzsche's goal then is to uncover the means by which moral ends are established in the first place as well as revealing that values are always supplemental—and hence changeable—in that they are only partial.

Nietzsche's archaeological project indicates the problematic 'nature' of religion: that is, religion is a particular form of sovereignty underpinning the world, which it engenders, through a field of technologies, beliefs, and figures. However, while his work examines the past in light of understanding the present (particularly with religion in mind), it only begins to comb the modern world. If we see his Anti-Christ as the spokesman of modernity, then we can already see a contradictory temporality: a will to be sovereign that does not have a supporting modality. In *The Outsider*, Richard Wright picks up this thread that Nietzsche leaves dangling—the possibility for a new time freed from religious authority. Thus, I read *The Outsider* as a modern sequel to *The Anti-Christ*. As I noted earlier, not only do they share a formal resemblance, but also they are akin intellectually.

First, note Wright's and Nietzsche's similar understandings of religion. Wright follows Nietzsche's analysis of religion as a way of veiling human nature:

> [Cross Damon] was convinced that an idiot could have conceived better ways of establishing emotional alliances, and men must have felt uneasy about it too, for they had sought to make desire and passion stable by subjecting them to legal contracts! If human emotions fail to remain constant, then draft laws stipulating that that which will not remain constant *must* remain constant... As he brooded over the meaning of desire a quiet sense of awe drenched him. Moods like these were the nearest he ever came to religious feelings.[31]

By having Cross approximate religious sentiments, Wright insinuates that the same sort of deliberation over human desire occurs in religion—this attempt to fix and distort that which cannot be fixed. Religion's crystallization of God—a translation of desire—occurs because there emerge men with a will-to-power who resemble

Nietzsche's priestly class. This class forces its flock to submit to "legal contracts" as a sign of their renunciation of asocial passions. Religion frees us from thinking about—let alone dealing with—the realities of human existence.

Wright continues his assessment of religion as ideology by examining the role of law. These men reign supreme because they can effectively control the discursive field and the objects produced in that field:

> It was an emotional compulsion, religious in its intensity, to feel and weigh the worth of himself that was pushing him into the arms of the one thing on earth that could transform his sense of dread, shape it, objectify it, and make it real and rational for him... His passion was to recast, reforge himself, and he was certain that Gil and Hilton had once in their lives felt what he was now feeling, that his reaching out for another pitch of consciousness had haunted them just as it now plagued him. But they had resolved their tangled emotions in the rigid disciplines of Communist politics, thereby ejecting from their hearts the pathos of living, purging their consciousness of that perilous subjective tension that spells the humanity of man. And now they were warring to slay in others that same agony of life that had driven them to the wall... It was [Communism's] believing that it *knew* life; its *conviction* that it had mastered the act of living; its *will* that it could define the ends of existence that fascinated [Cross] against his volition.[32]

In this lengthy passage, Wright affirms his connection with Nietzsche's stance on religion. Cross' sense of dread is linked to being confronted with human nature unmediated, desire in its gaping enormity. This encounter takes on sublime heights, the burden of which is too great for Cross to maintain. He himself has the urge to follow the way of religion and Communism, a modern religion in Wright's estimate. Like Nietzsche's priest, the Communist party leaders Gil and Hilton 'transvalue' and reduce unmitigated nature into manageable icons. These icons, as Wright describes, are purged; in order to arrive at their respective religion, their conviction, means both a fabrication and a negation of nature (the process of reforging and ejecting the self). Wright parallels Nietzsche's understanding of religion by linking Cross' agonies to ideology and its religiosity that promote suffering and internal violence.

Not only does Wright parallel Nietzsche's critique of religion, he parallels his form, namely, the use of metaphor to reveal his thought. That is, in following Nietzsche, Wright also is aware of the proliferating chain of reductive metaphors. Wright compares and remaps modern feelings onto religious feelings; yet, his claim of religion is in turn only a reduction of another reduction (Nietzsche's religion), which was itself a reduction (religion's nature). In these metaphoric domains, Wright does two different things. On the one hand, he strictly follows Nietzsche's understanding of religion as a reduction of nature. On the other hand, Wright deviates from Nietzsche as metaphors do in their comparative manner. Basing his theory loosely on Nietzsche's model, Wright suggests that modernity creates the idea of religion in order to define its difference. In other words, the difference in Wright's employment of Nietzsche is to equate directly what only Nietzsche implies: that is, Wright's interest in religion is to show its persistence in modernity by modernity's attempts to surpass and break from religion. The twin processes of reducing and negating only bind modernity further to religion—not simply to remove them as different.

Already within Wright's remapping of religion, we see that religion is transferred into modern domains. The questions become, "How does Wright show that modernity is like religion?" and "How exactly is Nietzsche's understanding of religion translated?" As Wright begins to demonstrate, modernity has specific interests in religion's demarcated claims on the human. Cross Damon established how this interest arises:

> One way, among a variety of others, of looking at what is happening in the world today is to view all modern history as being tied together by one over-all meaning. And that meaning is this: to a greater or lesser degree, all human life on this earth today *can* be described as moving away from traditional, agrarian, simple handicraft ways of life toward modern industrialization. Some nations, owing mainly to historical backgrounds, have already made this journey, have already, so to speak, exhausted their industrial potentials of development and are industrially overripe. I'm referring to nations like England, Japan, Germany, etc. In these nations the problem of the future structure of society and the question of what kind of faith will sustain the individual in his daily life constitute a kind a chronic spiritual terror.[33]

The fact of industrialization in Cross's view creates a rupture in the existential ethos of the community, a spiritual crisis. This rupture echoes Nietzsche's account of religion's origins, its rupture from initial harmony with nature. Instead of nature per se being used to figure religion, what Cross suggests is that now these encounters with religion and tradition are "discovered" and used in such a way to differentiate the modern.[34] As a result of this anxiety, the West uses this "discovery" for its own benefit and detriment: "These non-Western nations and peoples were retarded inasmuch as they were captive peoples in the hands of the industrialists of the West who needed their physical labor and the natural riches of their soil to keep the giant industrial machines of the West going."[35]

Just as Nietzsche demonstrated how Judaism used foreign gods to define itself, the West is seen in light of its use of non-Western people. They literally become the work that is covered and negated in service of the West. These tactics allow Wright to link modernity with religion:

> Modern man still believes in magic; he lives in a rational world but insists on interpreting the events of the world in terms of mystical forces.[36]
>
> They helped undermine the old world, but they discovered that the building of a new world was much harder than they had suspected. They called for new systems of ideas, and when no valid ones were forthcoming, they demanded new myths that would not insult the intelligence of man.[37]

These magical myths of rationality and progress work through the denial of peoples violently expunged in support of this god-like ideal of industrialization and capitalism. Modernity's attempts at proclaiming its supremacy over and against that which it encounters position it in the role of religion. Furthermore, it is ironic that modernity speaks one thing, while doing another: modernity is religious as it reduces what it names religion to a relic of the past, a meaningless tradition. Precisely in its attempts at reduction, modernity is made in the image of what it negates.

Modernity's use of religion is not a simple or even innocent use; similar to Christianity's use of human nature, modernity deploys its sovereignty in insidious ways. *The Outsider* specifically unearths the subterranean stratum by following its protagonist on his quest for freedom. Cross Damon is both the product and process whereby modernity defines itself: he is the *Outsider* that simultaneously defines the inside from which he is excluded. He is explicitly set up as the "negative" in this work. Cross bears his namesake as he bemoans, "I've made no promises...I'm nobody...I'm responsible for nothing." What best characterizes his negative place in the book occurs relatively early; Cross is involved in a subway crash from which he miraculously survives, but also which enables his "escape": he is presumed dead. These events take on a more significant meaning; they suggest what has already occurred, an already given fact of modernity. To the modern world, not only is Cross dead, but he along with his "tradition" have already ceased to exist. Wright returns to this "accident" near the book's close: "To be sure, the accident had made possible his decision to dishonor those unwritten vows that he had been made to promise, but his eagerness in embracing the opportunities presented by that subway accident had robbed that accident of its element of contingency, and the rest had flowed naturally and inevitably."[38]

Wright confirms this view that the accident had already occurred in the sense that the continual enslavement and negation of Cross and his race is far from arbitrary or accidental. The denial of race is part of the process of defining itself. Cross'—and consequentially his race's—accident is planned all along. As Cross himself realizes, it did not matter what the ideological content and group; such a space would have been constructed and filled eventually.[39]

Wright thus is able to write a novel that demonstrates how modernity buries its dead—represented by Cross, but, more generally, by that which is negated—yet, the novel is not just a work of mourning. Wright also writes a novel where the dead return, where they are never really done with once and for all: they uncannily return. Cross even remarks: "... since religion is dead, religion is everywhere... Religion was once an affair of the church; it is now in the streets in each man's heart. Once there were priests; now every man's a priest."[40]

What Wright is doing then is writing into modernity its own transgression. It is by having modernity explicitly participate in the transgression of its own values that Wright catches modernity guilty. In short, by blurring the difference between Cross and Ely Houston—the district attorney and the embodiment of the law—Wright undermines modernity's claims of sovereignty and implicates it in that which modernity attempts to surpass: religion and tradition:

> Cross, in his first encounter with Houston, recognizes their mutual entanglement: And, intuitively, he knew that Houston was caught in the same psychological trap. That was why Houston was a District Attorney sworn to uphold a system of law in which he did not really believe. Houston was an impulsive criminal who protected himself against himself by hunting down other criminals![41]

Cross realizes that Houston and he belongs to the same time of modernity. He points to an uncanny awareness that Houston himself will have to achieve, if he is to track down Cross for the crimes he commits: namely, that this shared temporality also produces the same psychology, the same set of desires, that is outlawed, monitored, and punished. What Cross fails to realize is that this knowledge does not liberate himself—rather this knowledge blinds him from his eventual entrapment and from his temporal alienation. Houston speculates:

> This man would be hard to detect, because he'd act like other men, maybe a little more subdued. But above all, this man must feel that he *knows* what's right. Then, all at once, he sees something that violates that sense of right and he strikes out to set it as it ought to be. Then he resumes his cool and correct manner of living.[42]

Houston accurately assesses Cross in his personality sketch. Like Cross' sketch of Houston, Houston here begins to formulate the underside of Cross. Cross acts like the law in his ability to suspend ethics and to define the criteria of right and wrong. However, just because Cross can act in this way that formally resembles the law does not lead to them being interchangeable. After all, a difference exists between an act that is legal and one that is criminal. Yet Cross' actions suggest that such a difference is suspect and potentially only formal and arbitrary—since the law can be just as violent as crime. Houston's sketch is Cross' sketch—both are sketches of the other. The law and the crime are inversions of each other; their respective value depends on each other. Cross himself realizes this relational quality involved in maintaining modernity when he realizes that he needs the law's recognition. Since Houston cannot prosecute (or persecute) Cross for his crimes, he abandons Cross in his anonymity to brood over his unacknowledged crimes. Both realize that they are each other's double that, although separated, are linked.

The implication that this revelation spells is that modernity is not free from its religious past, rather it is bound to its other. Cross' girlfriend, Eva, also bears witness to this temporal entrapment and impossible notion of freedom. In her diary, she writes:

"I've begun to notice things that I've noticed before. Is this because of what Gil and the Party have done to me? I feel like a victim and everywhere I look I think I can see other victims in the making."[43] Eva is brought to the harsh realization that modernity is just a repetition of what came prior. The Party, industrialization, capitalism all begin to blur together not only in Eva's paintings but the whole novel. Eva captures the awareness that these institutions are just variations on a theme found in religion; their claims at progress and advancement actually turn out to be threats. Although the Party has threatened to expel her from its ranks, the threat itself testifies to the price of freedom. In this sense, modernity is like religion in Wright's crafting of his metaphoric parallel: religion like modernity enslaves. Modernity incorporates this aim of religion, but transfers the psychological effects to racism. Eva's epiphany of her doubledness, her implication in this hypocritical practice, tragically attests that the only way out—the only absolute freedom—always comes with a price, in this case, her own suicide.

Wright's title suggests an extreme, almost impossible condition: namely, the state of getting "outside" and dwelling there. The work is a problematization of this limit; neither Cross nor Eva are able to get outside without some price deeply qualifying this attempt. I suggest two interrelated reasons. Wright is concerned with modernity's claims and scope; there are both promises and failures in his eyes. He grants modernity the boon of devaluing some myths; its corollary problem is that new myths have arisen. Like Nietzsche then, Wright questions notions of origins and rebirths. But Wright critiques modernity's belief in its absolute break with what came prior. Modernity, like religion, tries to speak on behalf of the universal by means of negation. Wright ties this concern with the insider/outsider problem: he links not only racism but also any oppression to systematic reduction and negation. Wright undoes modernity's claim by revealing its violent underside and questioning its sovereignty (or the lack thereof). In order to challenge effectively modernity's claims to be a superior time, Wright proceeds to uncover the impossibility of freedom. He is worried that even those marginalized will reemploy the same dangerous, dominant terms and strategies. Wright's understanding of this dynamic allows for a more complex view of historicization and temporality as not entailing distinct phases. Wright's vested interests in theorizing modernity's time implicate both dominant and oppressed groups, blur constructed differences, and work toward reconfiguring modernity. The dominant group has always already constructed the other, but to leave both groups unchallenged is for Wright a crime. Following Nietzsche's idea of transvaluation, Wright's work is an exemplification of this challenge against the repetition of the same phenomena and temporality.

Wright's epic challenge to modernity affords a few closing remarks. His novel's analysis of the techniques used to characterize the modern as present and religion as past challenges and confounds such a periodization—exemplified in the secularization thesis—which I set out to interrogate. The religious past still exists in consequential ways: it haunts the very enforcement and modality of modernity. The spirit of modernity is betrayed by its own failures to be free from its "past." This means that modern time is still religious. This present religiosity then makes the modern a future temporality. Ironically then, Wright's critique constitutes the modern as an event yet to come. On the one hand, Wright suggests that this messianic hope will arrest this ongoing repetition. Yet, on the other hand, he forecloses this hope from arriving in his indictment of Cross' hope and search for meaning and freedom. It is perhaps here where Wright and Nietzsche are in the most visible alignment. Nietzsche's Anti-Christ too is an impossible figure because his manifesto relies on religion. His form deconstructs his content—his sovereignty and freedom. Together, both Cross Damon and the Anti-Christ sacrifice the belief in a future hope and bear witness to modernity's impossibility. Instead, their acts leave the event of transvaluation as a possibility existing outside their texts and temporalities.

Notes

[1] These components are culled from Talal Asad's "Religion, Nation-State, Secularism" and Peter van der Veer's "Religious Nationalism." See Talal Asad, *Formations of the Secular: Christianity, Islam, Modernity* (Stanford: Stanford University Press, 2003), 13 and Peter van der Veer, *Religious Nationalism: Hindus and Muslims in India* (Berkeley: University of California Press, 1994), 178.

[2] Peter van der Veer, "War Propaganda and the Liberal Public Sphere," *Media, War, and Terrorism*, eds. Peter van der Veer and Shoma Munshi (London: Routledge, 2004), 9–21.

[3] One such postmodern example, which critiques the belief in the religion/modernity split, comes from Paul Gilroy, whose work *The Black Atlantic* offers a characterization of Richard Wright as subscribing to this naïve view. Gilroy writes, "For Wright the decisive break in western consciousness which modernity identifies was defined by the collapse of religious understanding of the world." (160) Gilroy's understanding of Wright establishes an inverse relationship between religion and modernity; modernity is not what religion is. To be sure, Gilroy locates loss, anxiety, and uncertainty in modernity. But, Gilroy uses "religion" without defining religion and modernity as well as their necessary connection and rupture. While Gilroy correctly notes Nietzsche's hand in Wright's work, he does not incorporate how Nietzsche's understanding of religion is translated into Wright's own understanding of religion and, consequentially, modernity—or in Gilroy's words the "metaphysical condition of the modern's world existence." Paul Gilroy, *The Black Atlantic: Modernity and Double Consciousness* (Cambridge: Harvard University Press, 1993), 160–61.

[4] Such a view would ironically contribute to and further support Gilroy's theory of "doubledness," which is the subject of his work.

[5] Wright was an acquaintance of the public intellectual, H. L. Mencken, who was also the translator of Nietzsche's *The Anti-Christ* from which I quote. They met in Wright's hometown of Memphis in 1927.

[6] Gilroy notes that this name "...combines an acknowledgment of Judeo-Christian morality with its Nietzschean overcoming" (*Black Atlantic*), 171.

[7] Sovereignty has a long history that can be traced from Nietzsche back to Hegel and even the Greek's anxiety over the tyrant, and a complicated genealogy that can be traced to Bataille and further to "poststructuralism's" concern with the subject. The concept usually has two senses, which are integrally related. The first is the idea of exercising supreme authority over a *limited* territory. The second prominent sense of sovereignty is that it is a state of *unlimited* autonomy and independence; it is the essence of the absolute because it has brought about perfection with-

in itself, while retaining its purity. I want to note that I am concerned with sovereignty both as *a power over* and *an absolute freedom*. These definitions are based on the entries in *The Oxford English Dictionary*.

8 Tim Murphy, *Religion, Metaphor, and Hermeneutics: A Study of Friedrich Nietzsche's Der AntiChrist* (Ph. D. Dissertation: University of California, Santa Cruz, 1997), 30.

9 Friedrich Nietzsche, *The Anti-Christ*, H. L. Mencken, trans. (Tucson: See Sharp Press, 1999), Chp. 7, 25. Hereafter cited as AC.

10 AC 57, 82.

11 AC 24, 40.

12 Nietzsche writes, "But every hope remained unfulfilled. The old god no longer *could* do what he used to do. He ought to have been abandoned. But what actually happened? Simply this: the conception of him was *changed*—the conception of him was *denaturalized*..." AC 25, 41.

13 AC 25, 41.

14 AC 26, 44; bold mine.

15 Murphy also appreciates this proliferating multiplicity; although he derives this through an analysis of metaphor, namely, how Nietzsche's early concept of metaphor informs Nietzsche's reading of religion, especially by remapping religion onto other metaphoric domains.

16 See Murphy again, whose findings would also support this section because Nietzsche's metaphors like language involve substitutions, supplements, and partial negations (*Religion, Metaphor, and Hermeneutics*), 124–25.

17 For example, in the next quotation, the "abortion" best captures this sense. The good god of the inferior tribe fails to provide either rain or victory; hence, this god is removed from the natural abundance of the life instinct. The devil of the superior folk provides food and booty; hence, this devil, although labeled as anti-nature, bears what the good god is supposed to secure.

18 AC 9, 26.

19 AC 57, 82.

20 AC 29, 47.

21 Nietzsche writes, "In order to be able to say Nay to everything representing an *ascending* evolution of life—that is, to well-being, to power, to beauty, to self-approval—the instincts of [*resentment*] here become downright genius, had to invent an *other* world in which the *acceptance of life* appeared as the most evil and abominable thing imaginable." AC 24, 40.

22 AC 26, 44.

23 AC 3, 22.

24 AC 58, 84.

25 Cf. Nietzsche: "We have just studied a code of religious legislation whose object it was to convert the conditions which cause life to *flourish* into an "eternal" social organization—Christianity founds its mission in putting an end to such an organization, *because life flourished under it*"

(AC 58, 84). Religious value arises since it separates life's essential non-utility from its supposed utility by the familiar tool of negation.

[26] AC 57, 82.

[27] AC 57, 83.

[28] AC 62, 90.

[29] AC 12, 29.

[30] AC 13, 29.

[31] Richard Wright, *The Outsider* (New York: Harper, 1953), 23. Hereafter cited as TO.

[32] TO 174–75.

[33] TO 324.

[34] In *White Man, Listen!*, Wright especially links religion and tradition (New York: Harper, 1957), 3.

[35] TO 325.

[36] TO 326.

[37] TO 330.

[38] TO 347.

[39] TO 325.

[40] TO 331.

[41] TO 126.

[42] TO 260.

[43] TO 192.

Hetero-Temporalities of Post-Socialism

Lisa Rofel

Time has, if you will forgive the paradoxical phrasing, long been a terrain for projects of social justice and utopian dreams. As Susan Buck-Morss has recently pointed out, this function of time is self-consciously true in historically sedimented ways for the praxis of formerly socialist nation-states.[1] What has become of time in the aftermath of socialism? My answer to this question will focus on China, a country whose continuously radical transformations have coursed through the twentieth century and now into the twenty-first. Unlike in the countries of eastern Europe, China's Communist Party still holds the reins of state power. Yet its policy of "socialist market economy with Chinese characteristics" is a clear abandonment of the heart of China's former socialist practices. Perhaps the most telling sign is the state's official retreat to a strict adherence to evolutionary stages of history—that China needs to pass through capitalism before reaching socialism. Maoist socialism, by contrast, believed in jumping stages to bypass the history of capitalism in the West. One symbolic sign of this abandonment is the decision made recently by the Central Committee of the Communist Party that capitalists and entrepreneurs are welcome to join the Party. In the aftermath of socialism in China, in the development of capitalist relations of production, what has happened to time? Does time still function as the terrain for utopian dreams? Of what sort?

In her recent compilation of essays, *Dreamworld and Catastrophe,* Susan Buck-Morss offers a provocative argument. She argues that liberal capitalist and socialist countries historically converged in their construction of a terrain of power—what she terms "the wild zone of absolute power"—that stands above the law and out of the control of the masses in whose name these modern sovereignties claim legitimacy to rule. This terrain reflects the intimate connection between the state and violence. The legitimation of this wild zone of absolute power derives, according to Buck-Morss, from the ability of the state to define its purpose as protecting citizens from the possibility of war and the threat of a common enemy. Its legitimation further derives from the ability of both socialist and liberal democratic states to appear as representatives of the popular will. Citing the Russian philosopher Valerii Podoroga, Buck-Morss argues that socialist and capitalist states construct a political imaginary in which the designation of putative enemies enables a collective "we" represented by the state. However, she points out that capitalist and socialist states build this political imaginary on different terrains. The most striking difference, she argues, is that liberal democracies envision a world of nation-states at war and thus their political imagery is dominated by the dimension of Space. Socialist states, on the other hand, engage with war as class struggle, and thus their political imagery is dominated by the dimension of Time. Socialist states, according to Buck-Morss, held themselves up as the vanguard of history, as having the ability to vanquish the capitalist world. One of their inner contradictions was that, by their own estimation, they repeatedly fell behind the industrial development of the West. Yet throughout their existence socialist states continued to produce utopian dreams in this dimension of Time.

To argue that capitalist and socialist states resemble one another in fundamental ways is, indeed, a provocation. While I do not completely agree with this provocation, I would like to take the argument in a different direction. Buck-Morss recognizes only one mode of temporality for socialist states: the classic one of linear progress embedded in modernist notions of history. This cohesiveness and singularity of Time that Buck-Morss represents as having been true of socialist states undermines, I believe, her goal to release the dreamworlds embedded in those former utopias and allow them to live—again, forgive the inevitable temporal mode—in the future. Such unitariness and universalizing obscures a terrain of struggle over time, a terrain in which heterogeneous elements co-exist, though not on equal terms. This idea of unitariness, certainly a cultural dominant in China, and the heterogeneous elements of time that continue to exist on the margins of that cultural dominant are both haunted by the history of colonialism. Since the mid-nineteenth century, China has been violently forced to grapple with Eurocentric ideologies and practices of time: the belief that History (in the singular) first happens in Europe and then appears elsewhere.[2] This belief rests on several corollaries: that Europe represents the pinnacle of civilization while the rest of the world represents mere shadows; that Europe constructed itself wholly from within rather than through differences it imagined between itself and Asia, Africa and the Americas; and that European colonialism, while relevant to the history of Europe, did not define the fundamental "Europeanness" of Europe. As Dipesh Chakrabarty has argued, the non-West has been forced to read the history of the West while the reverse has never been true.[3] Chakrabarty reminds us that "Europe" operates as an imaginary figure that informs the historicism of the non-West. As Johannes Fabian famously argued, Eurocentric historicism denies co-evalness to non-Western cultures.[4]

This imaginary figure of Europe has deeply informed the notions and practices of time that subtend modern Chinese history. It continues to haunt China's post-socialist dreams. But China's need to grapple with Eurocentrism and the Enlightenment notion of progress masks how government institutions, through the marking out of time into programs and plans, create the metaphysical effect of the state. It masks how any cultural hegemony also always creates its exclusions—never more so than in a history of socialist class struggle, in which the struggle partially centered on precisely who could carry forward the banner of socialist progress and who would be designated as stuck in a feudal past. It masks the partiality of the official discourses of time that, while dominant, were not always hegemonic. It masks the work that it took to create articulations amongst differing approaches to time. Finding and releasing the dreamworlds necessitates this kind of theoretico-historical work. It necessitates asking after heterogeneity, hegemony—where the terrain of struggle is never fully won—and articulation. To think through these contingent, non-necessary connections between different notions and practices of time will enable us to break with the Eurocentric logic of necessity and inevitability, with colonial teleology and evolutionary thinking, with the notion of homogeneous, empty time and with the idea of a unitary state. Finally, it will enable a more expansive praxis that will recognize multiple dreams.

To take one example of the heterogeneity of time in China under socialism: Official time in China was marked, in Gail Hershatter's felicitous phrase, by "campaign time." "Campaign time," according to Hershatter, "cordon[ed] off the pre-1949 past 'before Liberation' and measur[ed] the present by state initiatives and popular participation in them. State-initiated campaigns were assumed to be more or less uniform."[5] But Hershatter turns the lens towards the margins of political power: poor women in fairly inaccessible rural areas. Hershatter reminds us that state edicts were carried out in widely varied environments, by local personnel according to particular circumstances. The women living in central and southern Shaanxi province, whom Hershatter came to know, spoke of their lives in other temporalities that overshadowed that of state-initiated campaign time. "Marital temporality divided the lives of women between early years in one's natal village and life in a new community as a married-in bride."[6] These women used the lunar calendar rather than calendar years used in official time. Their lives embraced multiple temporalities cross-cut by levels of political activism, generation, and geographical location. Hershatter calls some of their narratives "wrinkled time," after the children's author Madeline l'Engle, to signal the ways many of these women truncated stories about official campaign time into events marked by individual histories of childbearing and their sufferings under extreme poverty under socialism.

In the aftermath of socialism, the Chinese state continues to operate on a notion of time that reflects that which they embraced for the socialist revolution: linear, evolutionary time in which China must pass through established stages to reach communism. The aftermath of European colonialism remains embedded here. But, as under socialism, the state's propagation of this practice of time is uneven. Moreover, its propagation creates its own exclusions: those citizens were designated as off-the-track of progress, out of the line of appropriate development, behind-the-times, as it were. Among citizens who live in these other social sites, one finds representations and practices of time orthogonal to that of the state.

The three representations of time I address in the rest of this essay are located in quite disparate sites: a recent documentary about workers in the industrial Northeast, the experiences of young, professional middle-class urban residents, and that of a migrant worker living in the metropolis. All of them co-exist in China, though they each offer disparate and heterogeneous approaches to time. None of them reflects a one-to-one correspondence with the official time of linear progress as exemplified in evolutionary stages of history. Nonetheless, each is obviously in dialogue with, even as they counter or skew it, that official time.

West of the Tracks, a documentary by Wang Bing, reflects a new direction in film making in China: independent ethnographic documentary as a social practice grounded in the production process itself. Inspired by the Japanese documentarian Ogawa Shinsuke, who developed what is termed the engaged mode of documentary, this production process involves long-term relationships between filmmakers and subjects. Young filmmakers, mainly from Beijing, have set out to live for several years with those whom they film, more in the manner of a participant rather than an observer. The social and political commentary of the film develops organically out of this relationship, and itself participates in mobilizing the subjects of the film. Following Ogawa, these filmmakers understand documentary practice not as a representation separate from reality but as a part of life. Their films eschew conventional documentary genres with talking heads and linear stories. Instead, they let the camera—or digital video—run continuously while people go about their daily lives. These documentarians see their work as part of the lives of their subjects and they are concerned that their documentary making should be a social practice that helps the people they film.

In recent years, these documentarians have chosen to live amongst those who have experienced the underside of the post-socialist market economy. *West of the Tracks* films railway workers in the rustbelt of northeast China. Nine hours long, the film tracks the lives of the men who spend their days moving industrial goods on the train from one factory to another, back and forth, and back and forth. We see them in all the seasons—huddling over their small stove in the railway office during the winter, walking beside the train in the spring, suffering the stifling heat inside the box cars in the summer, and relaxing in the breezy fall. The Northeast of China long served as the industrial heart of the country. Industrialization began there early in the twentieth century under German, Russian and then Japanese colonialism. Coal-mining was vital. In the last decade, however, the Northeast has experienced severe unemployment, as workers have been laid-off in the thousands from unprofitable state-run enterprises.

This film tracks these experiences; we watch the few workers left who shuttle small amounts of materials to the remaining factories. In the manner of the documentary, this realization comes through only gradually, as the workers themselves talk about the fact that very little coal remains left to haul (and thus to warm their stove); that they have to scrounge illicitly for coal now that the old abundance is gone. Or they discuss among themselves the work of hauling the remains of a certain factory, as if they were preparing a burial.

Mostly, the men have little to do. We watch them playing cards on the train; simply trying to keep warm in the tiny office during their lunch break; taking naps while they rotate night shifts. Sometimes they tell stories of women. We watch them scraping together a livelihood out of the industrial wasteland that envelops their world. One long sequence follows an older man and his son, both long unemployed, who have squatted in a ramshackle hut by the tracks. The hut has just enough room for a bed that they share. The son stays in bed most of the day; he has nothing to do. The father has no formal employment with the railways. He is a temporary worker who helps the other workers and, in return, they look the other way as he scrounges for coal and other industrial scraps which he sells in small quantities. Since these scraps are state property, his activities are technically illegal. The old man sometimes hops on the train and rides for several hours with the other workers to chat and wile away the time and then hops off. Near the end of this sequence, government officials have decided—for reasons never made clear—that the old man and his son can no longer have the privilege of living in their hut. They are forced to move to an area near the airport, where the old man manages to scrounge other temporary work.

Time in this film goes nowhere. Literally. The back and forth movement of the train, or perhaps we should call it the side to side movement, simultaneously mimics and frames the lives of these men. The seasons might change, but the movement is the same. The train does not move forward; it does not have new destinations. Only the men's sensations of the weather in the different seasons indicates that time passes. But even this kind of passing of time is ambiguous. Much of the work takes place on the night shift. The nights might be colder or warmer, but they are always dark. It is difficult to tell time in the dark. And, besides, there does not seem to be much point in marking the time too closely. The men take breaks to eat, though some of them do not have full meals. But otherwise, they are marking time as it moves back and forth and goes nowhere.

Time also moves backwards in the film. It is as if the filmmaker had taken a socialist era film about the glorious lives of the proletariat and the future utopias to come and fed the reel backwards into the projector. Everything moves backwards: the factories fall apart and close, the poverty comes to the foreground, the railway workers are so marginalized that the government barely acknowledges that they exist. This tale conveys linear regress.

In stark contrast to industrial workers' time in *West of the Tracks*, the rising young urban middle class provides a wholly other experience of time. In the major cosmopolitan cities of China, educated youth are entering a world of work non-existent for their parents' generation. They seek jobs in multi-national corporations, high-tech companies, and the diverse information economy, including a vastly expanded media world. They are entering an economy that, until the late 1990s, was the fastest growing economy of any country in the world. Such dizzying growth produced contradictory affective energies. A belief that anything was possible mingled with anxiety about the meanings of such rapid transformations, even as the material environment reflecting the rapid transition transformed the senses as it folded past and future into one another.

This neo-liberal market socialism is accompanied by fundamental changes in moral evaluations of social life. As I argued in my book, what I term the allegory of postsocialism tells a story of how communism repressed human nature.[7] According to this allegory, Maoism deferred China's reach of modernity, against the specter of colonial representations of Chinese culture as that which suppresses individuation and therefore the capacity to produce modern selves. And it did so by impeding Chinese people's ability to express their natural humanity that, all along, lay beneath the cultural politics of socialism. Diverse public discourses put forth the view that the end of socialism means that human nature—whether this is the human nature of the neo-liberal free market economy or the human nature of gender traditionalism—has emerged to find its freedom of expression. This allegory of emancipation holds out the promise that people can unshackle their innate human selves by emancipating themselves from the socialist state. The articulation of this allegory with a transnational so-called free market economy is one of the key motivating forces leading people to embrace a different practice of time.

Yet, as we will see, this does not lead this rising urban middle class to embrace a linear notion of progress.

This generation of youth who have come of age in the aftermath of socialism imagine themselves as having within their grasp the possibility to become free of all constraints. Young women and men repeatedly described to me the need to cultivate a self that lacked constraints in order to express itself properly. Tao Ming, a 34-year old who works for the *Workers Daily* explained, as we sat in an American fast-food restaurant, Roaster's: "My mother is very frugal (*jiejian, jiesun*). That whole generation was. Very diligent (*qinlao*). They didn't treat themselves very well. They would sacrifice for others before doing something for themselves. Our generation is more selfish." By selfish, she meant literally centering on a self that is allowed to desire. But this focus on the self engenders unresolvable contradictions, as it leads to conflicts between a focus on the self and a focus on relations with others.

This post-socialist generation expresses this yearning to be free in three arenas—jobs, consumption, and private life. They self-consciously distinguish themselves from their parents, who relied on the state to provide them with a lifetime job. Indeed, one of the boasts of the socialist state was that, unlike capitalism, socialism provides work for all. This younger generation boasts of constant and rapid job change, what several theorists have called flexible accumulation under late capitalism. It was commonplace for these urban youth to remark casually to me about a new job they had found. I no longer was surprised when those I knew over the course of a year, mentioned two or three jobs they had in the course of our acquaintance. One young woman, who works for the Xinhua News Agency, the Chinese equivalent of the Associated Press, told me that her brother had changed jobs eleven times. These young people might move within a certain job world, such as computers or news media, but they might also move across it. They might work for a multinational office, especially if they hoped to work some day outside of China, in North America or Europe. Or they might have just returned from working in the West, convinced that working in China was better after all. They criticized others who were either unemployed or remained "immobile" in a lifetime job, arguing that there existed a sea of jobs out there for anyone with initiative. Indeed, to a striking degree, China's economy depends on youth. Official and popular discourse castigates the socialist system for its doddering cadres, who know nothing about market economics and who manage bankrupt state work units. Official policy thus promotes hiring young people with advanced training to, literally, run the economy. In a vast array of Chinese companies, people under thirty-five are in charge.

Perhaps it is not surprising that a popular phrase for this kind of rapid job hopping is called "jumping into the sea." "Jumping into the sea" means leaving the state-run sector, which is continually shrinking, in any case, for more lucrative but more uncertain prospects in the market economy of jobs. Jumping into the sea can be liberating, but also, of course, unsettling. These young people live with a paradox: a simultaneous yearning for freedom coupled with a sense that life does not move along a smooth and rising path.

Such a life leads these young people to have a discontinuous sense of time. They do not portray jumping into the sea as exemplary of linear process. Though they view this life as relatively free of constraints (failing to envision the market as another world of constraints), it does not add up to a teleology with an ultimate goal.

This emergent bourgeoisie's structured experiences of time reflect most closely that which has been associated with postmodernism: the feeling that life is unstable, that it is made up of temporary, provisional, contingent moments; the belief that history is finished and there is no need to revisit the past, for only the present exists and all one can do is be immersed in it; that social life is full of diversity but also fragmentation, innovation and experimentation at an ever increasing speed with an emphasis on the ever changing experience of the new; the idea that all is on the surface of social life; and that one's life is a pastiche of beliefs, styles, moments.

One can see how these youthful bourgeoisie experience this kind of time not only in their approach to work but also in their consumption practices. Vast energies devoted to consumption traverse China's cosmopolitan cities. In a nearly Bahktinian carnivalesque (or perhaps Fellini would be a more apt auteur), young professionals with a great deal of disposable income—in large part because they live with their parents until marriage—consume, it seems, without restraint. Their consumption obviously crafts class distinction, in Bourdieu's sense of the term.[8] Their consumption, as they frequently mentioned to me in an offhand manner, is also predicated on the unpredictability of the future. They do not live according the idea that they should save up for a future which they see clearly. But they also struggle with the tension they feel between a search for pleasure and a sense of superficiality, between a desire for novel stimulation and a yearning for substantive experience.

Their consumption practices are linked to time in yet another sense. For consumption, one of their measures of freedom, is not just about pleasure. It is a post-socialist technology of the self by which these urban youth and, by metonymic association, the Chinese nation, enable themselves to transcend the specificities of place and identity and be part of the "world." Of course, in many respects, this "worlding" is about linking specifically to an ideology of the American middle-class and the domination of American consumer goods in the networks of globalization. Thus, the formation of a consumer identity in China is, at heart, about a search for a cosmopolitan subjectivity. This yearning for cosmopolitanism in China consists of two aspects that are in tension with each other: first, a self-conscious transcendence of locality, posited as a universal transcendence, and second, a domestication of cosmopolitanism by way of renegotiating China's place in the world. These youthful bourgeoisie are a critical site for the negotiation of this tension between transcending the local and renegotiating China's place in the world—one could say between transcendent desire and protective Chineseness. They are the mediators of cosmopolitan "desire," they are the site of negotiating the meaning of "Chineseness" and they are posed as the ultimate consumers. At the same time, the way they domesticate cosmopolitanism is through a series of structural dichotomies and structured forgettings that re-invent the past.

The three main structural dichotomies, and their attendant forgettings, that I heard most often were between: 1) consumerism and the search for wealth versus politics; 2) consumerism and the search for happiness versus kinship; and 3) consumerism and the search for freedom versus communalism.

Every single one of the young people I spoke with began our conversations with a striking, and unsolicited, statement that, at first, I thought had nothing to do with the questions I asked. They all began our conversations with the declaration: "I am completely uninterested in politics." As I heard that declaration over and over again, I began to ponder its significance. For I realized that this declaration was not simply a means of warning me off of dangerous topics. Everyone volubly, vociferously and satirically condemned the government's treatment of Falun Gong, for example. Rather, this was an assertion that the passions of the self were not in the least bit directed towards the state. This opposition between a past where politics equals constraint versus a future in which consumption and the pursuit of wealth means freedom is, as I said, a structural dichotomy that depends upon structured forgettings. It forgets that politics once inspired many in China with a passion that gave meaning to their whole lives as well as the most menial or trivial tasks they might be engaged in. It forgets that many of their parents, in their youth, thought of politics not as constraint but as the ultimate moment in taking control of their lives and giving them more freedom, as least for a brief moment.

This youthful bourgeoisie also structures an opposition between consumerism and the search for happiness versus kinship. When Tao Ming said that her generation is more selfish—and here she used the term *zisi*, meaning literally "self privacy" or "unto the self"—I wasn't sure whether being selfish, that is, focussed on the self, was a good thing or a bad thing for her. So I asked her. She implicitly acknowledged that both were true, in her answer: "Yes, but there have been good changes, too. My friend and all her sisters do not want to have any children. Their mother says they are selfish." Tao Ming, in other words, thought the decision not to have children was a positive sign that women, too, can have a self free of desire as opposed to a kinship-oriented self that seemingly characterizes her mother's generation. Young women placed their mother's generation in the category of "traditional culture," which supposedly forced them to marry only out of kinship obligations. Once again the opposition holds a structured forgetting. After all, their parents, and their parents' parents, grew up with ideologies of companionate marriage that were written into the first constitution of the People's Republic of China. The kind of revolution they fought for based itself on an ideology of gender equality that posed class identification over and against kinship identity. During the Cultural Revolution, moreover, their parents participated in movements in which many of them denounced their parents for having the wrong politics. Far from being mere subjects of a traditional Chinese culture, their parents also thought they were fighting the four "olds" of feudalism in China.

The final structural opposition, with its structured forgettings, that I encountered was consumerism and the search for freedom versus communalism. These young urban women refuse the space and time of nation building and imagine themselves as transcending city, country, class and, ultimately, China. For the young women I met, as well as for others in China, excitement about the possibilities of a cosmopolitan future includes a search for the freedom to move through space and time that their parents, they imagine, did not have. They imagine that the socialist state fixed people in place in order to regulate and stabilize social relations. Of course, there were efforts to fix people in place, such as the household registration system that drew a strict line between the countryside and the city. And people in the cities were organized into work units, often, though not always, including housing organized by work unit. This arrangement was meant to stabilize full employment and social welfare benefits. David Harvey (2000) has argued that utopias are usually spatialized, that "spatial form controls temporality, an imagined geography controls the possibility of social change and history." Unlike the utopias Harvey addresses, which, he claims, repress the dialectic of social process, the stabilized spatial arrangement under Maoism produced the opposite. That is, part of the structured forgetting entailed in the imaginations of these young women is the history of the social struggles under Maoism that precisely addressed the issue of space and place: should certain young people remain in the city or be sent to the countryside, should urban people remain in their work units or transfer out to other work units. It also forgets the experience of *chuanlian*, when young people during the Cultural Revolution enthusiastically travelled everywhere to experience revolution as Mao urged them to do and to learn from the peasants, and each other.

In contrast to the urban middle class experience of time as fragmentation, accompanied by these structured forgettings, and in contrast to the sense that time does not move at all, as experienced by male industrial workers in the Northeast rust belt, I offer one final example of an experience of time by a woman migrant worker. Ma Xiaoduo was in her late thirties when I met her. She had moved back and forth between her home in the countryside and various Chinese metropolises numerous times over twenty years. Ma Xiaoduo, in her narrative of her life, portrayed this movement as progress and in that sense, appears to cleave most closely to the official representation of time. But Ma Xiaoduo's story of her life is not one of progress built automatically into the scheme of things. She emphasizes the contingency of life and the inability of migrant workers to reach any kind of destination with a clear sense of progress over time as underlying their movement through space. The story she tells of her life does not rest on a teleology. Perhaps one could say she has a goal: stability, as symbolized by an urban residence permit. Her story is full of struggle, and, in this sense, her story haunts the post-socialist regime with reminders about class inequality, precisely the kind of class inequality that used to accompany socialist stories of progress since abandoned by the state.

Ma Xiaoduo grew up in the countryside in western China. She was unusual in that her father supported her efforts to get through school. Most young women in the countryside fail to obtain more than a few years of schooling before they are pulled out to do agricultural work. This problem is especially prevalent since the reversion to household farming with the end of socialist communes. Ma Xiaoduo graduated from high school, determined not to get stuck with agricultural work. She even took the national test to enter college, but given that only about one percent of high school graduates attend China's few universities and she attended a rural high school, she did not really have a prayer. This was the early 1980s, when universities were rebuilding themselves after the Cultural Revolution and many older urban youth who had been sent to live in the countryside were also desperately trying to attend university.

Ma Xiaoduo turned to what was then a risky path: illegal migration to the big city. She had an aunt who lived in Beijing and she also relied on one of her neighbors who had already migrated. Migration from the countryside to the city is illegal in China. It resembles illegal migration from Mexico and Central America to the United States. And the discourse about this migration among urban residents resembles the anxiety and hypocrisy of U.S. citizens. To live in China's major cities, one needs a residence permit. The permit allows one's children to attend city schools and one's family to receive medical care. It allows one to obtain a permanent job in the city. Thousands of rural residents have migrated illegally to Chinese metropolises on the eastern coast. They do construction work if they are men and service work, including sex work, if they are women.

Ma Xiaoduo went the route of domestic service. She landed a job through her neighbor, but lasted only twenty days, for, as she averred, she had no clue how to care for a child and could not stand the finicky crying baby of the middle-class family she was stuck with. The family never let her rest and never let her sit down. Fortunately, an older neighbor of this family urged her to call her aunt to get her out of that situation. She then moved on to become a cook for a large family. The only problem was that, although she told them she knew how to cook in order to land the job, she did not know anything about Beijing cooking. She had no clue how to make the most basic dish of all, northern dumplings. After the family scolded her, the grandmother taught her how to cook. After that family, Xiaoduo managed to move on to take care of an older woman, who turned out to be a rather prestigious communist party cadre and introduced her to several important cadres when they vacationed in the favorite spot of old party cadres, Beidaihe, northeast of Beijing. Xiaoduo stayed with this woman for several years until she saved enough money to go back to school. She struggled to earn the equivalent of a college trade school degree and hoped to go back to her countryside home to become a teacher. For a time, she lived with the old woman, working part-time and studying part-time. Then she lived in an abandoned building near the school, so that she could study full time. Afterwards, she returned to her family in the countryside. All through this period, she had lived illegally in the city. Though the movement in space, back to the countryside, might appear to be a movement backwards in time, the move from domestic servant to teacher was a move forward. When she arrived back in the countryside, however, Xiaoduo realized that rural living had become too harsh for her after her years of acclimating to the city. The dilemma that Ma Xiaoduo experienced is one that most migrants experience: the desire to be immersed in a network of family and familiarity, but the feeling of estrangement from that home due to the scales of development that pull migrants away from homes and into cities that treat them hypocritically as transgressive interlopers. This experience is one of internal exile. Internal exile means an experience of time marked by continual displacements and lines of escape, a jagged, uneven time that pushes towards progress but not through evolutionary scales.

Xiaoduo eventually left her rural home again, for she was unable to find a teaching job there. This time she went to the city that has a reputation of "the wild west": Shenzhen. Shenzhen is a new city created wholly by the state in the 1980s, as their first experiment with foreign capitalist investment. Located close to the border with Hong Kong, Shenzhen has expanded enormously since its inception. Cordoned off from the rest of China, even other urban residents are supposed to apply for official permits to enter Shenzhen. The off-shore factories established in Shenzhen by foreign investment, initially almost entirely from Hong Kong, beckoned women like Xiaoduo, who were unofficially allowed to enter Shenzhen. There, at last, Xiaoduo began to flourish. She managed to obtain an office job, in part through serendipity and the sympathy of one man in the company willing to take the risk of hiring her. But Xiaoduo still had no urban residence permit. The problem of residency became acute because she had become engaged to a young man she had met in Beijing. His parents were opposed to the match, because Xiaoduo did not have a residence permit, which meant that their child would be designated with the residence of the mother. Xiaoduo was determined to overcome this obstacle and, through struggling over several years in Shenzhen, managed to obtain a Shenzhen residence permit. Although she would still be illegal in Beijing if she returned, she decided to do so to marry her fiancé. Through the marriage, she was finally able to transfer her official residence to Beijing.

At the time that I met her, Xiaoduo worked as a staff person for the Rural China Knowing All organization established in Beijing by middle-class women who want to help rural women in Beijing learn skills they can take back with them to the countryside. In some respects, Xiaoduo could point to the successes of her life. But she did not do so in an easy way. For Xiaoduo, the Chinese economy is so uneven that she moves through different temporal modes of class within a seemingly single national space, a space that is itself transversed with transnational presence. What she pointed to instead were the fundamental divisions between herself and urban middle-class people. That division always made her feel like an outsider, an exile within. She does not like the way the city changes peasant women, teaching them, she says, not to care about one another's problems, not to feel any connection to their family in the countryside. On the other hand, she acknowledges that it seems inevitable, given the hierarchy of spatial scales of development in China, that moving forward for peasant women means moving to the city. But, this is a fraught movement forward in her eyes, one with losses as much as gains, and one where the continuous reproduction of the urban–rural divide within Beijing creates a class division that makes progress for rural migrants an elusive spectral presence.

These three disparate experiences of time are clearly related to the rapid, one might say torrential, sweep of capitalism in China. In the 1990s, China's rapid economic growth was subtended by China becoming one of the world's major labor workshops. While the fact of capitalism's restless search for profit might be predictable, however, the direction or form of that search is not. In each of these cases, micro-practices operate that appear through an unceasing, restless, but also unpredictable, capitalism. In the first case, the industrial workers of the Northeast experience the unmaking of Maoist socialism's class system. The Northeast, as the rustbelt of China, has experienced this displacement in the most dramatic manner. It is thus the site of numerous labor protests in the face of mass unemployment. Messianic movements like Falun Gong have their home here as well. Not only is the state steadily closing down all state-run work units, but the state has repressed the language of Maoist socialism. No one speaks of themselves anymore as "the proletariat." These workers face a stasis in their pensions, along with a government program that has subsidized their ability to buy their apartments. Their security is real and also fixed in time. Thus they experience time as moving from side to side rather than backwards and forwards.

In contrast, peasant migrant workers like Xiaoduo move through a network of time that is built on the state-constructed division between rural and urban. From the beginning of the socialist state, resources poured into the urban areas, under a plan for rapid industrial modernization. The rural areas were organized to feed the urban areas, but had to rely on their own resources for any improvements in their lives. The major micro-practice here that shapes their experience of time is the residence permit. It is nearly impossible for rural residents to get an urban residence permit. Without an urban residence permit, none of these workers can move in any kind of linear path towards progress, even as they try desperately to move forward in their lives. Migrant workers who move to the cities work in the absence of law and contracts to protect them. Yet peasant workers, unlike the urban workers of the Northeast, still have access to material resources in the form of land. Thus, the urban workers like those in *West of the Tracks*, whose experience of time was first shaped by Maoist socialism and then reshaped by post-Mao capitalism, often see themselves as worse off than these migrant workers who can, in their view, leave urban work at anytime.

Finally, urban professionals experience a fragmented time marked by the urban turn to the proliferation of middle-class consumerism, the official desire for youth to lead the country to wealth and power, and the blurred state-private investment in an information economy.

Capitalism has shaped these heterogeneous representations and experiences of time. Yet capitalism is not simply a universal process with universal actors with universal needs.[9] Its differential struggles with historically situated workers make for uncertain outcomes and constant upheavals. Time is the arena in which these power relations are contested and negotiated. Time is therefore both the result and the outcome of social struggles. The project of articulation of these heterogeneous representations and experiences of time is not already given, but has to be crafted, and the kind of futures such an articulation will produce will be contingent on exactly how people's interests are made to speak to or past one another. This heterogeneity will define how China grapples with postcolonial hauntings of its future.

Notes

[1]Susan Buck-Morss, *Dreamworld and Catastrophe: The Passing of Mass Utopia in East and West* (Cambridge, Mass: The MIT Press, 2000).

[2] Some might object that "Europe" is not a singular entity. Certainly, in the empirical history of European nations, the diversity of experiences is evident. Nonetheless, in their colonial endeavors, Europeans tended to represent themselves as a unified entity in contrast to the natives they ruled over. See Ann Laura Stoler, *Carnal Knowledge and Imperial Power: Race and the Intimate in Colonial Rule* (Berkeley: University of California Press, 2002). Moreover, this imagined Europe, constructed in the colonies, has also haunted European history, especially those nations that seem to be within Europe but are deemed to be on its margins.

[3] Dipesh Chakrabarty, *Provincializing Europe: Postcolonial Thought and Historical Difference* (Princeton: Princeton University Press, 2000).

[4] Johannes Fabian, *Time and the Other: How Anthropology Makes Its Object* (New York: Columbia University Press, 1983).

[5] Gail Hershatter, "The Gender of Memory: Rural Chinese Women and the 1950s," *Signs* vol. 28, no. 1 (2002): 44.

[6] Hershatter, Ibid., 45

[7] Lisa Rofel, *Other Modernities: Gendered Yearnings in China After Socialism* (Berkeley: University of California Press, 1999).

[8] Pierre Bourdieu, *Distinction: A Social Critique of the Judgment of Taste*, Richard Nice, trans. (Cambridge, Mass: Harvard University Press, 1984).

[9] Sylvia Junko Yanagisako, *Producing Culture and Capital: Family Firms in Italy* (Princeton: Princeton University Press, 2000).

The Politics of Temporality

Heidegger, Bourdieu, Benjamin, Derrida

David Couzens Hoy

"The time of our lives"—this expression condenses into one phrase a series of questions that could require much more than one lifetime to answer. Is the time of our lives a function of a life as a whole, a lifetime, or can it be condensed into a single moment of vision? Does a life have a unity that runs through it, or is the unity of time, and of a life, a narrative, a story, a fiction, or even an illusion? In this essay the question that particularly interests me is, what notion of time is the time that we encounter in our experience of our world and ourselves? Is the time of our lives the same as the time of nature or of history? In particular, if time runs through our lives, in which direction does it run? Does time come toward us from the future, as Martin Heidegger maintained, from behind us through the past, as Pierre Bourdieu asserted, or from the present, cycling perhaps in an eternal recurrence, as Friedrich Nietzsche speculated? Then there is Immanuel Kant's question, is temporality a feature of us or of the world? That is, is the time of our lives subjective or objective, or is there a third possibility? I begin with Heidegger's phenomenology of temporality in the writings of the late 1920s before moving on to the temporal theories of Pierre Bourdieu, Walter Benjamin, and Jacques Derrida.[1]

Dasein and Time: Martin Heidegger

Heidegger's intention is to show that Kant's way of thinking about time and space is derived from what Heidegger calls a more "primordial" level of questioning. In contrast to Kant's transcendental arguments, which show that if something is required for knowledge, something else is also required, Heidegger's derivation arguments try to reverse the ontological ordering of the terms of analysis. Kant thinks that time in the objective Newtonian sense of the present-at-hand universe comes before the human, qualitative experience of temporal moments. Heidegger inverts that order, arguing not that objective, clock time is invalid, but that without Dasein's prior qualitative temporality objective time is unintelligible. Heidegger's project in *Being and Time* (1927) is an attempt to show that starting from objective time, the philosopher will not be able to explain qualitative temporality, but starting from qualitative temporality, the philosopher can successfully explain objective time.

This question of priority has produced a debate about whether Heidegger is a temporal idealist or a temporal realist. For instance, William Blattner reads *Being and Time* as a version of temporal idealism because he understands Heidegger to be saying that time follows from Dasein, rather than Dasein following from time. The latter would be "temporal realism." Is Heidegger a temporal idealist who believes that time is dependent on subjectivity? Blattner thinks so, and says: "If Dasein (Kant: the subject) did not exist, time would not obtain."[2] As I read Heidegger, however, if Dasein is not construed as a subject and time is not a present-at-hand object, then neither idealism nor realism is the right category for thinking about temporality. The contrast between idealism and realism depends on being able to draw a sharp distinction between subject and object, or inside and outside of the mind. Heidegger is questioning this philosophical picture when, instead of saying that temporality is either subjective or objective in origin, he asserts the third possibility that temporality itself temporalizes.[3] For Heidegger, "Time is primordial as the temporalizing of temporality,"[4] and this in turn constitutes "the basic possibility of authentic or inauthentic existence."[5]

But then the greater mystery becomes, what does it mean to say that temporality temporalizes? In *The Fundamental Concepts of Metaphysics: World, Finitude, Solitude* (1929) Heidegger adds analyses that indirectly clarify how temporality temporalizes. Using boredom as the paradigm instead of anxiety, Heidegger maintains that time is *lengthened* by boredom. The German word for boredom is *Langeweile*, or literally a "long while." What is profound boredom about? Heidegger suggests ironically that what is boring is neither objects nor subjects, but time, or more precisely, temporality. In boredom, Heidegger says, "one feels timeless, one feels removed from the flow of time."[6] Boredom is oppressive because time weighs heavily.

Profound boredom is ontological, and it reveals an emptiness, into which everything withdraws. This withdrawal of everything makes Dasein aware of the whole of its existence. Boredom is thus as ontologically revelatory as anxiety is in *Being and Time*. Neither boredom nor anxiety is about any particular thing, but each is about everything (and nothing). Heidegger's 1929 lecture, "What is Metaphysics?" confirms this ontological status of profound boredom. There too he distinguishes between ordinary boredom, as when a lecture is dragging on, and genuine or profound boredom. Of the latter he writes, "Profound boredom, drifting here and there in the abysses of our existence like a muffling fog, removes all things and human beings and oneself along with them into a remarkable indifference. This boredom reveals beings as a whole."[7]

As I understand the purposes of the argument in *Being and Time*, the reason Heidegger chooses *Angst* rather than profound boredom is that boredom's indifference to everything cannot be complete. Dasein cares about this indifference and presumably it does not want to be bored to this extent. Dasein cannot settle for saying, "Nothing matters, so that does not matter either." Simply shrugging one's shoulders and muttering "whatever" will only aggravate the problem. Kierkegaard's aesthete is an example where profound boredom tries to exist as a contradiction by preoccupying itself with its inner life.

In my interpretation, though, anxiety and boredom have different effects on our self-understandings. Anxiety *individuates* Dasein by making Dasein confront its unique fate and destiny. Individuation is an encounter with what is meaningful about the world, and it still involves what Heidegger calls Care. Rather than *individuate* Dasein, however, I read Heidegger as suggesting that boredom *subjectivizes* Dasein.

To say that *Angst* individuates is to say that each Dasein finds out what it cares about, and what makes it "in each case its own." To say that boredom subjectivizes is to maintain that Dasein becomes indifferent to all meaningfulness and ceases to care about the world. Because of this extreme degree of indifference to all meaningful interactions, Dasein is left merely with its inner life. Heidegger says that the temporality that is profoundly boring "constitutes the ground of the possibility of the subjectivity of subjects, and indeed in such a way that the *essence of subjects* consists precisely in *having Dasein,* i.e., in always already enveloping beings as a whole in advance."[8] To be a subject is not to be an individual who is engaged in a determinate way in the world and who has an identity. The subject is an undifferentiated "no one" who is precisely not engaged with the world. Heidegger says that when we say that we are ourselves bored, we do not mean our individuated selves: "Yet we are now no longer speaking of *ourselves* being bored with…, but are saying: It is boring for one. It—for one—not for me as me, not for you as you, not for us as us, but *for one*. Name, standing, vocation, role, age and fate as mine and yours disappear…. That is what is decisive: that here we become an undifferentiated no one."[9]

In other words, one becomes a *subject*. However, insofar as one *becomes* a subject, that is already a temporal process. Temporality is thus a condition for the emergence of subjectivity, and the reverse is not the case: subjectivity could not be a condition for temporality. Temporality could not be something produced by the subject and therefore could not be subjective. Hence, if idealism means dependence on subjectivity, Heidegger could not be a temporal idealist. My hypothesis is that for Heidegger, becoming subjective is one way that temporality temporalizes, and this is a different temporalization from becoming individuated.

The next step is to show that this temporalization can then be either authentic or inauthentic, depending on whether we become genuine individuals or mere subjects. At the end of Division 1 of *Being and Time* Heidegger has given a complete, synchronic account of what it is to be a human being at a particular moment of time. In Division 2, however, he wants to describe Dasein as a being whose life is "stretched out" between birth and death. His goal is to account for the diachronic "connectedness" of Dasein's life. This is both an ontological and a normative task. As an ontological task, he needs to describe how Dasein can be the same being at different times of life. As a normative task, he wants to show how it is possible for Dasein to fail to connect its life, on the one hand, and to connect the various moments in a cohesive manner, on the other. To fail to connect one's life is to be inauthentic, which means that one does not think or act on one's own accord. To succeed in connecting one's life cohesively is to be authentic, i.e., to act on one's own accord.

Heidegger's account of authenticity is tied to his critical understanding of the flow of time. Some philosophers see time as a river flowing from the past, to the present, and into the future. For Heidegger, in contrast, *authentic* temporality experiences time as coming from the future. Thus, the inauthentic relation to the future is simply to sit back and "wait" for it, in the belief that it will come along anyway. Instead of experiencing the future as a "going-into," in a way that one can only passively await, one can encounter it actively as a "coming-towards." The authentic relation to the future is thus to "anticipate" it by actively resolving to make the future be as significant or meaningful as possible. This judgment of significance is not simply subjective, because significance is a feature of the world for Heidegger. The world presents itself, and this presence of the world thus becomes the difference between an authentic and an inauthentic present.

The inauthentic sense of the present, which Heidegger refers to as the "lost Present," is illustrated by Heidegger's analysis of people who never have enough time. These people have lost or wasted time because they have lost themselves in the busyness of everydayness. In contrast, authentic Dasein always has time. Instead of letting the past determine the present, the authentic relation to the present involves the *Augenblick,* or "moment of vision," wherein Dasein projects a sense of the connectedness of its past, present, and future into a unified vision that it then resolves itself to act on consistently.

The moment of vision alters Dasein's relation to its past as well. Instead of viewing the past as over and done with, Dasein can see the past as an ongoing influence on the present and thus on the future. The relation to the past is not one of "forgetting." Forgetting is not simply a failure of memory but a "backing away" from the recognition of what one has been. In contrast to this "backing away," Dasein has the possibility of "repetition." In repetition, Dasein can "take over resolutely that entity which it already is."[10] As authentic, Dasein thus resolves the apparent paradox and "becomes what it is."

How Temporality Temporalizes: Pierre Bourdieu

For Pierre Bourdieu, Heidegger's privileging of the future is a sign not of authenticity, but of "inauthenticity," or what Bourdieu would identify as "misrecognition." Where Heidegger sees possibilities coming into the present from the future, Bourdieu will see the reproduction in the present of past objective structures. These structures are built into the body, or more precisely, into the set of bodily dispositions that Bourdieu calls the *habitus*. The *habitus* is situated in a social field and participates in it in much the same manner as one participates in a game. To give one of Bourdieu's favorite examples, in games such as football or soccer or basketball, the player sends the ball not to where the other player is at the moment, but to where the other player soon will be. A successful anticipation depends, however, on long hours of practice that enable the teammates to anticipate the bodily capabilities of one another. What seems like a successful instance of looking ahead is really, then, a case of implicit looking-back. In *The Logic of Practice,* Bourdieu inverts Heideggerian temporality by seeing "the presence of the past in this kind of false anticipation of the future performed by the *habitus*."[11]

Although Bourdieu differs with Heidegger's account of the direction of time, I see Bourdieu as providing a concrete (although ontic) way of understanding Heidegger's gnomic injunction that temporality temporalizes. Bourdieu argues that time is not to be thought of as a thing or an object that a subject "has." Bourdieu wants to describe the actual practices of "agents," not "subjects." He therefore says that he is reconstructing the point of view of the acting agent, which is the point of view of "practice as 'temporalization', thereby revealing that practice is not *in* time but *makes* time (human time, as opposed to biological or astronomical time)."[12] Bourdieu distinguishes further between the conscious project of the future (*avenir*) and the prereflexive protention of the "forth-coming" (*à venir*). Bourdieu is drawing on Husserl's notion of the protention of, for instance, the hidden faces of a cube. One sees a cube, but one does not see all the faces of the cube. Nevertheless, one could not be seeing a cube as a cube unless one sensed the hidden faces, which are said to be "quasi-present." Similarly, unlike the future in the sense of *l'avenir,* which is a conscious projection of a future present, Bourdieu thinks that in the *à venir* time is not noticed even though it is there. In my view, this is not quite the correct phenomenology, for the players certainly have a sense (often more implicit than explicit) that the "clock is running." Bourdieu is right, however, when he says that time is first noticed consciously only when there is a breakdown: "Time (or at least what we call time) is really experienced only when the quasi-automatic coincidence between expectations and chances, *illusio* and *lusiones,* expectations and the world which is there to fulfil them is broken."[13] How does a temporal breakdown occur? For Bourdieu, a breakdown in the *à venir* can be the result of a disruption of either the *illusio*, the expectations that are built into the *habitus*, or the *lusiones*, the probabilities that are built into the social field.

An example of such a breakdown is unemployment. Unemployment suddenly leaves one with time on one's hands. In contrast to the busy person who never has enough time, the unemployed become conscious of dead time, where time hardly seems to move at all. (Benjamin would call this "empty time.") Bourdieu thus adds a social explanation to Heidegger's distinction between the inauthentic person who never has enough time and the authentic person who always has time. Bourdieu sees this surplus of time as a sign not necessarily of authenticity, but more often of powerlessness. Consider the case of waiting. The powerful, whose time is valuable, make the powerless wait. In Bourdieu's analysis, waiting is a form of submission.

Even if Heidegger is right in suggesting that to have control of one's life is to have control of one's time, such that one can "take one's time," Bourdieu believes that there is more to say about how personal temporality derives from social temporalization. For example, the academic and the artist live privileged lives where the line between leisure and work is blurred and time seems to be bracketed. The quasi-free time of the academic resembles the negated time of permanently unemployed subproletarians in that both have time on their hands, but the subproletarian pays the price for the academic. For Bourdieu, the fact that both cases would satisfy Heidegger's criterion for authentic temporality amounts to a refutation of Heidegger's account by *reductio ad absurdum*. More generally, this social contradiction casts doubt on any analysis of individual authenticity that ignores the social conditions of temporality that are built into the body. Bourdieu therefore concludes that "time is indeed, as Kant maintained, the product of an act of construction, but it is the work not of the thinking consciousness but of the dispositions and practices."[14]

Angelus Novus: Walter Benjamin

In much the same manner that Bourdieu undermines Heidegger's argument for the priority of the temporal future, Walter Benjamin undercuts the priority that Heidegger gives to the future in history. Benjamin's famous interpretation of Paul Klee's painting, *Angelus Novus* (which Benjamin owned), resembles Heidegger's account of time as coming from the future, except for one feature: the angel is going into the future facing *backward*. Unlike Heidegger's account of "projection" as consciously and resolutely positing a *telos*, Benjamin suggests that what we really see is not purpose and meaning in our lives, but contingency and confabulation. As the angel is propelled into history, it looks *back*. In looking back, the angel does not see the connected and sequential chain of events that a forward-looking agent would see, but rather, it sees "one single catastrophe which keeps piling wreckage upon wreckage and hurls it in front of his feet."[15] The time of our lives, viewed backward, is disjointed, out of joint, discontinuous, a series of fragments.

I pause to ask three questions about Benjamin's allegory. First, is there not some tension between the directionality of the storm, which is blowing the angel into the future, and the pile of debris, which builds up at his feet, growing skyward, as if he were not moving away from each bit of wreckage? The answer requires us to consider the temporality of the debris. Presumably the debris is not left behind as the angel is blown into the future. Instead, the debris goes along with the angel, piling up at his feet.

The second question is, given that the meaning of history has crumbled into a pile of debris, what gives Benjamin the right to speak of "one single" catastrophe? What is one and single is not the debris, but what it is that has been wrecked, namely, universal history and the very idea of global progress. Kant and Hegel see history as the story of the development of universal reason and freedom. In contrast, Benjamin's angel sees that this story of the progress of civilization is an ideological shambles that distorts and enervates the present. We are at the mercy of the storm, and the message is that our sense of ourselves as moving forward is an ideology that ignores the victims of history and the reality of barbarism.

The third question concerns the directionality of time, and our own temporal orientation. Unlike the famous critique of Hegel by Marx, Benjamin neither stands us on our heads nor puts us back on our feet. Instead, he turns us around so that we are facing backward. The point of the angel's facing backward is that history has no *telos*. Unlike Marx's spatial metaphor, which has the Hegelian seeing the world upside down, Benjamin's temporal metaphor implies that we cannot see where we are going. Are we in fact going backward? No, because we are moving away from where we have been, not back to where we were before. The story thus is still linear. However, it is difficult to say that we are moving forward. There are no signposts, no indications of an increase in freedom. The wreckage just piles up and apparently leaves no basis for teleology. The debris does not contain the continuity and coherence of a narrative that would allow us to think of ourselves as approaching a *telos*. "There is no document of civilization," says Benjamin in the seventh thesis, "which is not at the same time a document of barbarism."[16] Does he thus deprive us of any sense of the time of our lives, a sense of how a life is connected between birth and death?

Benjamin suggests that past generations provide the present with a "*weak* Messianic power."[17] What is the basis for this messianic hope, given the starkness of the figure of universal wreckage? The answer is similar to the account that I just gave of the temporality of the debris of universal history. As the storm blows the angel backward, the debris is not strewn out in the receding distance, but accompanies him, piling up at his feet. The present is not "empty," homogeneous time, but rather is what Benjamin calls *Jetztzeit:* "The present, which, as a model of Messianic time, comprises the entire history of mankind in an enormous abridgment, coincides exactly with the stature which the history of mankind has in the universe."[18] The messianic moment does not come from knowing where we are going, but from seeing where we have been. Citing Nietzsche, Benjamin says that the image of enslaved ancestors provides more motivation than that of liberated grandchildren.[19] This vision of past enslavements is not the beginning of knowledge of how things could be better, although it does lead to the knowledge that universal history is untenable.[20] Universal history, which tells the story of the continuum of progress toward universal freedom, must be abandoned because it empties human freedom of concrete content. Universal history also leads to fatalism insofar as progress occurs automatically, whether mechanistically in a linear direction (Kant) or dialectically in a spiral one (Hegel).[21] The point of writing history against the grain is not to reinforce this continuum, but to blast it open.[22]

As a side note, I will point out that Benjamin prefers the temporality of the calendar to that of the clock because, instead of the smooth flow of uninterrupted transition, the calendar suggests a more punctuated sense of time. The calendar permits the recurrence of days of remembrance. The sense of time conveyed by a clock is continuous transition, whereas the calendar allows for a sense of time as coming to a stop and standing still. The calendar marks the possibility of the day when class action can explode the continuum of history.[23]

Messianicity without Messianism: Jacques Derrida

Jacques Derrida distinguishes two senses of the future.[24] There is the predictable, foreseeable future, *le futur,* and there is the unpredictable, unexpected future, *l'avenir,* that can break into the present unexpectedly at any moment. There is thus a close connection between this latter sense of the future and the idea of a Messiah, who is always about to come into the present. Even if one met the Messiah, the story goes, one could only ask the Messiah when he will be coming. That is the eschatological sense of the future, an unexpected eruption of the future in the present, in contrast to the teleological sense whereby the present unfolds or develops into the future in foreseeable ways. Derrida believes that even a *weak* Messianic power such as Benjamin posited still involves a vestige of teleology. In "Marx & Sons,"[25] Derrida calls messianicity a "universal structure of experience"[26] that cannot be deconstructed.[27] However, Derrida's notion of messianicity has no religious connotations. His notion is of a messianicity "without messianism,"[28] that is, without a Messiah. Messianicity therefore ceases to be a matter of *theoretical* knowledge, and it becomes instead a part of *practical* activity. All action has consequences that cannot be predicted. Derrida is suggesting that it is even a condition of praxis that one not know how to calculate outcomes. This unknowability and incalculability make action an open-ended matter. They make action itself messianic. In my interpretation of what Derrida is up to, I suggest that this open-endedness of practical action is the universal structure that Derrida calls messianicity.

The contrast here is between an account of historicity and an account of temporality. Messianicity as Derrida uses the term is not tied to a conception of universal history, and it does not rely on notions such as progress, or decline, or cycles. Instead, as a basic condition of temporality temporalizing, messianicity is prior to the whole enterprise of the philosophy of history. Messianicity is built into temporality, and temporality is a condition for the possibility of history. By attaching messianicity to temporality rather than to historicity,

Derrida corrects Fredric Jameson who thinks that there is a "utopian" hope at stake in Derrida's account. Derrida contests this attribution of utopianism because it puts Derrida's account of messianicity on the same plane as Walter Benjamin's utopian notion of the weak messianic force of history. Derrida maintains that Jameson is thereby making a category mistake. Neither a form of optimism nor a form of pessimism (although Derrida says that it is more optimistic than pessimistic), Derrida's notion of messianicity makes both optimism and pessimism possible. It is therefore neither utopian nor dystopian, because it is a quasi-transcendental condition for the possibility of *both* utopian and dystopian politics.

As I read Derrida, then, messianicity is a fundamental condition of action in relation to time. Messianicity involves a "waiting without waiting."[29] This means that on the one hand, we have to wait for our actions to have consequences, but on the other hand, we cannot simply wait, for we have to act. The messianic element thus prioritizes the present, but not in the manner of the typically conservative way of prioritizing of the present. A conservative interest in preserving the status quo could be simply the result of a desire not to think about how things could be different. Rather than try to conserve the status quo, messianicity looks to the future, that is, to the unexpected that is yet to come (the *à venir*).

Messianicity differs, however, from the Kantian and Hegelian stories whereby the end of history is achieved when the development of freedom is complete as a result of predictable advances. In contrast to these stories, messianicity separates teleology from eschatology. Messianicity avoids the teleological story of gradual progress toward the *telos* of universal freedom. However, it retains the eschatological aspect whereby a breakthrough event can interrupt at any moment. As Derrida explains, "*Anything but Utopian,* messianicity mandates that we interrupt the ordinary course of things, time and history *here-now*; it is inseparable from an affirmation of otherness and justice."[30] As an affirmation of the inevitability of otherness, messianicity must recognize both that the future is always unexpected, and that it cannot be calculated, predicted, or programmed. As an affirmation of justice, messianicity also recognizes that the future is now. As a theory of action, it represents an injunction and a responsibility to act now, without delay. The messianic event cannot be simply willed to occur. Nevertheless, its constant possibility precludes passivity or abstention. Derrida's sense of temporality is thus intended, I believe, to emphasize the necessity of commitment to the present and the need for action.

The Politics of Temporality

Let me now conclude with some reflections on the return of the present in the recent thinking about the "politics of temporality." In contrast to Heidegger, who prioritizes the future, the other theorists whom I have described here place more emphasis on the present. This return of the present may be the first sign that Continental philosophy is moving out of the period of the "post." Terms such as postmodernism, poststructuralism, post-Marxism, and post-historical make philosophy into a "late" or a "belated" social phenomenon. In contrast, emphasizing the present suggests that philosophy is moving into a new historical moment—one that it is still too soon to label.

Of these theorists of temporality, Bourdieu's theory might appear to put the most emphasis on the past. The bodily *habitus* incorporates dispositions that then are projected as expectations for the future. The *habitus* is thus a strongly conservative force. The *habitus* explains why we find certain patterns of action intelligible and why only specific actions seem plausible given the social field. However, to say that the *habitus* is strongly conservative is not to say that Bourdieu's theory of the *habitus* is conservative. Although we are deeply entrenched in our *habitus* and thus in the past, Bourdieu thinks that reflective sociology can contribute to active social change by letting the appearance of social necessity become apparent as just that: appearance. Bourdieu depicts necessity so strongly because he knows that we resist the appearance of necessity and that once sociology reveals it, we will take action against it. The point is not simply to become reflectively *self*-conscious or *self*-critical. We "become who we are" in the present not so much by changing our *selves* as by changing our *world*. I therefore read Bourdieu as prioritizing the present.

By removing the vestiges of teleology, Derrida too can be read as emphasizing the present as the time of our lives. At the same time, he does not advocate the inauthentic present with its attitude of "wait and see" [*voir venir*]. As he explains in *A Taste for the Secret*, the future erupts in the present unexpectedly: "the future rushes onto me, comes onto me, precisely where I don't even expect it, don't anticipate it, don't 'see it coming.'"[31] This analysis of temporality temporalizing itself should help to rebut criticisms of deconstruction for political quietism. One version of quietism is the reactive resistance of the sort that is labeled as "reformist." Reformists are accused of being averse to the prospects of revolution. Derrida insists, however, that just as deconstruction is not utopian, it is also not anti-revolutionary, and it can invoke the rhetoric of revolution.

These remarks show either that it is a misconstrual of poststructuralists and postmodernists to think that they reject the rhetoric of social progress entirely, or that it is incorrect to label Derrida as a poststructuralist or a postmodern. Critics tend to think that anyone who is designated as "post" should reject the Marxian story of class struggle and revolution. In "Marx & Sons" Derrida says that he does not reject either the idea of class, however problematic it is, or the figure of revolution, however complicated he finds it. To label him as either a poststructuralist or a postmodern tends to oversimplify his theory of temporality and historicity.[32]

Like Derrida, Benjamin also emphasizes the present, and thus the need for action, but with a more normative bite. In contrast to Heidegger, Benjamin does not tie his account of the connectedness of life to the future or to death. The messianic moment can erupt at any point, but it is motivated more by past enslavement than by future liberation. The priority is on the possibility of action in the present, and as in the case of Derrida, there is a suspicion of putting off until the future the immediate need for social change.

How, then, should the phenomenology of temporality be understood? At the risk of sounding elliptical, let me sum up these views on the politics of temporality with an American expression, "There is no time like the present." As a cliché, this truism means that it is better not to put off for later what one should really be doing right now. But in addition to this practical advice against procrastination, it also makes the valid theoretical point that "There is no time *but* the present." That is to say, from the point of view of the metaphysics of time, the past and the future do not seem really to exist: the past because it is always over and done with, and the future because by definition it always has not occurred. However, the past and the future in fact exist phenomenologically, precisely as features of what could be called either the lived present or the time of our lives.

The past can be changed, for instance, by reinterpreting the present. Moreover, the future is always the future of a past that has not occurred. Just as Disney's Tomorrowland does not really express today's sense of the future, but is obviously only the future as foreseen in the 1950s, the future is the projection of a present that is always already past. The future is not really ahead of us, but behind us, as possibilities that once were to be realized, but that exceed what was once present. The directionality of time, as exemplified by Benjamin's angel, is perhaps best expressed in the phrase, "*Back* to the future."

If this mapping of the terrain of temporality is on the right track, it helps to clarify both the genealogical and the political features of the time of our lives. Genealogically, the door is opened to a dimension of temporality that is real, insofar as it characterizes the temporality of our lives, even if it is not the same as either the objective time of the physical universe or the subjective sense of time passing within consciousness. Politically, each particular theory of how temporality temporalizes itself, whether from the future or from the past, will have consequences for an understanding of agency and action. Of course, we should not forget the primary sense of the expression, "we are having the time of our lives." In ordinary parlance, this expression means that we are enjoying ourselves, that we are having a good time, perhaps the best time, of all the times we have ever had. Enjoyment is itself a temporal dimension, and it too should be included in any analysis of the connectedness of life. After all, having the time of our lives goes a long way toward making life worth living.

Notes

[1] This paper was first given at the conference in Budapest in November 2003. It incorporates excerpts from the forthcoming book-length study, *The Time of Our Lives: A Critical History of Temporality.*

[2] William D. Blattner, *Heidegger's Temporal Idealism* (Cambridge, Mass: Cambridge University Press, 1999), 246.

[3] Martin Heidegger, *Being and Time*, trans. John Macquarrie and Edward Robinson (New York: Harper & Row, 1962), 377; cited as BT for the English pagination and as SZ for the German pagination. The German phrase is *Zeitlichkeit zeitigt* and appears on page 328 of the German text (BT 377).

[4] BT 380; SZ 331.

[5] BT 377; SZ 328.

[6] Martin Heidegger, *The Fundamental Concepts of Metaphysics: World, Finitude, Solitude*, trans. William McNeill and Nicholas Walker (Bloomington: Indiana University Press, 1995), 141; cited as FCM.

[7] Martin Heidegger, "What Is Metaphysics?" in *Basic Writings*, ed. David Farrell Krell (San Francisco: Harper, 1993), 99. Joy is said to offer a comparable revelation.

[8] FCM 158.

[9] FCM 134–135.

[10] BT 388: SZ 339.

[11] Pierre Bourdieu, *The Logic of Practice*, trans. Richard Nice (Stanford: Stanford University Press), 62.

[12] Pierre Bourdieu, *Pascalian Meditations*, trans. Richard Nice (Stanford: Stanford University Press), 206; cited as PM.

[13] PM 208.

[14] PM 213.

[15] Walter Benjamin, *Illuminations*, trans. Harry Zohn (New York: Schocken, 1969) 257; cited as I.

[16] I 256.

[17] I 254.

[18] I 263.

[19] I 260.

[20] I 257.

[21] I 260.

[22] I 262.

[23] I 261–262.

[24] The following discussion of Derrida reflects an earlier discussion of material in a different context in chapter 4 of my book, *Critical Resistance: From Poststructuralism to Post-Critique* (Cambridge, Mass: MIT Press, 2004).

[25] Jacques Derrida, "Marx & Sons," in *Ghostly Demarcations: A Symposium on Jacques Derrida's* Specters of Marx, ed. Michael Sprinker (New York: Verso, 1999); cited as GD.

[26] GD 248.

[27] GD 253.

[28] GD 242.

[29] GD 251.

[30] GD 249.

[31] Jacques Derrida & Maurizio Ferraris, *A Taste for the Secret*, trans. Giacomo Donis (Cambridge UK: Polity Press, 2001), 84.

[32] See GD 229.

"To the Planetarium": From Cosmos to History and Back

Eternity No More

Walter Benjamin on the Eternal Return

Tyrus Miller

I

> "Men of the nineteenth century, the hour of our apparitions is fixed forever, and always brings us back to the very same ones."—August Blanqui, 1872

On January 6, 1938, Walter Benjamin wrote to Max Horkheimer from San Remo to report on a remarkable development in his thinking about his Baudelaire studies and about the larger framework of the *Passagenwerk*, Benjamin's decade-long historical research about nineteenth-century Paris, a project that he described as an *Urgeschichte der Moderne* (an archaic history of modernity). The occasion of this development was his encounter with a largely forgotten text by the famous insurrectionist Auguste Blanqui, entitled *L'éternité par les astres* (Eternity According to the Stars). This short book comprised a set of cosmological speculations written in prison by the old revolutionary near the end of his life, and coupled arguments from the popular science of Blanqui's day with a remarkable vision of infinite repetitions of the same in an indefinite series of parallel worlds. Here is Benjamin's report to Horkheimer:

> In the last weeks I have made a strange find that will decisively influence the work: I came upon the text, Blanqui's last, which he wrote in his final prison, in the fortress of Taureau. It is a cosmological speculation. … and is, as far as I can tell, till now as good as unknown. … Admittedly, at first glance the text is tasteless and banal. Whereas what constitutes its main portion are the clumsy meditations of an autodidact, these prepare for a speculation about the universe that could be provided by no one lesser than this revolutionary. If Hell is a theological object, one could call these speculations infernal. The world view that Blanqui sketches here, while taking its data from the mechanistic natural sciences of his day, is in fact infernal—but is at the same time, in the shape of something natural, the complement of the social order that Blanqui must have recognized in the evening of his life to be the victor. What is astonishing is

> that this sketch is completely without irony. It represents an unconditional surrender, but at the same time the most terrible lament against a society that projects this image of the cosmos against the sky. The piece has as its theme, the eternal recurrence, the most remarkable relation to Nietzsche; and a more hidden and deeper one to Baudelaire, with whom in a few of its magnificent points it resonates almost word-for-word.[1]

Although this was only two years from the close of Benjamin's own defeated life, this seemingly minor discovery of a truly marginal text of Blanqui crystallized a whole new set of motives in the *Passagenwerk* studies and seemed to offer Benjamin a conceptual hinge for the juncture of modernity and myth he intuited in the culture of nineteenth-century Paris. In fact, in the final, 1939 version of his exposé of the *Passagenwerk* he granted the last word—in his text and about the nineteenth century—to the messengers of the eternal recurrence. In his final paragraph, following a quote from Blanqui's *L'eternité par les astres*, Benjamin concludes:

> The century was incapable of responding to the new technological possibilities with a new social order. That is why the last word was left to the errant negotiators between old and new who are at the heart of these phantasmagorias. The world dominated by its phantasmagorias—this, to make use of Baudelaire's term, is "modernity." Blanqui's vision has the entire universe entering the modernity of which Baudelaire's seven old men are the heralds. In the end, Blanqui views novelty as an attribute of all that is under sentence of damnation. Likewise in *Ciel et enfer*, a vaudeville piece that slightly predates the book: in this piece the torments of hell figure as the latest novelty of all time, as "pains eternal and always new." The people of the nineteenth century, whom Blanqui addresses as if they were apparitions, are natives of this region.[2]

Though perhaps not the shattering experience of being "6000 feet above man" that Nietzsche reported when he first envisioned the eternal return, nor perhaps the grim testament of a revolutionary activist doomed to live year-after-year in prison, it might be said that Benjamin's discovery of Blanqui's late text hit him with the force of a revelation. The content of that revelation was that the eternal recurrence—the utmost cancellation of modern historicity—was itself a historically determinate form of experience that had emerged, or punctually reemerged, at the acme of modernity in late nineteenth-century Europe.

In focusing on this motif in Benjamin, I myself will be pursuing three simultaneous paths of argumentation. The first, and narrowest one, will be philological and explicatory. I am using the motif to map out some potential interconnections between fragmentary points in Benjamin's oeuvre, to suggest how such themes as the new, the same, fashion, hell, myth, and phantasmagoria can be read in light of a temporal form of nineteenth-century metropolitan experience that Benjamin was at pains to disclose in its material context. More generally, I will also use this focus on the cosmological motif of the eternal recurrence to suggest the complex concatenation of time structures in Benjamin's thinking, one that is not reducible to such simple dichotomies as, say, the opposition of a Jewish and Marxist messianic time to a cyclical time of myth.

Second, according to Benjamin's own interpretative and historiographic method, there are particular "horizons of readability" that are time-bound and ephemeral, in which certain aspects of the historical past become discernable, because they are charged with the special intensity of a transhistorical configuration of past and present. Clearly, he believed that in his own discerning of a shared experience of the eternal recurrence by the poet Baudelaire, the philosopher Nietzsche, and the revolutionary Blanqui, that he stood in such a momentary horizon. On the other hand, at the moment of his writing, he could not really have the perspicacity to describe his own situation historically. From our later horizon, however, Benjamin's fragmented meditations on eternal recurrence become legible within a cultural context shared with such contemporaries as Georg Simmel, Karl Jaspers, Martin Heidegger, Karl Löwith, Mircea Eliade, and Pierre Klossowski, all participating in a serious contemplation of the meaning of the thought of eternal recurrence, especially in the work of Nietzsche. Moreover, this concern has not ceased to be important, but rather has persisted up to the present in the work of post-structuralist thinkers such as Gilles Deleuze and Jacques Derrida.

Finally, I want to suggest how Benjamin's focus on the modernity of the eternal recurrence highlights its general significance as a topic for meta-historical reflection. Namely, what Benjamin discerns in the nineteenth-century thought of eternal recurrence—and here he is preceded, even exceeded by the magisterial Nietzsche criticism of Karl

Löwith—is the extreme, forced conjunction of an archaizing cosmology with a modern psychology. Yet if, as Paul Ricoeur has argued, narrative offers a way of reconciling the opposition of cosmological time ("the time of the world") and lived time ("the time of the soul"), the maximum polarization of cosmos and experience that the eternal recurrence implies also marks the point at which historical narrativity is stressed to its limit and begins to unravel. In short, I will argue what seems to be at issue for Benjamin in the emergence of the eternal recurrence is the very narratibility of history and hence also the narrative transmission of culture. At the same time, however, this crisis of narrative representation of history may be less the end of history as its turn towards what Hayden White has suggestively called the "modernist event." Historical writing following this turn will need to be more conscious of how temporal concepts such as eternity, succession, causation, and recurrence were tied to concepts of meaning, mimesis, and generic form. Moreover, it will need to be more reflexively constructivist in its conception of the historical, in an awareness that—as David Wood has written—"concepts, theories, and representations 'of' time [form] part of the temporal itself."[3]

II

My discussion will center on a specifically modern formulation of the notion of "eternal recurrence"; however, it is useful to consider this concept more generally, before considering the texts and quotations that Benjamin assembled around it. Mircea Eliade's *Cosmos and History: The Myth of the Eternal Return*—begun just after the end of World War II and published originally in 1949—remains a key source for its bold theoretical stance and forceful interpretation of the significance of this concept for religious and historical thought. According to Eliade, eternal return is the characteristic underpinning of mythic thinking and arises in a plethora of variations out of a single fundamental, existential need of human societies: to master the despair and fear that comes from historical change, death, contingency, and transience. Out of this basic anthropological motivation, in Eliade's view, archaic societies, and even, to a large extent, postarchaic and modern societies as well, project an ontology that limits or excludes change by mapping it on to repeating patterns and archetypes. On the one hand, the macro-structure of natural and social reality appears already to be a repetition of a cosmic archetype, and at each level, from the cosmic to the political to the everyday, the cosmogonic act of world-foundation must be repeated and renewed on a periodic basis. The prescribed rituals and profane gestures, in turn, gain their place in a cosmos of social and religious meanings in so far as "they deliberately repeat such and such acts posited *ab origine* by gods, heroes, or ancestors."[4]

Eliade's conceptualization had the virtue of allowing him to define, with a single clear criterion, an essential difference between mythic thinking and historical thought, between archaic and modern societies, and between the foundations of pagan and non-European civilization in which cyclical cosmologies predominate and the historical orientation of Judaic and Christian civilizations along with their modern European successors. Moreover, it can be said that Eliade's discussion of the mythic thought of the eternal return has a special object that appears in stark relief against the background of this nearly universal myth: the case of Jewish civilization and its incomplete, hybrid confluence into Christian and post-Christian modern Europe. If the eternal return represents the symbolic means through which archaic and pagan societies disavow history, the Jews are singular in having broken with such archetypical, recurring cosmoi in favor of a God who reveals himself in one-time historical moments. With Jewish civilization—and in the Jewish legacy within Christianity—genuinely historical time emerges. For Eliade, then, myth and history, and the phenomenologies of time that underlie them, are pivotal terms in constituting distinct "ethnic ontologies."[5] The status of the Jews, built into the structure of Eliade's interpretation of eternal return, is at once that of privilege and exclusion, privilege from the point of view of historical thought, exclusion from the perspective of mythic thought. The relative valuation of these two interpretative postures is precisely what is at stake in Eliade's short, programmatic study.

It would be an easy extrapolation, and one which I in fact will venture here, to say that Eliade, writing in 1945, could view these civilizational tendencies at war in his own contemporary context. European civilization, at its present extremity of crisis, represented a wavering, unresolved amalgam of mythic and historical components, or to put it in baldly ethnic terms, of gentile and Judaic elements. European anti-Semitism, seen in this light, would mark a return of myth and an anti-modern disavowal of history, but history precisely as ethnically personified by that singularly historical people, the Jews. Insofar as history is the domain of terror and insofar as the Jews are seen from the mythic perspective as the bearers of history as such, they are subject, recurrently, to the violence of the mythic mentality, which seeks to impose a cosmic cancellation of history through recurrence. We might say that Eliade poses himself ambigiously between an historical view of the recent tragedy of the Jews as a singular horror of his time and a mythic view that sees anti-Semitism as the expression of basic anthropological forces that are not going to go away any time soon and are rooted in the very deep-seated existential revulsion against history. The *meaning* of the war and the Nazi genocide is not simply a matter of deciding between two historical actors, the Nazis and their opponents, but rather also between a historical and an anti-historical, mythic, anthropological framework of interpretation of these events. Or put otherwise, the war, in the most stark and virulent way, sets Europe before a decision—existential, political, even theological—as to whether history or the myth of eternal return gets the final say about who and what is European. Thus Europeans do not just have a particular historical decision to make; rather they have to decide for or against historical being as such, for or against the recurrence of myth. In arriving at this decisionistic conclusion, Eliade thus agrees with Walter Benjamin, who viewed his present as a polarized, extreme confrontation of myth and messianic possibility, an emergency situation that required an unequivocal, punctual decision. Benjamin, however, leaned emphatically towards a critique of myth in favor of a messianic, theo-political Marxism.

Notably, too, as Karl Löwith had suggested in his 1935 book on *Nietzsche's Philosophy of the Eternal Recurrence*, of which Benjamin read Theodor Adorno's copy in 1938, this issue of decision was already built into Nietzsche's presentation of eternal recurrence in *Also sprach Zarathustra*. Löwith was, along with Hannah Arendt, a student of Martin Heidegger and a Jewish refugee from Nazism, first in Italy, later in the United States. Löwith notes that the title of *Zarathustra* was originally to have been "Noon and Eternity," a key image in the text. He notes the temporal ambiguity of Nietzsche's noontide, which is at once the eternal moment in which time stands still and the instant in which the shattering insight that everything recurs can take place. Paradoxically, the crisis point in which one decides, in an instant, to affirm existence as a whole, in all its recurrences, is also the only moment in time in which eternity is manifest. As Löwith writes:

> A metamorphosis occurs unexpectedly in Nietzsche's presentation of the noon and of the eternal recurrence: out of what by nature comes *again and again*, develops something that is supposed to be decisive *once and forever*. Likewise, the "moment," too, is not eternal because what is forever shows itself in the moment, but because the moment (as a decisive moment) determines in advance what will be in the future.[6]

Löwith's analysis of this paradox in Nietzsche's noon becomes the central point of his critique: for all of its appeal to cyclical recurrence, myth, and the return to pre-Christian and pre-Socratic antiquity, the temporality of crisis, a typically *modern* time-structure predominates in Nietzsche's thought. Löwith concludes:

> Zarathustra thus teaches two things: as the redeeming man of the future, he proclaims a new way of existence; and as the teacher of the eternal recurrence of the same, he teaches... the highest law of the whole life of the world. But as a critical center, the noon is not the always recurring time of something everlasting, but rather the moment of decision at which previous history parts from the history of the future.[7]

III

I now turn to the constellation of materials that Benjamin assembles around the topic of the eternal recurrence, in order to consider how this concept offers a perspective on the associated themes of Benjamin's writing, especially in the *Passagenwerk*. First, a word about the texts in question. The first notes for the *Passagenwerk* date between 1927 and 1930, and Benjamin worked periodically on his research throughout the 1930s, but especially intensely following his definitive departure from Germany with the rise to power of the Nazis. Out of the *Passagenwerk* research notes, he intended to publish a book that would place Baudelaire in his material and intellectual context; the various drafts and reworkings of these Baudelaire writings represent that part of the research most near to some form of publishable completion. Thus, the main texts that I will be discussing include a 1935 and a 1939 exposé of his research, at least partly intended to win financial support from the *Institut für Sozialforschung* directed by Max Horkheimer; his first attempt at a draft of the Baudelaire study, "The Paris of the Second Empire in Baudelaire"; his late truncation and revision of the former draft into a rounded, shorter essay entitled "On Some Motifs in Baudelaire"; a group of draft notes assembled under the title "Central Park"; and a few of the topically organized "convolutes," or note-files, of the larger *Passagenwerk* research, especially Convolute D [Boredom, Eternal Return] and Convolute J [Baudelaire].

The theme of the eternal recurrence emerges in Benjamin's work with the elevation to an allegorical centrality of Blanqui, who had previously appeared strictly as an empirical, historical figure up to Benjamin's discovery of *L'éternité par les astres*; the emergence of the previously marginal Nietzsche as an important voice and especially as mediated through Karl Löwith's study; and a reinterpretation of several temporal and phenomenological issues in Baudelaire. Moreover, Benjamin's historiographic collage method entailed a kind of hopeful, patient empiricism, in which the historian collects, sifts, and rearranges countless bits of data and discourse, until hidden nodal points begin to take on relief and consistency. The resonance of the poet Baudelaire, the philosopher Nietzsche, and the revolutionary Blanqui around the eternal recurrence convinced Benjamin that he had uncovered a phenomenological key to nineteenth-century experience, a conceptual node that pointed to something more than simply the subjective imagination of these thinkers.

I will begin with Blanqui, whose text is likely to be unfamiliar to many. As already mentioned, Blanqui's text is centered on the view that given a finite number of basic elements in the universe and an infinity of spatial extension and time, then every entity, every event must exist in countless copies dispersed over these spatial and temporal expanses. In this cosmos, we may find it consoling that our lost loves are being embraced by our doubles an infinite number of times, yet at the same time, every defeat and instance of suffering in history are also replicated throughout eternity. Contemporary commentators in newspaper notices and reviews of Blanqui's book did not fail to point out the logical inconsistencies in the argument. Two, in particular, stand out, since they anticipate almost word-for-word

Georg Simmel's classic refutation of Nietzsche's "proof" of the eternal recurrence and critique of its inconsistencies.[8] First, the commentators note, it is not logically necessary that even a finite number of elements generate recurrences, much less exact sequential recurrences, in the course of infinite time. Second, even if it were the case that the recurrence did take place, so that our doubles led the same life that we did elsewhere in the universe or in another time, "we" would know nothing of it as individuals. Without any perspective outside the recurrences from which to grasp them as a whole, the hypothetical fact of recurrence should be perfectly indifferent to "us," as singular identities here and now. The Blanqui reviewers tend to praise the vivid imagination and vigorous mind of the old revolutionary, while remaining ironical about its content. Analogously, Simmel rejected the "scientific" or ontological claims in Nietzsche's conception, while admitting its force as a moral and psychological ideal, a kind of post-Kantian categorical imperative that values individual authenticity and rectitude over the social generality. Benjamin, however, in keeping with his materialist interests in the structure of experience and the framing of historical time, takes a different tack. On the one hand, unlike previous commentators, he tends to take the idea of eternal return not as a psychological metaphor or regulative moral idea, but rather as a literal experience; accordingly, he seeks to understand what are the social preconditions and phenomenological characteristics of eternal return as an experience of reality. On the other hand, he interprets the eternal return meta-historically and theologically, as a critique of progress, a recurrence to myth within modernity, and a post-Christian secularization of the theo-political topos of hell.

Benjamin cites Blanqui secondhand from a late nineteenth-century work, Gustave Geffroy's *L'Enfermé*, and quotes Geffroy's commentary as well. Blanqui wrote, "What I write at this moment in a cell of the Fort du Tareau I have written and shall write throughout all eternity—at a table, with a pen, clothed as I am now, in circumstances like these," and Geffroy comments: "He thus inscribes his fate, at each instant of its duration, across the numberless stars. His prison cell is multiplied to infinity. Throughout the entire universe, he is the same confined man that he is on this earth, with his rebellious strength and his freedom of thought."[9] By the end of his short treatise, Blanqui has risen to a pessimistic hymn for astral necessity and the vanity of human hopes measured against the relentless course of the stars. Following Benjamin's quotation from Blanqui—firsthand this time—I will quote the final three paragraphs of the book. The first reads:

> At bottom, this eternity of the human being among the stars is a melancholy thing, and this sequestering of kindred worlds by the inexorable barrier of space is even more sad. So many identical populations pass away without suspecting one another's existence! But no—this has finally been discovered, in the nineteenth century. Yet who is inclined to believe it?[10]

Benjamin does not follow Löwith's lead at this point in noting the exact corollary of Blanqui with Nietzsche here, but as with Nietzsche's eternity at noon, Blanqui's discovery of "l'eternité par les astres" is an analogous revelation at a particular moment in history, "in the nineteenth century," of a cosmic structure of eternal recurrence. Hence, Blanqui conjoins a temporality of crisis with a temporality of repetition. Though all moments in history are implicitly rendered equal by the fact that each will recur in an infinite number of copies, so that none can be said to be more important than any other, there is one privileged crisis point in the entirety of eternity: the infinitely repeated nineteenth century, as the crisis moment in which this cosmic order becomes conscious to Auguste Blanqui. To state the paradox differently: because every historical moment repeats and the future is in no way a consequent of precedent events, futurity itself is an illusion; yet it is this crisis of the nineteenth century, in which we come to know this cosmic fact, that determines the future for all eternity as illusory.

Blanqui spells out these paradoxical consequences even more clearly in the last two paragraphs that follow. He writes:

> Until now, the past has, for us, meant barbarism, whereas the future has signified progress, science, happiness, illusion! This past, on all our counterpart worlds, has seen the most brilliant civilizations disappear without leaving a trace, and they will continue to disappear without leaving a trace. The future will witness yet again, on billions of worlds, the ignorance, folly, and cruelty of our bygone eras![11]

"Until now": thus speaks a classic locution in the rhetoric of crisis. But this locution functions as if it were a performative speech act, which, once pronounced by Blanqui at a punctual moment in the nineteenth century, would stamp eternity, backward and forward, with the laws of astral necessity. What follows upon this pronouncement is the rule of cosmic mechanism, which subsequently renders every gesture and act merely a mime-show of inexorable cosmic forces (and note the theatrical metaphors that pervade Blanqui's passage throughout):

> At the present time, the entire life of our planet, from birth to death, with all its crimes and miseries, is being lived partly here and partly there, day by day, on myriad kindred planets. What we call 'progress' is confined to each particular world, and vanishes with it. Always and everywhere in the terrestrial arena, the same drama, the same setting, on the same narrow stage—a noisy humanity infatuated with its own grandeur, believing itself to be the universe and living in its prison as though in some immense realm, only to founder at an early date along with its globe, which has borne with deepest disdain the burden of human arrogance. The same monotony, the same immobility, on other heavenly bodies. The universe repeats itself endlessly and paws the ground in place. In infinity, eternity performs—imperturbably—the same routines.[12]

Taking off from these passages in Blanqui, Benjamin offers a number of phenomenological remarks on eternal return that interpret it as an index of modern experience. Here, the key issue is the meaning of the element of the "new"—of novelty—in modernity. This problem posed by Baudelaire in his stark conjunction of ephemeral, epiphanic flashes of beauty in the metropolitan environment with the experience of spleen, "I feel that I have lived a thousand years." Baudelaire, according to Benjamin, experienced spleen as the potential for devaluation that exists in any commodity or object of fashion, once its moment of newness had past. What once seemed new, now is just one more inert, material shell, to be accumulated in drawers and on shelves, while its purchaser seeks the new once again. Spleen results from a backward turned glance that perceives and suffers the indifferent weight of obsolete objects, and its experience may even by turned forward again and felt at the heart of the new. Already, at the glint of fashionable beauty, the object breathes the staleness that will soon set in around it. Thus, even in the kaleidoscopic procession of new objects, spectacles, fashions, there is the return of the same: the obsolete, the tasteless, the faded. As Benjamin writes in a fragment of the Baudelaire study, "The devaluation of the human environment by the commodity economy penetrates deeply into the poet's historical environment. What results is the 'ever-selfsame.' Spleen is nothing other than the quintessence of historical experience."[13] Against the overwhelming and repeated experience of spleen, Baudelaire seeks to seize the new, a struggle that elevate the artist and poet into the hero of modern life. The heroism of Baudelaire's "New," Benjamin emphasizes, is not a formal or objective newness, but an abstract, subjective temporal difference that intensifies the moment to free it from the melancholy sameness of spleen. "Baudelaire's work," Benjamin writes, "is not concerned with the attempt, decisive in all the arts, to engender new forms or to reveal new aspects of things; its interest is in the fundamentally new object, whose power resides solely in the fact that it is new, no matter how repulsive or bleak it may be."[14]

Baudelaire's hero of modern life struggles on the battlefield of abstract time, and thus, Benjamin concludes, is equivalent to those other heroes of time, Nietzsche and Blanqui, in their endurance of the ultimate intensification of nihilism in the thought of the eternal recurrence:

> The heroic bearing of Baudelaire is akin to that of Nietzsche. Though Baudelaire likes to appeal to Catholicism, his historical experience is nonetheless that which Nietzsche fixed in the phrase 'God is dead.' In Nietzsche's case, this experience is projected cosmologically in the thesis that nothing new occurs any more. In Nietzsche, the accent lies on eternal recurrence, which the human being has to face with heroic composure. For Baudelaire, it is more a matter of the 'new,' which must be wrested heroically from what is always again the same.[15]

Such heroic newness is the distillation of sheer temporal intensity from an object, which at the instant of newness falls away as the mere vehicle of this subjective experience and endures as a knick-knack of spleen. Subjective newness feeds off the new object, but in such a way as, in advance, to doom it to a spleenful afterlife among all the previous objects that were once-new. Thus, Benjamin argues, the Baudelairian new is not really a redemption of, or escape from, spleen, but rather its inseparable *semblable* and *frère*, the complement and perpetual reproduction of spleen in ever-expanded form. He writes: "Baudelaire's project takes on historical significance... only when the experience of the ever-same, which provides the standard for assessing that project, is given its historical signature. This happens in Nietzsche and in Blanqui. Here, the idea of the eternal return is the 'new', which breaks the cycle of the eternal return by confirming it."[16]

Benjamin tried to deepen this analysis by further exploring its phenomenological dimensions. These are encapsulated in a one sentence entry in the Convolute m [Idleness]: "Phantasmagoria is the intentional correlate of immediate experience."[17] That is to say, what appear to be extreme modes of thought and perception in Baudelaire, Nietzsche, and Blanqui are, in fact, rich descriptions of immediate experience. The hallucinatory characteristics of their texts—Baudelaire's negative and positive intensities, the coupling of cosmic pessimism and exaltation in Nietzsche, the inseparability of denunciation and resignation in Blanqui—are not symptoms of unbalanced minds, but rather of an objectively false appearance, a reality that disclosed itself to these thinkers in a systematically distorted way. In a letter to Adorno dated 23 February 1939, Benjamin tried to justify his latest attempts to organize his Baudelaire studies around a critique of the phenomenological pair "newness" and "sameness." He begins by noting that "sameness" is not, in fact, normally experienceable:

> Sameness is a category of cognition; strictly speaking, it is not to be found in plain, sober-minded perception. Even in extreme cases, perception that is sober in the in the strictest sense, free of all prejudgement, would at most encounter similarity. The degree of prejudgement which can normally accompany perception without detriment can be harmful in certain special cases. It can reveal the perceiver as one who is *not* sober-minded.[18]

Benjamin notes the example of Don Quixote, who recasts whatever he sees as part of his on-going romantic adventure as a knight-errant, and likens Cervantes's comic presentation to that of Daumier's drawings:

> Daumier constantly encounters sameness. In the faces of all the politicians, ministers, and lawyers, he perceives the same thing—

> namely, the meanness and mediocrity of the bourgeois class. But here, one thing is all-important: Daumier, like Cervantes, imparts a comic aspect to the hallucination of sameness. ... Daumier's laughter, however, is aimed at the bourgeoisie; it sees through the sameness this class prides itself on, revealing it as the fatuous *égalité* which flaunted itself in Louis Philippe's nickname.[19]

Benjamin here gives an original spin to the concept of ideology that he inherited from Marx and, more immediately, in terms of sources, from György Lukács's *History and Class Consciousness*. If ideology, in the classic Marxist view, is false consciousness, then in Benjamin, in contrast, it is false *experience*: that is, ideology is not just thought, but phenomenologically lived as a kind of intoxication of everyday life. One form in which ideology may be criticized, again in lived form, is through comic presentation and laughter.

This view of ideology and ideology criticism as experiential is fundamental to his interpretation of the eternal return as well, although the critical strategy of comedy seems, in Benjamin's view, to reach its limits here. Baudelaire—and by implication, the associated names of Nietzsche and Blanqui—does not follow Cervantes and Daumier in making fun of sameness, but rather takes up occupancy at the heart of sameness, heightening its intensity to the point of critical explosion:

> In laughter, both Cervantes and Daumier do away with sameness and thus fix it as a form of historical semblance [*Schein*]. Sameness has a very different stamp in Poe, to say nothing of Baudelaire. In "The Man of the Crowd," the possibility of comic exorcism may still flash up. In Baudelaire, this is out of the question. Instead, he gave artificial assistance to the historical hallucination of sameness which had taken root with the commodity economy. And the images in his work that relate to hashish are decipherable in this context.[20]

To put this in other words, the intoxication, exaltation, and hallucination that characterize the perception of the eternal return do not discredit this thought, but rather invest it with its significance as an historical index. The thought of the eternal return redoubles and heightens the very forms in which reality is already immediately experienced as an ideological phantom, and in so doing, it offers purchase on the phantasmatic character of the real.

This understanding is reflected in the plan that Benjamin presents to Adorno for the recasting of his Baudelaire essay: "From the standpoint of the Baudelaire study, the restructured text looks like this: The definition of flânerie as a state of intoxication is fully developed, together with its links to Baudelaire's drug experiences. The concept of the ever-selfsame is already introduced in the second book as ever-selfsame *appearance*, whereas in its definitive form as ever-selfsame *happening* it continues to be reserved for the third."[21] Benjamin also explores how this deforming of perception by sameness also breaks down the temporal fabric of experience and memory. If nineteenth-century lived time is increasingly marked by the repeated oscillation between intense punctual newness and the always-already imminent return of spleen, or by the eternal recurrence of the same, then the intervals between one point in time and another are no longer one of a genuine duration. This loss of duration has an individual lived dimension, but also a larger-scale, collective, historical face as well. In a dazzling passage, weaving together these two temporal frameworks, Benjamin writes:

> The idea of eternal recurrence transforms the historical event itself into a mass-produced article. But this conception also displays, in another respect—on its obverse side, one could say—a trace of the economic circumstances to which it owes its sudden topicality. This was manifest at the moment the security of the conditions of life was considerably diminished through an accelerated succession of crises. The idea of *eternal* recurrence derived its luster from the fact that it was no longer possible, in all circumstances, to expect a recurrence of conditions across any interval of time shorter than that provided by eternity. The quotidian constellations very gradually began to be less quotidian. Very gradually their recurrence became a little less frequent, and there could arise, in consequence, the obscure presentiment that henceforth one must rest content with cosmic-constellations. Habit, in short, made ready to surrender some of its prerogatives. Nietzsche says, 'I love short-lived habits,' and Baudelaire already, throughout his life, was incapable of developing regular habits. Habits are the armature of long experience [*Erfahrung*], whereas they are decomposed by individual experiences [*Erlebnisse*].[22]

Benjamin, in sum, utilizes the emergence of the concept of eternal recurrence as a meta-historical diagnostic of the condition of memory and experience in nineteenth-century modernity. In the heyday of historicism and visions of historical progress, the thought of the eternal recurrence—in the three very different heads of Baudelaire, Blanqui, and Nietzsche—casts a shadow of repetition and cosmic stasis over the bustle of nineteenth-century urban life, fashion, and commerce. For Benjamin, these two, antipodal modes of interpreting the historicity of experience in this period were, however, covertly interrelated. In their mutually canceling implications, they point towards a new, different form of historical thinking, writing, and acting, a practice of history that could shatter the continuity of historicist succession along with the continuum of mythic repetition. In the end, however, Benjamin's nineteenth-century thinkers only approached the threshold of this new form of historical practice. Its realization would require a new formal and conceptual vocabulary of montage that owed more to Aragon, Eisenstein, and Brecht than to Baudelaire, Nietzsche, and Blanqui.

Notes

[1] Walter Benjamin, *Briefe* II, eds. Gershom Scholem and Theodor W. Adorno (Frankfurt a/M: Suhrkamp Verlag, 1978), 741–742.

[2] Walter Benjamin, *The Arcades Project*, trans. Howard Eiland and Kevin McLaughlin (Cambridge, Mass.: The Belknap Press of Harvard University Press, 1999), 26.

[3] David Wood, *The Deconstruction of Time* (Atlantic Highlands: Humanities Press International, 1989), 11.

[4] Mircea Eliade, *Cosmos and History: The Myth of the Eternal Return*, trans. Williard R. Trask (New York: Harper Torchbooks, 1959), 5–6.

[5] I am indebted here to the work of Sorin Antohi for this conception.

[6] Karl Löwith, *Nietzsche's Philosophy of the Eternal Recurrence of the Same*, trans. J. Harvey Lomax (Berkeley: University of California Press, 1997), 103.

[7] Löwith, *Nietzsche's Philosophy*, 104.

[8] Georg Simmel, *Schopenhauer and Nietzsche*, trans. Helmut Loiskandl, Deena Weinstein, and Michael Weinstein (Amherst: University of Massachusetts Press, 1986).

[9] Benjamin, *The Arcades Project*. Convolute D 6, 2. Following Benjamin's system of organizing his notes into topical folders, "Convolutes," I cite the "Convolute" and note number in this and subsequent references to the *Arcades Project*.

[10] *The Arcades Project* D7; D7a.

[11] *The Arcades Project* D7; D7a.

[12] *The Arcades Project* D7; D7a.

[13] Walter Benjamin, *Selected Writings*, vol. 4, 1938–1940, eds. Howard Eiland and Michael W. Jennings (Cambridge, Mass.: The Belknap Press of the Harvard University Press, 2003), 96.

[14] *Selected Writings* IV, 96–97.

[15] *The Arcades Project* J60, 7.

[16] *Selected Writings* IV, 97.

[17] *The Arcades Project* m3a, 4.

[18] *Selected Writings* IV, 208.

[19] *Selected Writings* V, 208.

[20] *Selected Writings* IV, 208.

[21] *Selected Writings* IV, 208–209.

[22] *The Arcades Project* J 62a, 2.

A Microscope for Time

What Benjamin and Klages, Einstein and the Movies Owe to Distant Stars[1]

Karl Clausberg

In today's day and age, upheavals in human worlds of image and media are preferably traced back to technical achievements. The history of optical media—camera obscura, photography, the cinema, television, etc.—seems meanwhile firmly established as a prime example of such means of viewing. But how well do these perspectives of progress fit the "nature" of humankind, which has somehow struggled through the channels of anthropological predispositions, neurobiological influences, *et al.* to the peaks of civilization? Using a short and anonymously published text from the mid-nineteenth century by a Berlin lay-astronomer, whose considerable impact is suggestively illustrated in the title of this essay, I would like to demonstrate that the interaction between scientific discoveries, technical innovations and extrapolating fantasy was often very complex, and did not break down into simple sequences of cause and effect, with technology as the pace-setter, or genetics as the fine-tuning and brakes.

The presentation in extracts and commentary analysis of the previously almost disregarded writings by Felix Eberty, later Breslau professor of natural and criminal justice, enables one to merge the skeins of cultural and scientific history—previously regarded and researched separately—to produce surprising cross references. If the associations sketched out in the accompanying commentary are correctly understood, then one can still count on surprises, especially in the key areas of recent human feats of imagination. In the case presented here, common era-contingent views once inflamed a lay-astronomer's juristically augmented imagination, which then left behind culturally formative traces throughout different scientific-fiction domains. Hence, Benjamin and Klages, Einstein and the movies found their way into the title. The significance of this source reaches even further: In his short texts, Felix Eberty did not only describe a new way of reflecting on the light-image (*Lichtbild*), but also laid out important aspects for a future pictorial science.

Albert Einstein

In 1923 a small volume entitled *Die Gestirne und die Weltgeschichte/ Gedanken über Raum, Zeit und Ewigkeit*, authored by a certain Felix Eberty, was published.[2] As it could be assumed from the publisher Gregorius Itelson's preface and the editor's added foreword to the second edition in 1874, the short text, originally published in 1846–47, had a strange history behind it. The anonymous German first edition was distributed worldwide in an unofficial English translation, and the renowned English secretary of the British Royal Astronomical Society, Richard Proctor, adopted its speculative perspectives in his bestseller *Other Worlds than Ours* (1870). In addition, the Ebertian ideas were widely spun out in Camille Flammarion's "gossipy" book *Lumen* without crediting the source, according, somewhat acrimoniously, to the publisher Itelson. Eberty's astronomical fantasies were circulated by other means for about seventy-five years. The Berlin new edition was a latecomer; it presented once again the original chunks of text, which, because of the avalanche they triggered, ultimately drew more and more attention.

The name of the author was well-known among educated readers in the 1920s: Eberty had not only edited the biographies of Walter Scott and Lord Byron, but also a seven-volume *Geschichte des Preußischen Staats*. The small text on astronomy must have met with additional interest; in any case, the volume was reprinted[3] in the original version—along with a likewise reprinted autobiography.[4] The subtitle also drew much attention at the time, as it indicated that no less than Albert Einstein, the Nobel prize winner for Physics in 1921, had written the introduction. "This small book, written by a unique and witty man, is not outside of contemporary interests," according to Einstein's words on June 5, 1923, "because it, on the one hand, presents a critical intellect in contrast to a traditional concept of time, yet on the other presents those characteristic conclusions, from which the theory of relativity—still so often reproached for the bizarre character of its own conclusions—saves us."[5] Should it also be ultimately understood from this formulation, besides the reference to current factionalism in German or Jewish physics,[6] that Einstein was already familiar with the most important contents of the book and if anything, was inspired by these in his fundamental concept of the theory of relativity? Does it not sound as if he were referring to a personal pre-history? Why else would he have written such a foreword? And, if a déjà vu experience had not come into it, would he have been persuaded by the publisher Itelson to whom he felt indebted?[7] Or had he himself initiated the new edition? Speculations such as these are not just plucked out of thin air—Einstein in his youth memoirs specifically mentioned a popular science anthology of the nineteenth century, in which Eberty's work was cited page after page.

"I also had the good fortune," wrote Einstein in 1949 in his *Autobiographic Notes*, "of getting to know the essential results and methods of the entire field of natural sciences in an excellent popular exposition, which limited itself almost throughout to qualitative aspects (Bernstein's *Popular Books on Natural Science*, a work of five or six volumes), a work that I read with breathless attention."[8] This edition of Bernstein's *Volksbücher* was given to the eleven year old Einstein by a friend of the family, medicine student Max Talmud (after emigration: Talmey) in 1889, in order to encourage his thirst for knowledge and his independent studies. The pupil of the Munich Luitpold Secondary School was deeply impressed, as noted by Talmey in 1932 in his memoirs on the "formative period of the author of the theory of relativity."[9]

Aaron Bernstein

Aaron Bernstein (Aaron Rebenstein 1812–1884), in 1848 a barricade fighter, in 1849 a radical democratic publisher of the *Urwähler-Zeitung*, and in 1853 founder of the [Berliner] *Volkszeitung*[10] had been publishing his *Naturwissenschaftliche Volksbücher* since 1885 in innumerable extended editions. The five or six volume edition, given by Talmud to schoolboy Einstein, was most likely one of the newly improved and supplemented versions of the fourth edition published in 1873–74. In the twenty-first section, there was a chapter on "Die Unendlichkeit und die Naturwissenschaft" (Infinity and Science), which most certainly made an impression on young Einstein, as it addressed fundamental perspectives and problems of astronomy and cosmology: "The bold, long-range vision of William Herschel, who first entered the world of fixed stars (with his famous large telescope), opened up a new world view for mankind's astounded intellect," and further: "We are able to behold two stars, once they merely exist at different distances from us, essentially in different age-epochs, even if both were created at one and the same moment. Spatial difference, for our understanding, is also contingent upon the time difference. What we see of infinite space also shows us forms of infinite times." William Herschel did also not avoid, according to Bernstein, expressing the even sharper consequences of this philosophy: Allowing these observations to proceed even further, "then a panorama of the heavens will lead to a panorama of the history of cosmogonies and of the distant depth of space and time."[11]

After such premises, Bernstein sang the praises of an unknown individual, whom he did not hesitate to mention in the same breath as Herschel: the telescope pioneer's profound ideas had lain fallow for many decades without striking any fertile roots; but then

> In the year 1840, a free and logical mind produced short, anonymously published texts in Germany that followed on from Herschel's fundamental ideas, developing them into a witty play of thought. It is correct—he established—that light bears the message of distant events and conditions many years later, and in this way shows us images from the past—and thus the reverse is no less true. Also, what is currently happening on our Earth will be transmitted to distant places much later. … The events on our planet will first be seen ten years after the result on a fixed star, whose light takes ten years to travel to us. What happened a century ago is still the present in their view. If we think even further into space, then our distant and far-distant present transmits itself further and further into the distance. In one point in space, the light of the scenes of the French Revolution is just now coming into view. And even further away, the invasion of the barbarians has just become the order of the day, Alexander the Great is still conquering the world. … And even further away in space, the representation of Earth's past by way of light will just be advancing into the future, historical events that have long been dead for us will just be coming to life.[12]

As early as 1855, in volume seven of the first edition of the *Volksbücher*, Bernstein presented at length the insights of Herschel's anonymous counterpart and added bibliographic information. The speed of light, which can travel an astounding distance in an extremely short period of time, and the idea that this communication from afar had often been traveling for thousands of years, bringing with it messages from times long bygone, provoked a profound contemplation of space and time. "We are following a short text with these contemplations, which was published many years ago in Breslau by an anonymous astute thinker. A small piece of writing, which possesses such a great wealth of conciseness and thought, that we are obliged to hope its author might delight the reader with even more distant gifts." The following was added as a footnote: "The title of this text is *Der Mensch und die Sterne* von X.Y.Z. Breslau 1846 (Man and the Stars by X>Y>Z Breslau 1846)."[13] That was, half-correctly cited (unless there had been a previous, lost draft), the first anonymous edition of the Ebertian text. In a special chapter, "Über die Geschwindigkeit des Lichts" (On the Speed of Light) of his Bernstein edition, Einstein was also able to read this unaltered text, again quoted here, which movingly described Eberty's perspective of the universe—the past as photographic reality between the stars.[14] Everything that existed in the past, wrote Bernstein summarizing Eberty's ideas, "all that still exists somewhere, as soon as one directs their eye to the right place where the light first arrives."[15] Is it necessary to emphatically point out that such sentences must have stimulated an adolescent Einstein?

Felix Eberty

Georg Friedrich Felix Eberty[16] (1812–1884) belonged to the wealthy Berlin Ephraim family,[17] which embodied the fate and brilliance of Judaism in Germany like a Buddenbrook-microcosm: among them were well-known personalities such as psychologist William Stern and art historian Rudolf Wittkower. Eberty's great-grandfather, Nathan Veitel Chajim (Heine) ben Ephraim was a royal jeweler as well as Friedrich the Great's most important financier and played a significant role in financing the Seven Year War; his father converted to the Christian faith and adopted the Christian name Hermann Julius Eberty. Felix received a humanistic, classical language education at the well-known Cauerschen Anstalt; he loved Greek and Homeric mythology. Latin left him cold, although he mastered it perfectly, as his dissertations prove.[18] After his university entrance exams, he studied law theory and graduated as a Protestant Prussian of Jewish origins in Catholic Rheinland, in Bonn, a cheerful retrospect of which can be read in his memoirs *Jugenderinnerungen eines alten Berliners.*[19] Eberty entered the civil service as an assessor for the Superior Court of Justice, became a judge, and ultimately a town councilor and adjunct professor of natural and criminal law at the University of Breslau. His youth-era passions however had a broader range than his later occupational or educational fields: He was a talented draftsman and pursued additional studies in philosophy, literature history, mathematics and astronomy. His first writings he had published by the age of eighteen in 1830 in the *Crelleschen Journal für reine und angewandte Mathematik.*[20]

The text, *Die Gestirne und die Weltgeschichte* (The Stars and World History), which was first published in 1846–47 in two parts under the initials F.Y.—and therefore practically anonymous, was a direct product of these varied interests. In the two booklets, Eberty reacted, as it were, with a sharp forensic eye on the most recent achievements in astronomy: on the first precise parallax measurements of the closest fixed stars in the years 1837–1842. A mere decade earlier, John Herschel, famous son of the even more famous reflector telescope pioneer, Sir William Herschel, had still resignedly complained about the unfathomable distance of fixed stars:

> We have attained, by delicate observations and refined combinations of theoretical reasoning to a correct estimate, first, of the dimensions of the earth; then, taking that as a base, to a knowledge of those of its orbit about the sun; and again, by taking our stand, as it were, on the opposite borders of the circumference of this orbit, we have extended our measurements to the extreme verge of our own system, and by the aid of what we know of the excursions of comets, have felt our way, as it were, a step or two beyond the orbit of the remotest known planet. But between that remotest orb and the nearest star there is a gulf fixed, to whose extent no observations yet made have enabled us to assign any distinct approximation, or, to name any distance, however immense, which it may not, for any thing we can tell, surpass.[21]

Tycho Brahe had been the first to attempt—unsuccessfully—to measure, from the "baseline" of the diameter of the earth's orbit, that is, starting from the opposite half-year points of orbit, the small shift—

the parallax—of presumably nearby stars against the universe's background. Even two hundred years later, James Bradley, the famous English astronomer, despite much-improved instruments and observation methods, was not able to achieve any results. It was Friedrich Wilhelm Bessel (1784–1846), the Hipparch of the nineteenth century as he was called, who succeeded in 1837–1838. Using the new Fraunhofer heliometer at the Königsberg Observatory, he was able to get a reliable value for star Nr. 61 of the Swan constellation. At the same time in Dorpat/Tartu, another astronomical research center in the Baltic region, Friedrich Georg Wilhelm von Struve (1793–1864) also carried out successful parallax measurements. The parallax angles were tiny, the distances of the stars they yielded, massive, and only imaginable in light signal traveling-times of years instead of miles or kilometers. Behind the weakest, individually visible stars, whose distance was estimated up to four thousand light-years, there were expanses of yet more space and stars, whose spatiotemporal distances were estimated at ten-thousand or even hundred thousand light-years, according to Johann Heinrich von Mädler (1794–1874), observer at the Royal Observatory in Berlin, later director at the Observatory in Dorpat/Tartu and author of a repeatedly-published popular science astronomy: "Where our measuring standards might no longer comprise space, nor our history, time; where not only our Earth, but also the sun, even her whole system shrinks to an inconspicuous, insignificant point, that is where fantasy must be allowed freedom."[22]

The first exact measurements of star distances were sensational; a type of Columbus effect became noticeable: one had, if only with measuring instruments, crossed the previously unfathomable abyss between the Earth and the stars, and established first landing points for the imaginary foot of man. But gauging the depth of interstellar space led not only to a freedom of fantasy, but also to the staggering insight into the irrevocable past of the visible present. In 1845, Alexander von Humboldt concisely highlighted these now substantiated facts, surmised by Francis Bacon, in the first volume of his *Kosmos*. The light from the closest fixed stars aCentauri, 61 Swan and aLyra took, as one now knew from Bessel, MacLear and Struve, 3, 9.1/4 or 12 years to reach Earth; and so on. The view of the starry sky thus presented henceforth an evidential scalene; considering the speed of light (first measured in 1675 by Olaf Römer), it seemed more than likely to Humboldt that the light of distant heavenly bodies might deliver the oldest physical evidence of the existence of matter.[23]

Dr. Eberty, after 1840 at the Hirschberg District Court and temporarily a judge in Lübben in Silesia, and before he began his university career in Breslau around 1850, exceeded in 1846 the words of the doyen of German Natural Science in the first booklet on *Die Gestirne und die Weltgeschichte*. He applied his juristic experience of partisan view-point change to astronomy, and thus arrived at the equally plausible as sensational thought experiment (*Gedankenexperiment*), to have fictive witnesses observe Earth's past from fixed stars at different distances and at accordingly different times. If mankind could travel to other stars in any short amount of time, to aim telescopes of enormous capacity at the happenings on Earth, then it would also in principle "be possible to recall sensibly to our very eyes, an actual and true representation of every moment of history that has passed." Thus, each and every moment might be recalled throughout the course of the centuries back until the most recent times, to the present.[24]

> Eberty's conclusion:
> Thus the universe encloses the pictures of the past, like an indestructible and incorruptible record containing the purest and clearest truth. And as sound propagates itself in the air, wave after wave ... so, in like manner, according to our ideas, the pictures of every occurrence propagate themselves into the distant æther, upon the wings of the ray of light ... Thus, that record which spreads itself out further and further in the universe, by the vibration of the light, really and actually exists and is visible; but to eyes more powerful than those of man. The pictures of all secret deeds, which have ever been transacted, remain indissolubly and indelibly forever, reaching from one sun beyond another. Not only upon the floor of the chamber is the blood-spot of murder indelibly fixed, but the deed glances further and further into the spacious heaven.[25]

As examples of unavenged and yet conclusive crimes, documented in celestial light-images (*Lichtbilder*), Eberty had in mind the abduction of Caspar Hauser and the mysterious murder of Carl XII of Sweden.

Charles Babbage

Eberty's "light archive" in space was obviously modeled after the lingering (criminal) archive of atoms stimulated by sonic waves, which Charles Babbage (1791–1871) had presented shortly before—1837–38—in his famous Bridgewater book:

> The pulsations of the air, once set in motion by the human voice, cease not to exist with the sounds to which they gave rise. … The waves of air thus raised, perambulate the earth and ocean's surface, and in less than twenty hours every atom of its atmosphere takes up the altered movement due to that infinitesimal portion of the primitive motion which has been conveyed to it through countless channels, and which must continue to influence its path throughout its future existence. … Thus considered, what a strange chaos is this wide atmosphere we breathe! Every atom, impressed with good and with ill, retains at once the motions which philosophers and sages have imparted to it, mixed and combined in ten thousand ways with all that is worthless and base. The air itself is one vast library, on whose pages are forever written all that man has ever said or woman whispered. …

Eberty even adopted the moral criminologist impulse from Babbage:

> If the Almighty stamped on the brow of the earliest murderer—the indelible and visible mark of his guilt,—he has also established laws by which every succeeding criminal is not less irrevocably chained to the testimony of his crime; for every atom of his mortal frame, through whatever changes its severed particles may migrate, will still retain, adhering to it through every combination, some movement derived from that very muscular effort, by which the crime itself was perpetrated. … When man and all his race shall have disappeared from the face of our planet, ask every particle of air still floating over the unpeopled earth, and it will record the cruel mandate of the tyrant. Interrogate every wave which breaks unimpeded on ten thousand desolate shores, and it will give evidence of the last gurgle of the waters which closed over the head of his dying victim: confront the murderer with every corporeal atom of his immolated slave, and in its still quivering movements he will read the prophet's denunciation of the prophet king.[26]

Yet there were also significant differences between Babbage's hypothetical criminal archive and Eberty's. While the English mathematician effectively allowed his archive material to emerge from the victims and criminals themselves, and stored it inside Earth's atmosphere so that even a time-inverted reconstruction of the protagonists could be connected to the evidential proceedings, the Prussian jurist saw space filled with apportioned, immaterial piles of light-images (*Lichtbilder*) that only allowed for insight and overview, and not for reconstruction and intervention. This applied to even the highest of all witnesses that Eberty could call upon: God;—even if His intervention potential was not of particular thematic concern to Eberty.

> We have here a perfectly intelligible perception of the idea of the omniscience of God with relation to past events. If we imagine the Deity as a man with human powers, but in a far superior degree, it will be easy for us to attribute to Him the faculty and power of really overlooking and discerning, even in the most minute particulars, every thing which may be sensibly and actually overlooked and seen from a real point of observation. Thus, if we wish to comprehend how any past earthly deed or occurrence, even after thousands of years, is as distinctly and immediately in God's presence, as if it were actually taking place before his eyes, it is sufficient for our purpose to imagine Him present at a certain point, at which the light and the reflection of the circumstance is just arriving. Supposing that this result is established; Omniscience, with respect to the past, becomes identical and one and the same thing with actual Omnipresence with regard to space. For, if we imagine the eye of God present at every point of space, the whole course of the history of the world appears to Him immediately and at once.[27]

Omnipresence

The visual omniscience bestowed onto the godly eye led Eberty to remarkable theoretical considerations concerning representation.

> Thus we have here the extension of Time, which corresponds with that of Space, brought so near to our sensible perception, that time and space cannot be considered as at all different from one another. For, those things, which are consecutive one to the other in point of time, lie next to one another in space.

> The effect does not follow after the cause, but it exists visibly in space near it, and a picture has spread itself out before us, embracing space and time together, and representing both so entirely and at once, that we are no longer able to separate, or distinguish the extension of space from that of time.[28]

If one bears in mind the tenaciousness with which the established contemporary aesthetic—Lessing's categories and his followers' intensified standards of norm—were prescribing deployment of either pictorially simultaneous co-existence or linguistically serial consecution, then Eberty's daring views become all the more evident. It took another half a century until Viennese art historian Franz Wickhoff, in his explorations of the Vienna Genesis, conveyed the paradigm of Succession into the world of visible presentation. He coined the term "continuous form of narration" (*kontinuierende Bilderzählform*). "This type of narrative is very striking", wrote Wickhoff in 1895 in his facsimile-commentary,

> ...it completely diverges from what we commonly see in art now. It is not a decisive moment being chosen here, which unites the most important characters of the text in one of the mutually momentous actions, in order to show them on a second screen in a different, not less significant situation, while continuing the narrative again in a third and fourth specific scene. It is not individual images of outstanding, epoch-making moments that merge into a cycle ..., but rather how the text flows, accompanies one, softly floating and continuous, just as a coastal landscape streams by the eye while on a boat ride, the heroes of the story are involved in a continuous succession of states.[29]

Wickhoff only had the narrative style of a certain epoch in mind, that of late Roman Empire. Eberty on the other hand was concerned with truly universal characteristics of hypothetical "light-images" (*Lichtbilder*) and pace-setting modes of perception. Nonetheless, the commensurability of the viewer's prospects—here interstellar streams of light-images, there temporally stretched coastal landscapes—is astounding.

Thought Experiments

Eberty not only laid out optic principles for an imaginary retrospective of multiple yet fixed star viewpoints, which corresponded with the new precise distance measurements. He also made the fictive inspection system dynamic as a continuum:

> Let us imagine an observer, with infinite powers of vision, in a star of the 12th magnitude. He would see the earth at this moment as it existed at the time of Abraham. Let us, moreover, imagine him moved forwards in the direction of our earth, with such speed that in a short time (say in an hour), he comes to within the distance of a hundred millions of miles, being then as near to us as the sun is, whence the earth is seen as it was eight minutes before; let us imagine all this, quite apart from any claims of possibility or reality, and then we have indubitably the following result: that before the eye of this observer the entire history of the world, from the time of Abraham to the present day, passes by in the space of an hour. ... We want no further proof, and it is evident, beyond the possibility of contradiction, that if an observer were able to comprehend with his eye the whirling procession of these consecutive images, he would have lived through the entire history of the world, with all the events and transactions which have happened in the hemisphere of the globe, turned towards him, in a single hour.[30]

From this description of a 4000-year world history compressed into one hour, the "small" technical detail of the rotating Earth ultimately springs into view. In order to provide a site-specific and even an individual-specific continuity of image sequences, Earth had to turn its same side again and again toward the onrushing viewer. Apparently the principle of the stroboscopic disks and cylinders, discovered by mathematician Simon von Stampfer in Vienna and Belgian physicist Joseph Anton Ferdinand Plateau in 1832, served as a template. Such instruments were called phenakistoscope or phantascope at the time, in other words "deception seer," in order to indicate the illusionist character of the discernible pseudo-movement. Did Eberty's fantasy thus turn the Earth into a huge mechanism for stroboscopic observation of global pictorial stories? The description referred very much to such illusionist character, but also subsequently directly asserted the seamless continuity of the image sequence:

"everything entire and unshortened, but only in the quickest succession." A latent contradiction between the implicit model and its fictionalization was articulated in further specific details:

> If we give the observer power also to halt at pleasure in his path, as he is flying through the æther, he will be able to represent to himself, as rapidly as he pleases, that moment in the world's history which he wishes to observe at leisure; provided he remains at a distance when this moment of history appears to have just arrived; allowing for the time which the light consumes in traveling to the position of the observer.[31]

Ideas such as these, about the ability to completely pause or stop next to, or in front of images transported with the speed of light, corresponded to faster or slower rotation of stroboscopic disks and cylinders; up to a complete standstill, which however would result in a breakdown of the illusion of movement. Even if Eberty did not take such a change into consideration, are we not clearly dealing here with a secondary fiction, derived from experimental technology?

Despite some connections, there are obvious differences: "Light-images" were no longer materially fixed to canvas or paper; and the fictional, moving beholder hovering beside them essentially took over the active part of viewing. It was not the rotation mechanism of the stroboscope disks and cylinders that determined either the pace or motionlessness of images, but rather volition: the imaginary witnesses' voluntary lingering or hurrying away. Conclusion: Eberty's fantasies were not second-rate adaptations of an existing illusion technology, but an ambitious *Gedankenexperiment*—a term coined by Hans Christian Oersted in the same decade.[32] Eberty's fiction made use of single elements of instrumentally created real experience, in order to re-interpret them and re-stage them anew. The scientific validity here must seem dubious by today's standards; but the scope of its ideas historically was extraordinary.

Microscope for Time

In the second "booklet" with the same title, and published again under the initials F.Y in 1847, Eberty expanded on his idea of light-images escorted between fixed locations. He plotted his interstellar eyewitnesses in an even more vividly and conceptually concise manner:

> Exactly in the same way in which an infinitely quick passage from a fixed star to the earth crowds together the images of all earthly events into a single moment, so by reversing the process, the succession of these pictures may, in the following way, be indefinitely deferred. Let us suppose that the light, and with it the reflection of some earthly occurrence arrives at a fixed star of the second magnitude in about twenty years. Let us also suppose, that the observer mounts to this star in the space of twenty years and one day, starting at the moment when, for example, the blossom of a flower was beginning to unfold itself: he will there find the image of this flower in that stage of development in which it was one day after the commencement of its blooming. If he was endowed with infinite powers of sight and observation, and had been able to follow the development of the blossom throughout his entire journey, he would have had time and opportunity of studying for twenty years the changes which occurred to the flower upon earth in a single day. The successive changes in its form are, as it were, fixed before his eyes. As it is scarcely possible to catch with the eye, a butterfly which flits past us, so as to detect the colouring of its wings, and, on the contrary, if we could follow and observe it in its flight, we might count out and separate even the minute grains of colored dust upon its wings, so would the observer, who had the power of following the reflection of a transitory event upon the wings of the light, be enabled to distinguish the most sudden changes with the greatest accuracy and leisure. In this way we have, to a certain extent, a Microscope for time.[33]

That which the later technical film term time loop had substantiated, was in Eberty's case attributed to the standard optical instrument of modernity, the microscope. But the microscope was supposed to be directed at "light-images" in space, in other words, it was to infiltrate the domain of the other master instrument, the telescope—an unnerving exchange at first. The traditional allocation of instrumental task areas had been outlined in 1802 by Joseph Görres, the Koblenz natural history teacher of the "great physiologist" Johannes Müller, with the following aphorism:

> The attribute of the idealist is the telescope, with it, he transpires into infinity, the threads of his optical nerves extend into threads of light, and with these gentle feelers, which flow together in the eye and from there flow into all of space, he can feel out the most remote worlds as if he had them in the palm of his hand ... The attribute of the realist is the microscope, ... he dives down into a water drop and eavesdrops on its pluvial inhabitants; all of nature except its field of view is lost to him, in exchange, he sees a new form emerging from the old, and looks to bring this to manifestation as well ...[34]

Carl Gustav Carus

Carl Gustav Carus was probably first to come up with the idea of admixing or even merging the microscope and telescope's scope of activities. In 1841, in the first of his *Briefe über das Erdleben* (Letters about life on earth), he described a remarkable microscope fantasy: he often planned to versify a dream in which a transfixed advocator of the separation of animate and inanimate nature receives the command

> from the *weltgeist*, to look through a floating microscope into the distant æther. He soon sees luminous and enlightened small globes circulating around each other in regular rotation, ... he sees how they attract each other, how here and there a small globe crumbles into dust, while elsewhere new particles form from misty substances, and after more eccentric tumbling about, are drawn together by suction and increase in size, and through separation decrease in size again, etc., and everything reminds him of his previous microscopic contemplations about luminous lake infusorians and the regularly tumbling, mouthless, bowelless global beings, so exactly, that he readies himself to incorporate these æther infusorians as a new and special subspecies into his own system. But there, strange words resounded without abashment: "What you saw was the movement of sun systems, and while you thought you spent an hour observing them, an entire world year has gone by!"[35]

Eberty could thus have adopted acceleration effects of orchestrated perception from Carus' text. If that had been the case, his principle speculations on time experience could have also been inspired by the model's total deceptive character via the equalization of the micro and macrocosmos: If you imagine the course of the stars and our Earth accelerating to twice its speed, wrote Eberty in 1847, so that all human life processes would be also reduced to half their duration; "the drawing of our breath, and the stroke of the pulse would proceed with double their usual rapidity, and our new period of life would appear to us as the normal length." Would we notice such an acceleration? Eberty's answer: We would not notice any change, because we are only able to determine the course of time by comparison with other time measurements. Our forty years would seem like eighty. But that was by far the last word: The same would be true, according to Eberty, if one thought of time being accelerated to four times, and not twice, its normal speed; and further: "For the same reasons, if the period and processes of life, and the course of events in the world around us, were accelerated, a thousand or a million times, or, in short, if they were infinitely shortened, we should obtain a similar result; and we can in this way imagine the entire course of the history of the world compressed into a single immeasurably short space of time, without our being able to perceive the change."[36]

Karl Ernst von Baer

The idea of accelerated or delayed time was accompanied by the question about absolute standards of the subjective experience. Eberty argued that a doubled, a halved, a thousand or million times quicker or slower life tempo in correspondingly moving environments would not be noticed, that in other words, time as well as space should ultimately be only a human concept. This set of ideas was taken up—just to mention one of the most striking lines of propagation—thirteen years later by the renowned Estonian biologist, Karl Ernst von Baer (1792–1876). He developed it into a playful conceptual exercise envisioning twenty-nine-day and forty-minute men, accelerated a thousand times, or accordingly delayed millennium or billennium men, but entered a different setting: he impressively described their radically changed perception of their environments that were not likewise accelerated or slowed. This in turn led to the bottom line of his argument.

"The pulse in general seems to relate in some respects to the speed of perception and movement." With this observation, von Baer—in a lecture entitled *Welche Auffassung der lebenden Natur ist die Richtige?* (Which conception of living nature is the right one?)—set out the specific embodiment of his then often quoted philosophies. Eberty's hypothetical doubling of the heart frequency delivered the prelude: "Rabbits have a pulse beat that is twice as quick as man's, and cows have one almost twice as slow. Surely, that small animal's perception and movements proceed much quicker than the latter's. In other words, rabbits experience more within the same time frame as cows." Using this assumption von Baer emphasized, that inner life in man and different animals could proceed much quicker or slower within the same external period of time, and that according to this, the perception of their environment—von Baer did not yet use the Uexküllian keyword *Umwelt*, but he implied it—would also prove to be very different.[37]

Assuming the lives of man could progress much quicker than they actually do, thus goes von Baer's follow-up argument, all natural proportions would appear completely different to him. Let us imagine his life would be limited to one thousandth of its normal course. He would already be frail if he were twenty-nine days old, but also very able, within one pulse beat, to register six or ten sensual perceptions. He would even see some things that we cannot see, for example, a rifle bullet flying by; and so on:

> If we were to imagine man's life being even more shortened, and right down to one thousandth of the above mentioned measurement, then its duration would be 40, at the most 42 minutes. If the surrounding nature remained completely unchanged, it would still appear totally different to us. ... Every sound we hear would surely be inaudible for such people if their ear was still organized in the same way as ours, but they might instead hear sounds that we cannot, maybe they would actually hear the light that we see.[38]

And more to that effect.

Just as von Baer envisioned human life shortened and at the same condensed in relationship to the external world, it could also be imagined one thousand times slower. The sun's apparent speed of flight would then leave behind a fiery trail; and so on.

> If we were to assume a human life slowed one thousand times its normal duration, and imagine it slowed a further one thousand times, then external nature would be seen again very differently. … We would not be able to register the change from day to night. We would not even register the sun, but perceive it like a glowing piece of coal spinning around and around in a bright ring, we would see it only as an illuminated arc in the sky.[39]

Von Baer's conclusion: Man can only measure nature in space and time in relationship to himself, that is, by his sensual organization, "because an absolute measurement standard does not exist." Which view of nature was then closest to the truth? "Without a doubt, the one that comes from the greatest [more time comprehensive] standard. … The standard of its effectiveness can never be too great, but is rather always too small."[40] What was then the greatest thinkable standard? Eberty's world history in a single blink of the eye? Von Baer did not comment on such extremes, nor did he mention in the printed text any source of inspiration for his analyses.

That von Baer could have developed his biological theories without any knowledge of the Ebertian text seems—on the one hand the accurate matching of acceleration or delay factors of two, four, one thousand, one million—extremely improbable. On the other hand, by bestowing subjective time perception with the status of a measuring standard turned von Baer's senso-psychological into an antithesis to Eberty's purely speculative "light-image" impressions. But Eberty had a catchword ready for this twist as well: "A tune may be performed in different times, either quicker or slower, without altering thereby in any way its nature. The intervals, the succession of the tones, and the proportionate length of one note to another remain unchanged, but the impression which it makes upon the hearer will be different, if his entire life has not undergone a corresponding and proportionate change."[41] Combined result: Comparisons of time frames could provide reliable data for objective research from the different inner worlds of diverse life forms, and not only psychological speculation. Jakob von Uexküll later made this principle the basis of his theoretical biology.[42] In the long run he owed this concept, passed on by Karl Ernst von Baer, to the Berlin jurist and lover of astronomy, Felix Eberty. The role of his short text can once more be assessed as accordingly significant, since it triggered a true avalanche of scientific speculation and idiosyncratic ways of observing and comprehending the world.

Christian Doppler

Like von Baer's accelerated or delayed 40-minute or millennium men, Eberty's hypothetical "light-image" inspectors should have encountered similar radical shifts of sensorial impressions during their rapid travels in space. And there were even more fundamental physical limitations there. Already in 1842, Austrian physicist Christian Doppler (1803–1853) had presented for the first time the effect which eventually took his name. He had addressed the variability of sensorial impressions subject to movement from the physical side. But his initial point was not the acoustic Doppler-effect mainly connected with his name today, but the colorful light of binary stars.[43]

According to the original undulation (vibration) hypothesis, it was generally known that perception of color is a direct result, per Doppler, of regular sequential waves of æther at specific time intervals. The intensity of the colored light is dependent on the size of the excursions (oscillatory movements) of the individual æther particles that touch the retina of the eye. Everything that thus changes the time interval between the single impacts of æther would inevitably bring a shift of color, while the intensity of light is dependent on the energy of the waves. But up until then, questioning how sensorial organs actually record and perceive æther or air oscillations had been completely neglected. As long as one assumes that both the viewer and the sources of the waves maintain their original positions, the subjective and objective conditions will be in complete harmony. But what if either the viewer or the source, or even both, changed their position—moved away from or closer to each other, and this at velocities approximating that of the wave's propagation? Doppler's initially graphic, then mathematical answer: Like a ship sailing straight into oncoming sea, sustaining at the same time an increasing number and an increasing force of water waves, this is how an accumulating and culminating of air or æther waves moving in opposite directions could be calculated, and so forth. Such visually fleshed out pre-considerations brought Doppler to the assumption that the light waves from binary stars speeding around each other, as well as from the so-called periodic variables, could be so tightly compacted or stretched that they would not only radically change in color for the human eye, but would also become invisible;—because Doppler thought even star movement with superluminal velocity possible.

Since the discovery of infrared and ultraviolet radiation (1800–01), the world became increasingly aware that man's sensorial organs could open only small windows to the worlds of mechanical, sonic, heat and light oscillations. The reason for Eberty not including the stellar-related Doppler-effect into his speculations, which could have been of fundamental relevance for his theories, might be due to the remote place of its publication. Doppler's first article was printed in the *Abhandlungen der Böhmischen Gesellschaft der Wissenschaften* (Essays of the royal bohemian society of sciences). But there were critical reviews from, for example, the highly esteemed Berlin Hofrath and Dorpat observatory director, Mädler, and Doppler's touchy replies to these, about which Eberty may have heard.[44] Nor did Eberty consider the impact of the Doppler-effect on his swiftly moving space viewers. Would not the extremely squished or stretched æther waves cause their perception to fail them? What would "light-images" look like if one moved along with them at the speed of light? Yet for fundamental thought experiments (*Gedankenexperimente*), which in the most extreme case, even endeavored to have the Almighty Himself be an inspector of images, questions such as these may have appeared rather second-rate. However it may have then been, it can be established for now that Eberty's astronomical fantasies were already jeopardized on the level of physics at the time of their publication. But in no respect did it lessen their tremendous impact.

H. Baillière, Publisher

The first published reactions to Eberty's innovative image theory came immediately after the first booklet's publication. London publishing house Baillière released an anonymous translation entitled *The Stars and the Earth; Thoughts upon Space, Time and Eternity*. In 1847, a definitive unauthorized translation of the second part followed, as appears in a printer's note in the German original, whereby Eberty obviously signed a license agreement with Baillière, yet reserved the rights to translate his work into English himself.[45] However it came to pass, Baillière released also the second part without crediting a name, and the two part booklet, soon to be published as one, became a huge success. In America and England, thousands of copies were bought of the many reprints and new editions; the total number could easily be estimated at several ten thousand.

"Forty-eight small pages, suggesting food for a life of thought," was the quintessence of the numerous laudatory press commentaries that were already quoted at the time the second part was published in 1847. The second complete American edition of 1850[46] of the reprinted third English edition contained a readers' recommendation, which was from then on continually added to the reprints. It was written by a certain Reverend Thomas Hill in June 24, 1849; the later twenty-first president of Harvard was also an excellent astronomer; and through him the reference to Babbage's Bridgewater Treatise and its atmospheric criminal archive was already established. In 1874, Hill himself published the fourth and, in 1882, the fifth, American edition of the still anonymous Eberty texts.[47] The astounding small book was seen very much within the context of Anglo-American science theology. Yet, its German author profited nothing from this nameless bestseller success abroad. Not even his name, whose beginning and end initials F.Y. had adorned the original version, was able to penetrate the English speaking educated society when Eberty managed to regain the German publishing rights in 1874 and publish a slightly reworked new edition under his full name.[48]

Camille Flammarion

Fame and profit from Eberty's ideas was garnered by another who was an exceptionally gifted property master and "popularizer." Camille Flammarion (1842–1925) published *Lumen, récit d'outre terre*[49] in 1867, the first of four narrations, which from 1872 onward were consolidated with two other texts in one book entitled *Récits de l'infini*, and in an extended, then illustrated version called *Lumen*,[50] reached a high number of editions. In these *récits*, a deceased person named Lumen answered to a curious living person: the "questioner" Querens. After his death in 1864, Lumen had found himself transported to a planet of the star Capella, and from there was able to see, thanks to his now extra-terrestrially enhanced vision, the Earth at the time of his birth and scenes from his youth; and so forth. In a long-winded, detailed yet exciting narrative, all the important moments in Eberty's fictions were adopted and further developed. Distributed among the four narratives written between 1865 and 1869, central aspects of Eberty's "light-image" narratives are repeated almost verbatim; especially the idea of the ubiquity of world history for the creator.

> All the events accomplished upon the earth's surface since its creation are visible in space at distances proportional to their remoteness in the past. The whole history of the globe, and the life of every one of its inhabitants, could thus be seen at a glance by an eye capable of embracing that space. Thus we understand, optically, that the eternal and omnipresent spirit can see the entirety of the past at one and the same moment. That which is true of our earth is true of all the worlds in space. The entire history of the universe can thus be present at once to the universal ubiquity of the Creator.[51]

That which Eberty had labeled "light-image" (*Lichtbild*) with juristic, linguistic precision,[52] Flammarion equated more succinctly, yet ambiguously to photography with the following statement:

> Here then, as I say, we have a series of terrestrial images imprinted in space, at corresponding distances, one after another. Let us now suppose that between each of these centennial pictures we should find annual pictures, between each of which the distance is preserved in accordance with the time that light travels in a year, which I have just given you. Then, between each of the annual pictures we have those of every day, and as each day contains the pictures of each hour, each hour the images of its minutes, and each minute of its seconds, all in succession according to their respective distances, we shall have in a ray of light—or, rather, in a flash of light—composed of a series of distinct images in juxtaposition, an aerial register (*inscription fluidique*) of the history of the earth.[53]

Thus a smoothly flowing inscription, a chronicle of Earth's history in a ray of light consisting of single images stacked onto each other. It was not unthinkable, according to Flammarion elsewhere, that the flow of images would fall on a dark star whose light sensitive iodine surface might act like a photographic darkroom to "fix" the radiations of distant Earth. If such an *astre obscur* also had a cylinder form and rotated accordingly, then an indestructible column of images would be created, a *colonne impérissable*, "around which would be engraved the great events of the world's history."[54] Such celestial darkroom fantasies were not a spin off of nineteenth-century apparatus technology alone, but also developed from the time-honored, column-based, picture display format, such as the Trajan column in Rome or the

Paris Vendôme column (that was torn down in 1871 and then erected again). This double reference also makes clear how Flammarion's repeated discussion on photographs oscillated between opto-physical image transmissions and chemical-material reproduction.

un microscope qui grossirait le temps

If elaborations like the idea of a illuminable celestial pillar of images, which beautifully anticipated Edison's cylinder phonograph of 1877, went in one direction, then other examples of "borrowing" from Eberty's original text were even more obvious. Thus, Flammarion did not only literally quote the microscope for time, but certainly sent ahead one of the accompanying demonstration examples. Eberty, in order to illustrate the advantages of the time-microscopically slowed perception of images, had brought the phenomenon of lightning to the witness stand.

> A flash of lightning, for example, appears as a momentary light, which blinds us for a time, without permitting us any power of distinguishing the causes which produce it. But if we could follow the image of such a flash, only up to the sun, i.e. for eight minutes, we should be able to unfold secrets respecting the nature of the phenomenon, which would not be less astounding of their kind, than the living worlds which the microscope exposes to our view in a drop of water.[55]

With Flammarion, the relative passages (here slightly abbreviated) read thus:

> If you set out from the earth at the moment that a flash of lightning bursts forth, and if you traveled for an hour or more with the light, you would see the lightning as long as you continued to look at it. … Now, if instead of flying with precisely the same velocity as light, you had flown a little less rapidly, and if you had employed a thousandth part of a second more to arrive at the same point, instead of always seeing the same moment of the lightning, you would have seen, successively, the different moments which constituted the total duration of the lightning, equal to the thousandth part of a second. In this entire minute you would have had time to see, first, the beginning of the lightning flash and could then analyze the development of it—its successive phases—to the very end. You may imagine what

> strange discoveries one could make in the secret nature of lightning, increased 60,000 times in the order of its duration. What frightful battles you would have time to discover in the flame! What pandemonium! What sinister atoms! What a world, hidden by its volatile nature from the imperfect eyes of mortals![56]

It was following this eloquent description of the most minute secrets of lightning that Flammarion made direct reference to the microscope for time; or, more exactly: his own suggestion for improvement from the mouth of Lumen, who documented the theoretical loan of goods even more concisely: "I see in your thoughts that you are comparing this effect to that of a microscope that could magnify time. That is exactly what it is; we thus see time amplified. Such a device could not, strictly speaking, be called a microscope, but rather a chronoscope or a chronotelescope (to see time from afar)."[57]

Hermann von Helmholtz

Chronoscope was the name given to instruments developed in the nineteenth century that recorded the exact beginning of actions or measured the duration of events; in the simplest case stopwatches included additional manually controlled seconds devices. Cylinder chronographs were one instrument that served to record measuring data and, in our context, they naturally allow for another reference to Flammarion's photographic celestial picture pillars. While Flammarion, by means of such apparatus, and with his chrono-télé-scope, focused again on the old allocation of domains meant for the master instruments of the microscope and telescope, he was also on the emergent path toward Chronophotography, which his compatriot Étienne Jules Marey has coined as a term and perfected as a photo-technical process around 1882.

In Germany, discussions touching on the "microscope for time" spread into various circles shortly after Eberty's texts were published. In December 1850, Hermann von Helmholtz gave a lecture at the society for entertainment in physics in Königsberg entitled, *Ueber die Methoden, kleinste Zeittheile zu messen, und ihre Anwendung für physiologische Zwecke* (On the methods of measuring small units of time and their application for physiological purposes).[58] His interests were directed at the measurement of "fast, temporary processes of the living body"—namely, the duration of nerve signals—that, due to their speed, had long escaped man's perception. Helmhotz had attentively followed metrological progress in the forties and surmised it in these words: "You see that the microscopy of time has far surpassed that of space."[59] Was the microscopy of time a dictum then, circulating amongst the German scientific set and picked up on early by the astronomy-lover Eberty? Or is it in fact conceivable that the Berlin jurist himself, with a wide range of social contacts at his disposal, was the originator of the term? Considering the uniqueness of his speculations, it is not all that strange a thought. In any case, it can be tentatively assumed that the Berlin-bred *bonmot* did not only fall on fertile earth in France, but also found worldwide distribution in the English-speaking realm.

William von Voigts-Rhetz

Back again to Flammarion's *Récits de l'infini*, or *Lumen* respectively, and the question of how he could have heard about the Ebertian texts. In his memoirs, Flammarion in 1911 claimed that this idea for a story of heavens suddenly came to him while looking out the window.[60] That may well be, but it could only have been a re-appearance. Previous to 1859, W(illiam) von Voigts-Rhetz—brother of the well-known Prussian general, by then military chevalier at the Paris Embassy, who like Flammarion hobnobbed with the spiritualist set—had gotten hold of Eberty's text in the anonymous English version and enthusiastically translated it (back) into German. This re-translation was published by the renowned Costenoble in Leipzig with the title *Die Sterne und die Erde. Gedanken über Raum Zeit und Ewigkeit*. The translator's foreword makes clear how emphatically the text must have been received in Paris: "When chance allowed us to encounter the small work presented here," wrote Voigts-Rhetz,

> we were so surprised by the boldness and originality of the ideas and consequences it contained, that an endeavor of gratitude occurred to us to make it accessible to a German public less familiar with English language and literature. Just how significant a sensation this pursuing text—its ideas so far beyond the common and beaten track of thought and consequences—and how substantial its unique and very singular path—through an area previously considered almost unapproachable—must

> have made in its unnamed author's fatherland, can be sufficiently proven by the number of published editions distributed in such a short time.

An additional footnote specified: "The translation here was done after the 6th in 11,000 copies of published editions."

A partial edition of Voigts-Rhetz's re-translation into German was printed with a reference to the actual author, Felix Eberty, and an accompanying note by publisher Costenoble:

> The remarkable addition to this small text's title can be explained by the fact that bookseller Baillière had the small work translated in London against the stated will of the author, without mentioning either on the cover or elsewhere in the book, that he was not presenting an English language original work to the public, but rather a translation from the German. The English publisher's approach was why neither the present publisher nor the translator became aware of the true facts, since the original German work, published thirteen years ago, had missed their attention. They believed they were presenting the German public with an interesting novelty—and it was only after this translation, which now turned out to be a re-translation into the original language version, was printed and distributed, that they obtained the name of the author …

and so forth. There followed an unctuous address to the German genius Eberty, in which Costenoble expressed his wish that its witty author would at least now, after the text's very unique fate, receive the same recognition in his fatherland as he had abroad.[61] The result, in any case, was that the re-translation was soon sold out.

Eberty was ultimately able in 1874 to publish a new edition that he himself reworked; and he had to report on further legal bizarre events in the foreword: "The original publisher emigrated to South America and due to forfeit and sale, the copyright passed from one hand to the other, and thus after many years of efforts courtesy of Leipzig book trader Mr. Gebhardt, the author has now been placed in the stand to have this new edition step into the wide world." The new version, now authorized by name, experienced three editions (five, including the text identical re-prints of 1923 and 1925), but it was then too late for any instantaneous worldwide recognition. Flammarion's books had long since re-directed the proceeds of the Ebertian ideas into French accounts, and when Richard Proctor, renowned astronomer and non-fiction author was asked in 1879 to prepare a revised edition of the English version, it remained anonymous, since it did not occur to Proctor to look for the original author on the continent, even though he was full of praise: "It is one of the most strikingly suggestive books, and small though it is, one of the most remarkable of the present century."[62]

Richard Proctor

Whether Proctor (1837–1888), who died four years after Eberty, ever discovered the identity of the author he so revered and perhaps left behind notes or even an exchange of correspondence to that effect, remains to been seen. In any case, Baillière Publishing House, which also arranged the 1880 new edition, apparently left Proctor in the dark concerning the circumstances of the first English publication. Proctor on his part considered it best to make as few changes as possible and only add footnotes to bring the edition up to date. He had reason enough for this respectful way of dealing with what he still believed to be an anonymous text:

> I remember well the interest with which I read it, when my esteemed friend the late Mr. William Longman called my attention to the work as one I ought to study before I completed my treatise *Other Worlds than Ours*. It was the study of this charming little book which induced me to add to that work of mine the closing chapter on the '*Control and Supervision*' of worlds, wherein I used (with due acknowledgement) some of the thoughts suggested by the anonymous writer of *The Stars and the Earth*.[63]

The said closing chapter of Proctor's book—numerous reprints followed its first publication in 1870, and revised versions were still being reprinted well into the 1930s—contained, after a preliminary aforementioned reference to *The Stars and the Earth*, the following "beautiful and striking conception" by the author he was referring to:

> Suppose that a being, armed with such powers of vision as we have imagined, should watch from the neighborhood of our earth the progress of some interesting event. If he then began to travel from the earth at a rate equal to that at which light travels, he would see one phase of the event continually present before him, because he would always be where the light-mes-

> sage recording that event was actually traveling. By passing somewhat less swiftly away, he would see the event taking place with singular slowness; while by passing away more swiftly he would see the event occurring in inverted order. Suppose, for example, he were watching the battle of Waterloo, he could gaze on the fine picture presented by the Imperial Guard as they advanced upon the English army, for hours, years, nay, for centuries or cycles; or he might watch the whole progress of the charge occurring so slowly that years might elapse between each step of the advancing column, and the bullets which mowed down their ranks might either seem unmoving, or else appear to wend their way with scarcely perceptible motion through the air; or, finally, he might so wing his flight through space that the Guard would seem to retreat, their dead men coming to life as the bullets passed from their wounds, until at length the Old Guard would be seen as it was when it began its advance, in the assured hope of deciding Waterloo, as it had decided so many hard-fought battles for its Imperial Chief.[64]

The time loop effect mentioned at the beginning in fact corresponded, as we know, to the Ebertian text. But Proctor could not have taken the passage about an inverted Battle at Waterloo from *The Stars and the Earth*; it is found as little in this English version as in the German original. Did he perhaps add it afterward to provide his source's plot with even more lurid detail? The issue is curious, as are many details of the above outlined publication history of Eberty's text. The precise counterpart can be found in Flammarion's *Récits de l'infini*, first collected and published in 1872, in the second *Récit* entitled *Refluum Temporis*, written according to a footnote in 1867; here an excerpt of the text equivalent:

> It was really Waterloo, but a Waterloo of the afterlife, for the combatants were being raised from the dead. In this singular mirage, furthermore, they marched backward one against the other. Such a battle has a magical effect, and it impressed me all the more forcefully because I had seen the event before. It was strangely transformed into its counterpart. No less singular was the fact that the longer they fought, the more the number of combatants increased; each gap made in the serried ranks by the cannon was immediately filled up by a group of the resuscitated dead."[65]

Was Proctor already aware of the magazine versions of Flammarion's *Récits* and unintentionally credited the Waterloo episode to his older source? Or was Flammarion aware of Proctor's book, published in 1870, when he revised his *Récits* for the book edition in 1872? The issue will have to remain unresolved for the time being as only one thing is certain, both authors added a significant twist to Eberty's imaginary acceleration and dilation of time: the reversal of time, the "back-flow of time" as an inevitable outcome if hypothetical viewers were to overtake piles of images moving at the speed of light.

What Proctor—as Eberty—from the beginning characterized as an intangible appearance of images, Flammarion first presented as a narrative of a real 'reversed world' only to offer a solution ex post to the riddle in Eberty's style. Flammarion thus remained closer to the tangibility of fiction with his alleged paragons such as the humorous idea of inverted worlds described by Gustav Theodor Fechner under the pseudonym Dr. Mises, who had included even corpses leaving the grave, in 1824 and 1846.[66] In contrast, Proctor, as Eberty, stressed the intangibility of the images and thereby in essence, only made them accessible to observation, while denying any control. Inasmuch the title of the last chapter of Proctor's book, *Supervision and Control*, named a fundamental theological problem that had previously appeared in Eberty's design: God could see all; but how, with his mere insight into aloof images of world history, could he have exercised immediate control; in other words: how could he ever intervene? That fact remained unfathomable for man—and has become a source of deep and unsettling tragedy for less confident generations to come.

Carl du Prel

While a better known flood of rehashing and paraphrasing of Flammarion's *Lumen* material was looming in the Anglo-Franco world—the first English translation already appeared in 1873 —a somewhat less spectacular line of propagation of Eberty's text was unfolding in Germany, which, nevertheless, was not insignificant for the Berlin scene after the turn of the century, to which we will return at the conclusion.

In the second, expanded edition of his much-read book, *Der Kampf um's Dasein am Himmel. Versuch einer Philosophie der Astronomie* (The struggle of life in space. An essay towards a philosophy of astronomy), Carl du Prel (1839–1899)—before his "conversion" to spiritualism—reacted in 1876 very directly to Eberty's text of 1874. The important aspects were referred to in a six-page note and critically praised: A being, omnipresent in space, would also have to be omniscient in relation to past events. But if the creator's omniscience was to also sustain its scientific justification, then one was bound to ask what it would ultimately be worth, according to du Prel: because at the same time the wonder of omnipresence would have to be presumed. At any rate, another preconditioned characteristic of the Almighty, namely His knowledge of the future, remained unresolved. Yet despite such fundamental doubts, du Prel was nonetheless deeply impressed by Eberty's "image science," and considered the microscope for time worthy of a concise characterization: Just as an eye approaching a star with the appropriate speed would perceive the circumstances on that star extremely condensed, thus in reverse when swiftly moving away, the events could be infinitely extended. Intermediate stages, normally invisible, would be discovered within natural processes, and that could lead to new findings. An alleged omnipresence would well manage the same; "but the even discovery of all intermediate stages of changes would remain fruitless without an understanding of the fundamental powers."[67] This corresponded exactly to the rational argument method of up and coming monism, which du Prel, with his "celestial Darwinism," wanted to help gain cosmic recognition.

In the third, recently revised edition of the book, which in 1885 was published under the title *Entwicklungsgeschichte des Weltalls. Entwurf einer Philosophie der Astronomie* (The history of the development of the universe: Design for a philosophy of astronomy), du Prel attested additional exemplary capability features to the microscope for time, which again raised the notorious problem of the wandering motif: "A locomotive could be given the pace of a snail; by exactly matching the movement [of the viewer's], a meteorite hurtling through our atmosphere could be quasi brought to a standstill, yes, even be made to move in reverse. A cosmic being [with access to the possibilities of time microscopy] could thus have time flow quasi in reverse."[68] It was not Flammarion being quoted here, but shortly after, Karl von Baer, with his previously stated remarks on observing nature.

In the expanded volume on *Die Planetenbewohner und die Nebularhypothese. Neue Studien zur Entwicklungsgeschichte des Weltalls*

(The inhabitants of planets and the nebular hypothesis. New studies on the history of the development of the universe), published two years before, du Prel had already followed up on the exo-biological side of the Baer suppositions in a rather Ebertian perspective:

> Beings which have the ability to register within one second all objective changes as subjective conditions of consciousness, so that they, for example, perceive every single oscillation of æther, in which we need many millions within a second to receive a sensation of light and color, would be, so to say, in the possession of a microscope for time. ... Let us now assume there were beings, whose organization was equivalent to the functional properties of æther, which, for example, would have the powers of movement analogous to the speed of light, then those beings could also be able to create the above mentioned microscopic expansion or telescopic concentration of time-successive variations.[69]

That essentially led to the "Doppler Effect," without it being named directly:

> Since even the variations of things are based on the movements of æther, and it is simply conditional to the number of its oscillations, which in turn effect our senses—in other words, whether they are perceived as sound, warmth or light—in this case, a power of motion approaching or moving away from an object could also be used to successively effect the different senses by that same external object; because the appropriate speed of motion would increase the number of æther oscillations within one second while approaching, and decrease while moving away. Such a being could, if its sense of hearing were to react with a deep tone to an external process, transform this into increasingly higher tones by its motion, then experience an increasing sensation of warmth, finally transform the original tone into a sensation of color and then run through the entire color spectrum; or it could also be conducted through the reverse sequence of sensations and, if it possessed other senses for æther oscillations that we cannot register, through even more modulations of sensation conditional to the same identical external causes.[70]

This cycle of the senses on a single vibrating object was emulated after a well known thought experiment of Wilhelm Wundt, who had a rod run through every frequency, from the deepest acoustic vibrations to beyond visible light, in order to demonstrate the various limits of the human range of senses.[71]

Du Prel's interest in the Eberty text is not only remarkable due to its documented exo-biological physiologizing, but also because his own writings enjoyed a large distribution in his day, which is now only assessable in outlines. At the turn of the nineteenth century, a broad stratum of academics, artists and writers were influenced by the ideas of monistic spiritualism, etc, whose most established prophet after 1880 was Carl du Prel. Even the acknowledgement of the small Eberty text, which du Prel counted as being one of the most important of his time, was surely promoted by his publications.

Joseph Pohle

Direct evidence for such piggyback proliferation can also be found with concerted du Prel opponents in the ecclesiastic set. Thus, Joseph Pohle (1852–1922), theology professor in the seminary at Leeds, later astronomy professor in Breslau, took a decisive stand in 1884–85 against du Prel's Darwinist astronomy, yet adopted the Ebertian ideas and added his own glosses to them:

> Let us envision the not impossible case, that the creator had created an eye of such extraordinary focus, such enormous visual range and wonderful sensitivity, that how ever vast the distance, it would be impossible to baffle the visual perception of this marvelous eye. It could hence also observe and follow the objects and events on our planet from any distance with the same acuteness as ours. .... If this wonder eye were placed on the farthest planet Neptune ... it could exactly follow an historical fact from there, for example the richly eventful course of the battle of Waterloo.

And so on.[72]— The source provided in this case was clearly Proctor, however with some modifications, especially recognizable as well in the wonder-eye created by God becoming autonomous.

"Even more wonderful scenes than those depicted can be gained by slightly modifying our optical fiction," Pohle continued.

> Let us imagine this magnificent eye in movement and not in observant stillness, and assume that this eye would travel an unimaginable distance of several thousand light years at a racing speed straight toward our Earth: then it is clear that the first

> scene it would have been presented at the beginning of its journey, would have depicted an ancient event in Earth's history, for example the beginning and primitive stages of the first human beings, and the most recent scene the present state of man and Earth's point of development. ... During its great journey, the wonder eye would truly have had all of Earth and world's history, from its beginnings until today, rolled out before it in swiftly changing tableaux vivants, and in fact at a faster or slower velocity according to the speed with which it moves through space. If this journey took place in an instant, as short as a fleeting thought, then this would be a significant moment, like in a magical picture that would flash by the alleged eye with a long chain of the most significant or most minor events of every century ...[73]

Subsequent to Proctor, Pohle distinguished three cases in the wonder-eye's very quick movement: it could be the same as that of light, or quicker, or slower. In the first case, the image encountered by the eye on its journey would never change on the retina. Then du Prel's snail-paced locomotive and backward hurling meteorite was quoted. Then another named source reference appeared:

> An utterly paradoxical order of perceptions would apparently result if the eye were to move at a quicker pace than the light: then namely, the imaginary case would have to set in, that the event would run in reverse. The history of man and Earth would be turned upside down. People would first be seen on the deathbed, then the sick bed, then in the prime of life and finally as an infant in the cradle. The past would successively, but with reversed sequences, become present. Because the eye, owing to its swiftness, would even overtake all those series of light rays still en route carrying older communications. We would have a distortion of the events, an apparent reverse of time, in truth a "reversed world."[74]

For these passages of text, Pohle expressly specified Eberty's text of 1874 in the footnote. Yet the referenced contents prove: it was only quoted from memory and riddled with bits and bobs from other sources; obviously also from Flammarion's *Récits*, without however this having been named.

Pohle's analyses are also so remarkable because his modest scientific studies on the inhabitants of planetary bodies, originally published in 1884–85 in the *Jahresberichte der Görres-Gesellschaft* (Annual

Report of the Görres Society), later as an ongoing updated "Einführung in die moderne Astronomie" (Introduction to modern astronomy), reached a notable number of editions.[75] The seventh, much expanded new edition of the obviously powerful book was published as late as 1922. Yet the excerpts of the passages quoted here, which still existed under the chapter heading "Die Gestirne und die Weltgeschichte" (The Stars and World History), in other words, under the precise title of Eberty's book, remained completely unchanged.

Equally remarkable as the perpetual distribution of Pohle's book was a small yet significant paraphrasing of the "quoted" source: Pohle no longer referred to God himself or other cosmic observers, but rather to a separate wonder-eye, which could behold history flashing by in the form of *tableaux vivants*, in magical pictures.[76] Here, the psychic eye-witnessing became an almost mechanical one. One needed only to speak of a camera-eye, rather than a wonder-eye, to bring the phantasmagoric media stage fiction into focal range.

Berliner Urania

How the real phantasmagoria of an accelerated prehistoric history proceeded in the eighties and nineties can perhaps be illustrated for the intellectual eye using the Libretti of Berlin's Urania, an institution for the advancement of public education. Thus, the following enlightening prolog on *Wanderung durch vorsintflutliche Landschaften* (Wanderings through the Antidiluvian Landscape) can be found in the scores of slide lectures given by M. Wilhelm Meyer in the "scientific theatre" of the Urania Society from 1888 onward:

> (Stage direction: with curtain closed):
>
> Nothing is still in this world, everything is in motion, everything is developing, everything strives toward a higher, more beautiful aim. ... A spirit able to survey millions of years as quickly as a minute might pass for us, would see the stars, the powerful sun with the planets swarming around it like moths, all whirring through the sky like fireflies on a warm June evening. ... And thus, the artist, whose hand is led by the spirit of science, should place us back into the long forgotten past. ... We want to wander through the great history of our Earth, gallop in but one short hour through a period of millions of years with seven-league boots, experience the storm and stress time of Earthly nature."

> (Stage direction: The curtain opens) First scene. Chaos.
>
> (Stage direction: The scene is at first very dark and gloomy. The nebula glow more and more, are swiftly swirling and churning. Then the glowing watery surface of Earth can be seen cooling. Steam develops, and finally, Earth can be seen entirely flooded by oceans, its line of demarcation at the horizon emerging from the mist.)[77]

The last of the twelve scenes showed, after diverse prehistoric landscapes, the arrival at the cultural historical present at a "Mediterranean beach similar to Sorrento" with antique ruins. Text and images were carefully interlocked: The words of the lecturer bridged the change of scene behind closed curtains, and the images were purportedly dramatized by diorama-like lighting effects and/or dissolving views (with fog effect projectors). Put in a nut-shell: That was an accelerated one-hour vision of world history, and its pictorial sequencing and dramatic presentation very likely had been derived from Eberty's astronomical fantasies.

Interim Result

What can be considered a provisional result of the above outlined odyssey of the small Eberty text up to the performances at the Berlin Urania Science Theatre? One thing at least: The Berlin lay-astronomer's cosmic perspectives were not a secondary fall-out product of an amusement park technology merely interested in vulgar illustrations of motion. On the contrary: Eberty's intellectual lightening-bolt was ignited only in concurrence with a revolution in instrument technology—the first precise fixed star parallax measurements—in a milieu of conceivability already sensitized to shifts in time scales.

"Let us conceive," Alexander von Humboldt had written in 1845 in the first volume of *Kosmos*,

> If for a moment we could yield to the power of fancy, and imagine the acuteness of our visual organs to be made equal with the extremest bounds of telescopic vision, and bring together that which is now divided by long periods of time, the apparent rest that reigns in space would suddenly disappear. We should see the countless host of fixed stars moving in thronged groups in different directions; nebulae wandering through space, and becoming condensed and dissolved like cosmical

> clouds; the veil of the Milky Way separated and broken up in many parts, and *motion* ruling supreme in every portion of the vault of heaven, even as on the Earth's surface, where we see it unfolded in the germ, the leaf, and blossom the organisms of the vegetable world.[78]

Humboldt had thus adopted Carl Gustav Carus' dream image of the space microscope and with that directed the eye toward the celestial past; and such an accelerated perception introduced the reverse perspective, the microscopic dilation of time. The acceleration and deceleration of a thoroughly imaginary, fictionalized perception gained, however, its startling impressiveness only in view of the gigantic spatiotemporal abyss, which opened for the instrumentally enhanced vision.

Felix Eberty was the first—and of this he was clearly aware—to put the equally plausible and consequential idea of a momentous thought experiment into words, to have fictive witnesses, of whom the most noble could be none other than God, observe Earth's past from different fixed stars at different distances and at accordingly different times. But there is more: He had these more or less almighty eyewitnesses watch the events not only motionlessly, but thought of them moving either with or against the flow of these light-images, so that now, not only time acceleration but even time dilation could enter into the picture. He labeled this principle of inspecting light-images (*Lichtbilder*), which embraced the Babbageesque idea of an indestructible criminal archive of sonic waves broadcasted into the atmosphere, a microscope for time. On the other hand he adopted the metaphor of an elevated instrument, which Carl Gustav Carus shortly before had transferred into space. With this, a preliminary model of the "cosmic cinema" was born, a half a century before the technical realization of the cinema, far away from the standard categorical doctrines of the purely time-successive or spatial-coexistent arts (Lessing's hardy inheritance).

Celestial Cinema

In 1988, a special edition of the *Beiträge zur Film- und Fernsehwissenschaft* (Contributions to film and television science) was published, which focused on Berlin in the Twenties, Center of Theoretical Film Philosophy (Berlin – *zwanziger Jahre, Zentrum filmtheo-*

retischen Denkens). Among the essays: *Die Utopie vom kosmischen Schauspiel und der Kinematograph* (The utopia of the cosmic spectacle and the cinematograph) by the Russian media expert, Michail B. Jampolski.[79] The author stressed in his essay that the state of film consciousness could also be clarified explicitly by studying utopian film theory. With this, he introduced the concept of cosmic film, the so-called celestial cinema, which roused a lot of attention especially in the twenties. French film theorists such as André Imbert and Albert Bonneau, along with Carl Hauptmann and Rudolf Harms in Germany, were all talking about film that would roll across a stage that is the universe. That was, according to Jampolski, no exalted metaphor, but rather had a significant tradition behind it, a certain style of thought. A special film utopia came of this, which was to be traced back to German Romanticism, Franz Xaver Baader and Johann Wilhelm Ritter's *Siderismus* as well as Jean Paul's *Hieroastronomy*. In Germany, this complex of ideas was further proliferated by Carl du Prel's spiritualist para-astronomy and intensified by Camille Flammarion's *Lumen* on the grand idea of a cosmic cinematograph.

Six years previously, in 1982, Max Milner focused the tradition of celestial light-images in his book *La Fantasmagorie*.[80] To his mind—after Tiphaine de la Roche who, in 1760 had described a rather hypothetical method of fixing light-images (*Lichtbilder*)—Eugène Mouton (1823–1902), with his fantasy *Historioscope* (1883), was its most prominent mentor; Flammarion, at the same time for him, followed in second place as an idea-recipient. Meanwhile, the esteem together with the more precise appraisal of his work turned out completely in Flammarion's favor. "Perhaps the single most innovative work of the entire literature of the scientific imagination," according to the relevant overviews.[81] And the equivalent seems to apply to literature and film theory as well: Thus we find in Danielle Chaperon's 1998 portrayal: *Flammarion. Entre astronomie et literature* (Flammarion: Between astronomy and literature), a collection of significant key words and buzz phrases on the new pillars of science-fiction and media history: Not only "le cinématographie céleste," coined as a title of a chapter, but also *la mémoire cosmique* as well as *la tristesse des astronomes* and *la mélancholie scientifique*.[82]

The sadness of astronomers and scientific melancholy, which Danielle Chaperon especially ascribed to Flammarion and his times, prove to be the equivalents of an image-related visual urge and yearning, which Roland Barthes once again expressed in a poetically compact formula: "From a real body, which was there, proceed radiations which ultimately touch me, who am here; the duration of the transmission is insignificant; the photograph of the missing being ... will touch me like the delayed rays of a star. A sort of umbilical cord links the body of the photographed thing to my gaze ..." And a bit later in contrapuntal insistence on the present: "The Photograph does not call up the past (nothing Proustian in a photograph). The effect it produces upon me is not to restore what has been abolished (by time, by distance) but to attest that what I see has indeed existed."[83]

The inevitably painful and perpetual distance from the past, prescribed by Barthes, was, however, not exactly Flammarion's main objective in his *Récits de l'infini*. His *Lumen* narrations had the character of spiritual, but yet very bodily-motivated time-travels that even included intimate encounters with loved ones. In comparison, the aspect of star-columns capable of retaining photographic images was rather marginal (one half a page of two hundred thirty six in the 1873 edition). Nevertheless, Jampolski stressed that Flammarion had designed a material model of duration (*durée*) particularly with this cosmic recording instrument, and thus anticipated the significant philosophical ideas of Henri Bergson, namely the cinematograph model of consciousness.[84] Flammarion's Quærens/Lumen dialogs had spawned the most comprehensive and extensive, proto-typical utopia of cosmic film—according to Jampolski, who then rounded off his essay with an impressive number of successive examples such as Paul Scheerbart, El Lissitzky, Blaise Cendrars, Waleri Brjussow, to name a few.

Were then Flammarion's long-winded *Récits* and Eberty's concise flashes of insight really, in essence, early forms of cinematograph theory and with that, principle witnesses to a cinematic mentality-in-the-making that precipitated cinema itself? Can one therefore follow a line of thought implied by Jampolski in a reference to Bergson?—Instead of immediately focusing on the later 1907 cinematographic model of *Évolution créatrice*, one should better stick to Bergson's debut work *Matière et mémoire* of 1896, in order to track the astronomical backgrounds of his philosophy.

Henri Bergson

In *Matiére et mémoire*, Bergson described himself as an "image among images." He saw the other images as being under the influence of one certain image, namely one's own body. This belongs to a special sensorial–physiological tradition: For the consciousness, the body is also a part of the imagined, spatial outer world. This was propounded by Ewald Hering in 1862 in his ground-breaking, yet today hardly-known *Beiträge zur Physiologie* (Essays on physiology).[85] According to him the body is thus itself an imaginary image, which is continually being completed from memory. It is introduced to visual space, together with individual parts such as hands or feet, which are in actual view. And further: The extent to which the image of a spatial world we dream of with open eyes is determined by the respective retina image, was ultimately to be established empirically by physiology, according to Hering. In this sense, Bergson as well: As a realist, one would assume an entirety, a universe of images that has no center, but in which all images successively line up infinitely. If all of these images however referred to one specific image (that of one's own body), they would have to re-group with every movement of the body. As soon as one, however, would generally connect the present with the past and want to foretell the future, the egocentric composition would have to be abandoned once again. Only thus was a science of the universe possible, according to Bergson.[86] But how could human perception and memory function in such a world of images?

The entire difficulty of the problem, thus Bergson's diagnosis, would result from the fact that perception is seen as a type of photographic view of things, which is recorded from a certain point with a specific instrument—our organ of perception—in order to then be developed in the brain. Now it seemed that if a particular point in space would be assumed, the effect would pass through the entire matter without resistance or loss. The possible photography of the whole would remain [flowing] light, because a plate to capture this would be missing. The human brain with its "zones of indetermination" would thus have to somehow take over the role of the photographic plate, and transform according to its own methods. [87] Was Bergson then pitting moving pictures against fixed, photographic ones in his opening ideas in the first chapter of *Matiére et mémoire*?

And with this, was he considering the exact opposite of cinematographic models of consciousness? The pervasiveness of Hering's optics and Flammarion-Ebertian space perspectives—with noticeable reserve toward pure mechanical (photographic) reproduction techniques—seems characteristic in our context. In its approach however, this was a rather anti-rationalist concept as well as one skeptical toward science, as it has been so vehemently pursued by the controversial expression theorist, Ludwig Klages (1872–1956).

Ludwig Klages

Objects residing in distance could at least become visible images, wrote Klages in 1922 in the *Cosmogonic Eros*; as the reverse was also true, that from the infinity of boundless space, distant images could approach nearer and nearer. A characteristic of the difference between close pictures and distant objects would be inviolability (*Unantastbarkeit*). If it was thus part of the nature of Earthly bodies for both the distant and near to be tangible—then distance's essential feature would befit only pictures from the heavens and especially all bygone events. Elapsed temporal distance would appear in spatial distance or: that which appeared in the distance of space would be the distance of time. Here one could as an exception, according to Klages,

> ask science to assist; astronomy teaches us that even the light from Sirius, the closest fixed star [!], would take many years to reach us; hence we then, in the present act of perception, behold something, which, in a considerable amount of time, has actually already passed! Meanwhile, even that about which we could not yet be sure, and which the science of tomorrow might discard as a misguided popularized opinion, everything would be nevertheless confused if we were to switch the measurable distance from Earth to Sirius' body with the distant character of Sirius' image! The issue here, is not mileage or even light-year lengths of matter-separating distances, but of that distance (effectively, distance itself), which, in observing the image (in a pre-intellectual way) is immediately observed as well.[88]

Everything spatially distant, according to Klages,

> can approach closer—except unconditionally only the stars! Let our eyes, by means of honed glasses, permeate still even

> more light-years deeper into the abysses of space, and, in the darkroom of a myriad of stars, capture an image at places in the universe where hence, only an expanse of empty darkness existed, let us ... chemically analyze the matter of the suns and burning misty haze of other planets: we would still presently only have the appearance of the stars, never their (merely scientifically presumed) physicality. Hence then: the stars are not being sought ("*die begehrt man nicht*" [Goethe]) and their present state—if we are allowed a joke here, to illuminate the associated suspected paradox—shine in absence!—Could thus accordingly no more convincing allegory of the past be contrived than the star-lit splendor of the nighttime firmament, we subsequently understand not only the shower of grandeur which, while beholding the sparkling heavens, can pass through even the most cultured observer, rather we also understand why, according to the original meaning, stars were either souls of deceased or locations of their residence, yet always the past choirs of the previous world, shining on in eternity, remaining linked to present everchanging proceedings on Earth.[89]

Measured by the spirit of endeavor and literal confidence demonstrated by the texts of the nineteenth century, Klages' statements on astronomy sound strangely both resigned and exalted. In actuality, he projected a modern state of knowledge, of the distances of fixed stars measured in light-years, onto the alleged shower of grandeur of prehistoric cultures. Had the latter not always counted on the presence of their Gods and ancestors?

Eros of the Distance

In 1921, Klages had set an irrational point of conclusion and departure to such heavenly perspectives in his essay on *Vom Wesen des Bewußtseins* (The nature of consciousness'): "Not things, but images are ensouled: that is the key to all of life's teachings (*Lebenslehre*)." This key however could not be possessed by scientists, because they laid out, instead of the original reality of images, according to Klages, the derived reality of purely alleged things.[90] What then did images have to offer? Their origin appeared in Klages' early psycho-philosophical works heavily charged with word-figures of multi-modal sensations: "A devotion that melts any self-assertive resistance, far beyond all objects and their embeddedness in a stream of temporality that engulfs us as well: from these three threads the veil has been woven through which the awake spirit beholds the world 'like a dream'."[91] In another formulation, with the melting into the world of images, the cosmogonic eros' suffering from insurmountable distances and ensuing insatiable longing had already been announced: "From close-up physicality shifted and exposed to the attraction of distance, the ego would loose its foothold, even without the self-devotion that, acting from within, now places it completely among the images, to such an extent that it confuses itself with them."[92] Synaesthetic qualities were addressed directly, if he somewhat assumed "that however, the perception of spatial distance would have a singular 'color'."[93] Klages ultimately condensed the multimodal experience into the overall complex of an unappeasable longing for images from the past called "Eros of the Distance" (*Eros der Ferne*).— Veil of Isis and Nimbus were the two terms created from this sense of melancholy, which came closest to Benjamin's expression of aura.

Walter Benjamin

There are few, but revealing examinations of the relationship between the two very opposite characters, Benjamin and Klages, which do not need to be especially referenced here.[94] In any case, it is clear that Benjamin adopted very central terminologies for his observation of art from Klages. It is notable in our context, that Benjamin's famous definition of the aura as a "unique appearance of a distance, as close as it may be," does not reveal any direct traces of its astronomical contextual origins—on the contrary. The multimodal–synaesthetic quality of inviolability (*Untastbarkeit*), so expressly provided by Klages with the stars shining in absence, was quasi cut from the firmament by Benjamin and re-deposited onto cult objects of the history of mankind. And even other features of perception, which originally took the shape of imaginative excrescences of astronomy, were considered by Benjamin to be the entirely Earthly achievement of new technical media.

What Eberty had imagined as a possible observation of "light-images" (*Lichtbilder*) racing away between the stars, Benjamin—according to a note, following Arnheim's prosaic filmmaking-recipes—took as sturdy, down-to-Earth produced mechanical illusions and imbued them with their own inner profundity:

> With the close-up, space expands; with slow motion, movement is extended. And just as enlargement not merely clarifies what we see indistinctly 'in any case,' but brings to light entirely new structures of matter, slow motion not only reveals familiar aspects of movements, but discloses quite unknown aspects within them—aspects 'which do not appear as the retarding of natural movements but have a curious gliding, floating character of their own.' Clearly, it is another nature which speaks to the camera as compared to the eye. 'Other' above all in the sense that a space informed by human consciousness gives way to a space informed by the unconscious. ... This is where the camera comes into play, with all its resources for swooping and rising, disrupting and isolating, stretching or compressing a sequence, enlarging or reducing an object. It is through the camera that we first discover the instinctual unconscious through psychoanalysis.[95]

Could the idea not occur to one—considering the evidence collected here—that Benjamin's lecture on the Optical Unconscious also referred to a larger context, the previous history of a more extensive camera concept, which had escaped him, in which astronomy and the performance of the eye were still directly connected, and which, with the Microscope for Time, had been given a handy name?

Angel of History

> There is a picture by Klee called Angelus Novus. It shows an angel who seems about to move away from something he stares at. His eyes are wide, his mouth is open, his wings are spread. This is how the angel of history must look. His face is turned toward the past. Where a chain of events appears before us, he sees one single catastrophe, which keeps piling wreckage upon wreckage and hurls it at his feet. The angel would like to stay, awaken the dead, and make whole what has been smashed. But a storm is blowing from Paradise and has got caught in his wings; it is so strong that the angel can no longer close them. This storm drives him irresistibly into the future, to which his back is turned, while the pile of debris before him grows toward the sky. What we call progress is this storm.[96]

The reverse run of history strived for in vain by Benjamin's angel—Otto Karl Werckmeister wrote about this left-wing icon in 1997[97]—

recalls a text from 1919 by Blaise Cendrars: *The End of the World, Filmed by the Angel Notre Dame*. The fictive screenplay reaches its climax when the said angel blows a trumpet from the cathedral's tower in Paris, at the end of the world, so that Paris and every other city in the world collapses, burying mankind under rubble and ruins. But then the film runs in reverse, the dead stand, the cities rise again. It returns in fast-forward to the beginning of the world. In the last section of the screenplay entitled *À rebours* (Against the Grain), the film replays world history again in forward motion; until the scene with which it began: Our God the Father, a hectic cigar-smoking entrepreneur, sitting back in his office again. "ECT" comes at the end, and: "*C'est la banqueroute*."[98]

Just as Cendrars differentiated between the angel with a camera and man as projector, between filming and watching the film, according to Werckmeister, Benjamin in his thesis also differentiated between the angel of history and us. Yet Benjamin did not assume Cendrars' playful, anarchistic back and forth between world catastrophes before and after the present. Nor, may we add, could Benjamin have adopted from Cendrars the very remarkable physical turn against the direction of movement. Cendrars only described the mechanical switch of film direction in an interim scene from the projector room, but failed to mention the angel of Notre-Dame assuming any particular posture. What then moved Benjamin to make his unique interpretations of Klee's painting?

In Flammarion's Quærens/Lumen dialogs, the problem was directly addressed of which bodily orientation a space observer should assume when moving along with "light-images": "In order to observe events [on earth] while hurtling through space, do you fly backwards or are the spirits endowed with the capacity of a rear-view?" Answer: "What a question! If I tried to explain with which intimate senses the spirits can see, I would plunge you into a debate on a problem unsolvable for you. For your personal satisfaction just assume that I turn from time to time to inspect Earth; that idea should be much easier to accept."[99] Had Benjamin thus at one point read Flammarion's *Rècits*, and borrowed not only the angel's physical orientation, which was so central to the scene, but also very possibly the entire concept of observing history? He could have found an appropriate, yet more clearly formulated answer even in the many editions of Pohle's *Sternenwelten*, and this time, as we have seen, with an explicit footnote reference to Eberty: "The panorama for the feigned [wonder] eye would present itself most remarkably, if we were to have the same wonder-eye move away from our Earth towards space at an incalculable speed, yet always be turned toward it to record light impressions."[100]

Textual Evidence

Browsing step by step through Benjamin's pictorial descriptions of his Angel of History, it becomes obvious how clearly—through the historically conditioned discolorations and distortions—references to nineteenth-century texts take shape: The angel stares at something, from which he is moving away in backwards motion; he is facing the past. That is the variable position of the space witness accompanying "light-images," as first described by Eberty. The reference to Eberty's text becomes even more distinct, if one compares once again the perceptual form of coexistence he himself emphasized: "Thus we have here the extension of Time, which corresponds with that of Space, brought so near to our sensible perception, that time and space cannot be considered as at all different from one another. For, those things which are consecutive one to the other in point of time, lie next to one another in space. ..."[101] For Benjamin's angel the painting of which Eberty spoke, became a single catastrophe seen at one comprehensive glance, a heap of rubble that reached to the heavens. The observer's voluntary movement, which, according to Eberty, led to the painter's view, of time, became with Benjamin the whirl-wind of progress turned loose, which, now autonomous, would sweep up the powerless viewer. The futile wish of the angel to linger, is nothing other than the reverse power of the Ebertian eyewitness, "to halt at pleasure in his path," in order to study a moment of history at leisure.[102]

The angel would like to linger, to awaken the dead and amend the beaten, according to Benjamin's description. Here one might indeed recall Cendrar's backwards running film or Flammarion's Waterloo *d'outretombe*. But such an emotionalized borrowing was, in my opinion, less fundamental than the formal take-over of space-witnessing, which transformed the chain of events into a horror painting. With the powerlessness of the angel to halt or intervene in history, a further aspect, also found in the Ebertian observer status, was carried to the most extreme ramifications: The imaginary eye-witness—let it be even the Almighty Himself—can still only observe, inspect images, but can not retrospectively change anything. With this restrictive view, we are, so to speak, at the visual keynote of Benjamin's historical thesis, which further underscores the perspective depth of the motif's peregrinations pursued here.

If one considers the description in context with the rest of the series of Benjamin's historical theses, then a collection of epitomes on the complex of motifs discussed here almost merges together as one continuous prose; with Babbage's air-archive as prelude, which Benjamin could have found in Gregor Itelson's preface to the Eberty edition of 1923–25: The past carries with it a secret index, which points it toward redemption. Does not a breath of air touch us, which once was exhaled by beings in the past? Does not an echo of the silenced oscillate in voices to which we lend our ear? (II)—Nothing, which has ever occurred, is lost to history. But only a redeemed mankind's past was cited in each of its moments. (III)—Characteristics inflamed in the class struggle, such as courage, humor, cunning and steadfastness, retroact in the distance of time. The past turns its secret helio-tropical powers toward the sun that is rising in the sky of history. (IV)—The true image of history flits past. The past can only be captured as an image, flashing the never-to-be-seen-again at the moment of its discernibility. (V)—To historically articulate the past means taking possession of a memory as it flashes at the moment of danger. (VI)—It is effective to take possession of the true image of history that is but a brief flash. History is an object of constructions whose location is not in homogenous and empty time, but in the present. Thus ancient Rome for Robespierre was a past loaded with the here and now, which he sprung from the continuum of history. That is the tiger's leap into the past. Marx termed the same leap under the open sky of history as the dialectic of the revolution. (XIV)—To spring open the consciousness, the continuum of history, is idiosyncratic to the revolutionary classes at the moment of their campaign. Revolutionary reversals of the calendar function thereby like an historical accelerator of time. (XV)—The historical materialist could not dispense with the concept of a present that is not a passage, but that has emerged from time and was brought to a standstill. (XVI)—Thinking is not only about the movement of thoughts, but also their suspension. It is necessary to recognize the signs of a messianic suspension of events. (XVII)—The present, which, as messianic model, summarizes the history of the entire of mankind as one massive abbreviation, coincides precisely with that figure which establishes the history of mankind in the universe. (XVIII).

Glimpses and flashes on the one hand, suspensions and enforced openings on the other; these are literal opposite pairs, which Benjamin offered as the human methods of experience and dealings with history or images of history; they suggest both a spatial positioning of the time-witness and they refer directly to Eberty's and Flammarion's extensive flash observations. According to the figures of speech, the witness finds himself next to or opposite the flood of images from history, which flash by him and which he must either suspend or spring open in order to understand. That which the angel of history can view as an entirety, but can no longer change, can be entered as knowledge by mankind into the revolutionary moments of messianic suspension. This illuminating capability of insight, in which Benjamin formulated both personal tragedy and comforting, leads us once again finally to the question of how the relationship of eyewitnesses and time-flows in longitudinal section, and in the ramifications of the imaginary space-perspectives unfolded.

The Theory of Relativity

In 1949 Albert Einstein reported in his *Autobiographische Notizen* (Autobiographical notes) a paradox he encountered when he was sixteen, thus in 1894:

> If I pursue a beam of light with the velocity c (velocity of light in a vacuum), I should observe such a beam of light as an electromagnetic field at rest though spatially oscillating. There seems to be no such thing, however, neither on the basis of experience nor according to Maxwell's equations. From the very beginning it appeared to me intuitively clear that, judged from the standpoint of such an observer, everything would have to happen according to the same laws as for an observer who, relative to the earth, was at rest. For how should the first observer know, or be able to determine, that he is in a state of fast uniform motion? One sees that in this paradox the germ of the special relativity theory is already contained.[103]

Einstein's memoirs of his youth not only captivate us with their lively, incarnate metaphors, that one personally hurries after a ray of light in order to observe it "at leisure"; this imaginary excursion, with the feigned parallel movement of the eye-witness toward the ray of light, also recalls the flow of precursory texts here. The conclusion: Einstein was not only demonstrably able to adopt from Bernstein's *Volksbücher* the Ebertian idea of a witness who accompanies 'light-images'; he also took it on as a concept and then made it a springboard for his revolutionary physics contemplations.[104]

Here, a further thought experiment (*Gedankenexperiment*) of Einstein's might rudimentarily clarify the situation at the outset of his (special) Theory of Relativity of 1905—and the visual proximity to the earlier fictive perspectives in space: Let us assume a train traveling at a constant speed past an observer standing beside the tracks. A second observer is located exactly in the middle of the traveling train. At the precise moment at which the observer seated in the train passes the observer standing outside, both observe a simultaneous arrival of two signals of light, which come from the beginning and the end of the train. Which conclusions do the two observers now draw, regarding the points in time at which the light signals at the beginning and end of the train were transmitted? The observer in the train takes it into account that he is standing in the middle of the train and that the speed of light is independent of the state of movement. He concludes from this, that both light signals arriving simultaneously also emanated simultaneously from both endpoints of the train. Even the observer standing beside the tracks knows that the speed of light is constant and that in order to cover a finite path, a light signal needs a finite amount of time. From the simultaneity of the arrival of both light signals, he concludes however, that the signal at the end of the train was given earlier, as the train's end at the point of flash must have been further away from him than the train's front part. The signal transmitted from the head of the train, and which is distributed at a constant speed as well, can only arrive simultaneously with that transmitted from the end of the train if it is created at a later point in time. The observer at rest concludes from this that both events must have taken place at different points in time. Einstein repeatedly played with the train scenario paraphrased here in his first own depiction of the theory of relativity for the general public in 1917. The embankment of the railway tracks with the passing train formed the continuously used scene to depict classical deficits and specific relative effects.[105]

His essay, which substantiates the Theory of Relativity, focused initially "only" on moving bars, yet nevertheless on the elapsed time of light.[106] In principle, Einstein's minimal models thus remarkably corresponded to the great interstellar imagined stage set, as it was launched by Eberty and expanded on by Flammarion and others. One only has to replace the railway train with the fictive flow of "light-image"—the illuminated windows of the wagons now reduced to two light signals—in order to envision the situation depicting Einstein's developed theories. Even Einstein's later narrative style in his depictions "for the general public" of the Theory of Relativity of 1917, in which incidentally lightning-flashes stood in for light signals, recalls echoes of Ebertian eye-witnesses. The repeated travels of or views from the train recall the double-tracked quality of the observer's attitude, which Benjamin later characterized as a flash-like flitting-by or suspended burst. Because the bursting open of history can also be visualized in the leap onto a train.

Eyewitnesses

Not only proto-relativistic problems were addressed from the feigned position of the observer; there were also anthropological questions: stillness and movement, a forward, rear or side view; these are apparently the basic possible poses of the human direction of observation. For Eberty's pure imagination and that of his successors in the nineteenth century, the fictive adherence to a certain sectional field of vision with front, back and sideways movements, even in the cosmic dimension, was not a very pressing problem. Under sensorial–physiological aspects or considered from the camera viewpoint, one would be confronted with enormous zoom effects,[107] however: the perception of images rushes into detail or dwindles with the square of distance to invisibility. Even the side-view, and the hypothetical pile of images it inspired, could not actually remain unfazed by the fast-as-light interstellar observer movements. Nevertheless the optical unconscious avowed by Benjamin would remain ready in all these different attitudes of viewing to spawn distinctive emotional reactions.

The sideways view itself—similar to the backwards glance—must have been afflicted with a curious self-limitation regarding destination. The actual vanishing point of motion is, when adopting the position of the passengers who only have the possibility of a sideways view, at the extreme edge of, or beyond, the human field of vision. The rear view, if it excludes all other positions, was completely drenched with a resigned immersion in the past. One needs only to imagine the view from the last car of a traveling train in Benjamin's time, in order to empathize intensely with his angel syndrome, the journey backwards into the unknown. The conclusion: Space fantasies of eyewitnesses accompanying "light-images" were still modeled after human experience of earthly, ground-level travel. The straight-ahead forward-looking inspection movement never once pretended to be fixed primarily on an assumed destination (for example a star). It was to remain concentrated primarily on the archive of the images emitted from Earth. The all-encompassing, quasi-simultaneous overview, reserved initially for God as an other-worldly authority, was only waiting to be adopted by the new demigods of technically produced images—the film directors. Also for this reason, Eberty's imagination, turned loose in the world as a visual thought experiment (*Gedankenexperiment*), was able to trigger such a widely ramified impact.

Concluding Observations

The chain of proliferation of the Ebertian text summarized here is only the tip of an iceberg, whose gauge has still to be completely assessed. Other sources and quotes, which have not been mentioned here, point to significant material or even detailed original passages circulated in the Anglo-American and Baltic realms, in Switzerland and France, before Flammarion adopted the ideas as his own. Contrary to the well-entrenched literary opinion, it must be emphasized that Flammarion was merely an intelligent and imaginative, yet also an unacknowledging reproducer of the Eberty's space perspectives—on the way toward a landslide-like distribution. Ultimately they returned to Berlin in the twenties, then the center of theoretical film debates, spawning a new edition introduced by Einstein, which may have found its way into the hands of Walter Benjamin.

Much supports the idea that Benjamin—regardless of which source—adopted at least the core of the Ebertian perspectives that had been circulating awhile: Not only the concept of the angel of history sailing backward into time, while the (image) wreckage of history is flung at his feet and towers into the heavens, was borrowed from the principle of Eberty's "light-images" (*Lichtbilder*) accompanying witness and transformed into the tragic. Benjamin's entire historical concept is permeated by pictorial aspects, which were derived from Eberty's fiction. With the auratic distance (*auratische Ferne*), a further significant aspect of modernism can be developed from the Ebertian prospects. In his *Cosmogonic Eros*, Ludwig Klages underlined the intangibility of images with astronomy's evidence that fixed-stars shine in absence. Benjamin transformed this intangibility into an endangered cult-image aura (*Kultbild-Aura*). Thus, these two very contrasting personalities were united in a cosmic dimension, before a backdrop of shifting view points, triggered by modern astronomy and by Eberty's initial, essential focus on space-time speculations. They guided Benjamin into the desperate confidence of his history theses; for Klages, they were the basic tenor of his concept of the intellect as adversary of the soul. Yet with the tragic/anti-rationalist pitch and tendencies of the interim war era, the consequences of the Ebertian space perspectives can still not be completely gathered or totally paced. One could go even further and find to some extent, as quoted above, new reflexes of this mid nineteenth-century revolutionary insight in Roland Barthes' *Camera Lucida* in 1980.

The outlook into the world of nineteenth-century orthodox and spiritualist astronomy proves how strongly the effect of Eberty's text was in the entire realm of scientific speculations. The fact that even the most well known representative of modern physics, Albert Einstein, verifiably as early as in his youth—via Aaron Bernstein's *Naturwissenschaftliche Volksbücher*—encountered the Berlin lawyer's world of ideas, completes the astounding spectrum of the little booklet's impact. To repeat shortly Richard Proctor's assessment of the English version: "one of the most strikingly suggestive books, and small though it is, one of the most remarkable of the present century."

With the inspired effect it had on Karl Ernst von Baer's concept —and more indirectly, but no less profoundly, on Jakob von Uexküll's— the unusual role of a pace-maker in the bio-sciences can also be ascribed to Eberty's text. The dilations and expansions of image perception conducted by hypothetical inspectors of light-beams, had caused Eberty to consider objective variations in life's pace as well. With this, a basic philosophy of bio-semiotics was born, which was then systematically developed by Karl Ernst von Baer into a theory of "temporal signs" (*Theorie der Momentzeichen*). Comparisons of time scales provided the means to draw from different inner worlds of various beings genuine data for objective research; they were left not merely to psychological speculation, according to Uexküll's multifold experimentally confirmed axiomatic position. This conceptual seed of theoretical biology he had received, conveyed through Karl Ernst von Baer, from the Berlin lawyer and lay-astronomer, Felix Eberty.

The interminglings of technology, physiology and fantasy in the nineteenth century lead to a history of the visual perspective as Image Science, in which progress and speculations in astronomy played a fundamental role. They delivered the mutual background for such different achievements as Einstein's special Theory of Relativity, but also for Benjamin and Klages' unique melancholic rear and distant views. In addition, they also helped prepare the path for the concept of a ceaselessly running cosmic cinema, of space filled with archives of images between the stars. In the fantastic amalgamation of technical/scientific progress (the first exact star distance measurements in 1837–42) and speculative release of observers in space (simultaneously with Edgar Allan Poe's *Eureka* text, which right now inflames cosmologists), a highly significant, recent counterpole to age-old astrological views, that have been studied in the Warburg circle, was established.

The conclusion: For the understanding of cultural and media history, Eberty's text is a first rate source, which has long been completely overlooked due to its anonymity and academically non-specialist remoteness. It is an unusual stroke of luck that so many different constructions of ideas—the Theory of Relativity, the concept of a celestial cinema that left behind such a manifold of traces in the twenties, the cosmogonic irrationalism of Klages and Benjamin's transversally-sighted philosophy of history—can be linked back to one single triggering text. Eberty, by sidestepping the canons of discipline, was surely not the only one to supply ideas to better-known intellectual master achievements of the twentieth century; yet in 1846–47, with a sharp legally trained eye, he distilled the ethical essence, of a truly universal reciprocity of possible viewpoints, from the physical–astronomical insights of his time, and thus for the first time, designed the basic structure of light-image-based filmic fictionality.

One must attempt, mournfully, to grasp the irrevocably lost: Felix Eberty was a typical representative of a wide-awake Jewish–German intellect, which had its mind's fertile grounds in Hardenberg's better Prussia. It was not incidental that Eberty chose the proceeds of his writings to go to the needy in the Silesian Riesengebirge, and in a further anonymous text, focused on the state ideal of a sovereignty above confessional constrictions. All this did not only make him a historically remarkable figure, but considering today's multi-cultural debates, it also makes him surprisingly contemporary.

In 2002, my attention was brought to Felix Eberty the lay-astronomer either through coincidence or remote curiosity, within the context of a series of lectures given together with Hermann Schweppenhäuser at the university of Lüneburg. I then searched out every recorded edition, sometimes only available in single copies, and when possible purchased them in the original. The lucky find commented on here (and offered separately as book containing the original German/English texts of 1846–47) provides many further starting points for a comprehensive history of science and culture: From it, natural science's or humanities' related associations to the cosmological complex of sentiment over the last 150 years—astronomical depth of time, the big bang and expanding universe etc.—can be reconsidered. Such an enterprise offers in addition painfully obvious insights into the changing use of our language in the light of now omnipresent technical "light-images." Proximity and distance, as well as their according tenses: the past, present and future, which started out in human language as image-replacements (best to be re-read in Karl Bühler's books) have played a fundamental role. They arguably made the development of grammar possible in the first place. What do the Ebertian imaginings stand now for, and what is going to happen when more and more technically produced images replace them? But that is a different field and another subject.

Notes

[1] The text presented here is an adapted version of: "'Ein Mikroskop für die Zeit', Was Benjamin und Klages, Einstein und das Kino den fernen Sternen verdanken"; in: Ralf Schnell (ed.), *Wahrnehmung Kognition Ästhetik. Neurobiologie und Medienwissenschaften* (Bielefeld, 2005), 157–212; a first draft-translation was done by Laura Bruce/Berlin in cooperation with the author. For final refinements and insertion of English-language quotations I wish to thank the CEU Press editors. – A more comprehensive account accompanies the bilingual German/English full-text edition of Eberty's booklet: Karl Clausberg, *Zwischen den Sternen: Lichtbildarchive. Was Einstein und Uexküll, Benjamin und das Kino der Astronomie des 19. Jahrhunderts verdanken* (Berlin: Akademie-Verlag, 2006). A comprehensive bibliography of editions and related sources discussed in this essay appears in *Zwischen den Sternen*.

[2] Felix Eberty, *Die Gestirne und die Weltgeschichte. Gedanken über Zeit, Raum und Ewigkeit.* With an introduction by Albert Einstein, published by Gregorius Itelson, (Berlin: Gregor Rogoff Verlag, 1923).

[3] Felix Eberty, *Die Gestirne und die Weltgeschichte. Gedanken über Zeit, Raum und Ewigkeit.* With an introduction by Albert Einstein (Berlin: I.M. Spaeth-Verlag, 1925).

[4] Felix Eberty, *Jugenderinnerungen eines alten Berliners* (Berlin: Verlag für Kulturpolitik, 1925). After being amended by hand-written notes and republished by J. von Bulow [Enkel Ebertys]; Preface by Georg Hermann. The first edition was published in 1878 in Breslau.

[5] "Dies Büchlein, von einem originellen, geistreichen Menschen geschrieben, entbehrt nicht des aktuellen Interesses, denn es zeigt auf der einen Seite kritischen Geist gegenüber dem überkommenen Zeitbegriff; auf der anderen Seite zeigt es, vor welchen eigentümlichen Folgerungen uns die Relativitätstheorie rettet, der doch so vielfach gerade der bizarre Charakter ihrer Folgerungen zum Vorwurf gemacht wird."

[6] Charlotte Schönbeck, "Albert Einstein und Philipp Lenard. Antipoden im Spannungsfeld von Physik und Zeitgeschichte" in *Schriften der Mathematisch-Naturwissenschaftlichen Klasse der Heidelberger Akademie der Wissenschaften*, vol. 8 (Berlin: Springer, 2000).

[7] Itelson had already translated Einstein's book *Über die spezielle und die allgemeine Relativitätstheorie. Gemeinverständlich* (1917) into Russian. "I am happy my book is also being published in Russian, and all the more because the most venerable Mr. Itelson has vouched for such an excellent translation." Einstein, in the foreword. On Itelson see Otto Buek, "Gregorius Itelson U" *Kant-Studien* (1926): 428–430. Gideon Freudenthal has recently drawn my attention to a correspondence between Einstein and Paul Feldkeller, who eventually published a sharp review of the Eberty reprint in the *Kant-Studien* 34 (1929): 198–199. In 1927 he

had already contacted Einstein in order to get an obituary on Newton (by then 200 years dead). Einstein was, according to his answer in the Einstein-Archive (No. 25–054/56), reluctant and apparently avoided further discussion concerning Eberty by remarking that he had merely done Itelson a favor: "Ich schrieb das das kleine Vorwort dazu nur, um dem von mir geschätzten alten Itelson einen Herzenswunsch zu erfüllen."

8 Albert Einstein, *Autobiographical Notes* [1949], ed. and trans. Paul Arthur Schilpp (La Salle, Illinois, 1991), 13–15.

9 Max Talmey, *The Relativity Theory Simplified / And the Formative Period of Its Inventor* (New Bedford, Mass., 1932), 159ff., spec. 162.

10 *Die Urwähler-Zeitung. Organ für Jedermann aus dem Volke* began being published in Spring 1849. It was banned in 1853 after being confiscated several times and then continued as the *Volks-Zeitung*; in 1904 it was renamed the Berliner *Volkszeitung* and in 1944 consolidated with the *Berliner Morgenpost.*

11 Aaron Bernstein, *Naturwissenschaftliche Volksbücher*. Affordable complete edition. The fourth improved and augmented edition. Fourth print. (Berlin: Dümmler Verlags-Buchhandlung, no date) [1873–74], vol. 5, 21st part, 76–111, spec. 99.

12 Bernstein, *Naturwissenschaftliche Volksbücher* [1873–74], vol. 5, 21st Part, 100–101.

13 Aaron Bernstein, *Aus dem Reiche der Naturwissenschaft. Für Jedermann aus dem Volke* (Berlin: Verlag Franz Duncker, 1855), vol. 7, chapter 2, "Von der Geschwindigkeit des Lichtes," 78–108, spec. 102.

14 Bernstein, *Naturwissenschaftliche Volksbücher* [1873–74], vol. 2, 8th section, 124–159, spec. 152ff.

15 Bernstein, *Aus dem Reiche*, vol. 7, chapter 2, 1855, 104; Volksbücher [1873–74], vol. 2, 8th section, 155.

16 *Archiv Bibliographica Judaica* (Red. Renate Heuer), *Lexikon deutsch-jüdischer Autoren*, vol. 6, Munich 1998, 40–43. ADB volume 55, 473–476; NDB vol. 4, 546.

17 Dolf Michaelis, "The Ephraim Family," *Leo Baeck Inst. Year Book* XXI (1976) 201–228. 2nd section The Ephraim Family (II), *Leo Baeck Inst. Year Book* XXIV (1979), 225–246.

18 Georg. Fridericus Felix Eberty (Berolinensis): DE VERA UNIONIS PROLIUM NOTIONE. Bonnae MDCCCXXXIV.—Professorial dissertation: F. Eberty, DE RECEPTIONE LEGUM JULIARUM DE VI PUBLICA ET PRIVATA IN GERMANIA [...]. VRATISLAVIAE (without date).

19 Eberty, *Jugenderinnerungen*: see note # (4).

20 Felix Eberty "Beweis der Lehrsätze [...]," in *Journal für reine und angewandte Mathematik. In Zwanglose Heften*. Ed. A.L. Crelle. Volume 5 (Berlin Reimer, 1830), 107–112.

21 Sir John F.W. Herschel, *A Treatise on Astronomy* (London, 1833), spec. Chapter XII, "Distance of the Stars," 376–377ff.

[22] Johann Heinrich von Mädler, *Der Wunderbau des Weltalls*, 1841; 1876, 473.

[23] Alexander von Humboldt, *Kosmos. Entwurf einer physischen Weltbeschreibung*, vol. I (Stuttgart and Tübingen, 1845), 159–161.

[24] F.Y. [Felix Eberty], *Die Gestirne und die Weltgeschichte. Gedanken über Raum, Zeit und Ewigkeit* (Breslau: August Schulz Verlag, 1846), 19–20. Citation after: *The Stars and the Earth; or, Thoughts upon Space, Time, and Eternity* (London, 1846), 35.

[25] F.Y. [Felix Eberty], *Die Gestirne* 1846, 20–22. *The Stars and the Earth* 1846, 36–39.

[26] Charles Babbage, *The Ninth Bridgewater Treatise A Fragment.* First published 1837; second edition (London, 1838); second edition (Philadelphia,1841). Chapter IX, "On the Permanent Impression of our Words and Actions on the Globe we inhabit." Quoted from the second edition.

[27] F.Y. [Felix Eberty], *Die Gestirne* 1846, 14–15. *The Stars and the Earth* 1846, 26–27.

[28] F.Y. [Felix Eberty], *Die Gestirne* 1846, 15–16. *The Stars and the Earth* 1846, 28.

[29] Franz Wickhoff, *Römische Kunst* [New edition of the commentary on the facsimile edition of 1895] (Berlin, 1912), 9–10.

[30] F.Y. [Felix Eberty], *Die Gestirne* 1846, 23–25. *The Stars and the Earth* 1846, 40–42.

[31] F.Y. [Felix Eberty], *Die Gestirne* 1846, 23–25. *The Stars and the Earth* 1846, 42–43.

[32] Hans Christian Oersted, "Der allgemeinen Naturlehre Geist und Wesen," *Die Naturwissenschaft und die Geistesbildung*, 3rd edition (Leipzig, 1851), 18.

[33] F.Y. [Felix Eberty], *Die Gestirne und die Weltgeschichte. Gedanken über Raum, Zeit und Ewigkeit*, 2nd booklet, by Georg Philipp Aderholz (Breslau, 1847), 2–3. Cited after: *The Stars and the Earth; or, Thoughts upon Space, Time, and Eternity*, Part II (London 1847), 10–12.

[34] Joseph Görres, *Aphorismen über die Kunst* (1802), *Gesammelte Schriften*, vol. 2, no. 1. (Cologne, 1932), 148.

[35] Carl Gustav Carus, *Zwölf Briefe über das Erdleben.* After the first edition 1841 published by Christoph Bernoulli and Hans Kern (Celle, 1926), 19–20.

[36] F.Y. [Felix Eberty], *Die Gestirne* II, 1847, 15–16. *The Stars and the Earth* II, 1847, 37–38.

[37] Karl Ernst von Baer, "Welche Auffassung der lebenden Natur ist die richtige? Und wie ist diese Auffassung auf die Entomologie anzuwenden?" Spoken at the opening of the Russian entomological society in October 1860, *Reden gehalten in wissenschaftlichen Versammlungen.* Second edition (Braunschweig, 1886), 258.

[38] von Baer, "Welche Auffassung" [1860], 260–262.

[39] von Baer, "Welche Auffassung" [1860], 265–266.

[40] von Baer, "Welche Auffassung" [1860], 267–268.

[41] F.Y. [Felix Eberty], *Die Gestirne* II 1847, 16–17. *The Stars and the Earth* II, 1847, 39.

[42] Jakob von Uexküll, *Umwelt und Innenwelt* (Berlin: Springer, 1909), 6.

[43] Christian Doppler, "Ueber das farbige Licht der Doppelsterne und einiger anderer Gestirne des Himmels. Versuch einer das Bradley'sche Aberrationstheorem als integrirenden Theil in sich schliessenden allgemeineren Theorie" in: *Abhandlungen der Böhmischen Gesellschaft der Wissenschaften* 5th sequel, vol. 2, 1842. Reprinted in *Ostwalds Klassiker der exakten Wissenschaften* vol. 161. *Schriften aus der Frühzeit der Astrophysik von Christian Doppler* (Frankfurt/Main: Harri Deutsch, 2002), 5–7.

[44] Christian Doppler, "Ueber das farbige Licht der Doppelsterne, mit vorzüglicher Bezugnahme auf eine von Herrn Dr. Mädler im Stuttgarter Morgenblatte No. 51 erschienene Recension einer unter obigem Titel verfassten Druckschrift" in *Oesterr. Blätter für Literatur und Kunst* 1844, No. 15. Reprinted in *Ostwalds Klassiker,* vol. 161, 25–34.

[45] The printer's note reads: "The editor retains the right to deliver an English translation and has disposed this requested copy in London, in accord with the contract of 13 May 1846."

[46] *The Stars and the Earth; Thoughts upon Space, Time and Eternity* (Boston: W.M. Crosby and H.P. Nichols, 1850).

[47] Gregorius Itelson must have adopted his Babbage reference from one of the American editions in which Hill's comments can be found. Reception of the anonymous Eberty-booklet was widespread and lasting. In addition to evidence given in 2006 *Zwischen den Sternen*—E.A. Poe's case, paleontologist Edward Hitchcock, architect Claude Bragdon—Emily Dickinson should be mentioned. There was a volume in her family-library, which may have furnished all the impetus her poetry needed. See Charles R. Anderson, "The Trap of Time in Emily Dickinson's Poetry" *ELH* vol. 26, no. 3 (Sept. 1959): 402–424.

[48] Felix Eberty, *Die Gestirne und die Weltgeschichte / Gedanken über Zeit, Raum und Ewigkeit* (Breslau: J.U. Kern's Verlag, 1874).

[49] Camille Flammarion, *Lumen, récit d'outre-terre* in *Revue du xix° siècle,* no. 11, 1 (février 1867), 163–184ff. The continuation appeared in: *L'Artiste, revue dirigée par Arsène Houssaye.* Statements and sources cited from: Danielle Chaperon, *Camille Flammarion. Entre astronomie et littérature* (Paris: Editions Imago 1998).

[50] Camille Flammarion, *Lumen.* Illustrations by Lucien Rudaux (Paris, no date. [1887]).

[51] Camille Flammarion, *Récits de l'infini. Lumen. Histoire d'une comete. Dans l'infini* [1872] (Paris, 1873), 143. English translation taken from Brian Stableford, *Lumen. Camille Flammarion* (Middletown, Connecticut, 2002), 69.

[52] F.Y. [Felix Eberty], *Die Gestirne* 1846, 21. *The Stars and the Earth* 1846,

37. The term *Lichtbilder* was not translated as light-images, but circumscribed.

[53] Flammarion, *Récits* 1873, 108. Stableford, *Lumen* (2002), 52.

[54] Flammarion, *Récits* 1873, 137. Stableford, *Lumen* (2002), 65.

[55] F.Y. [Felix Eberty], *Die Gestirne* II 1847, 3. *The Stars and the Earth* II, 1847, 12–13.

[56] Flammarion, *Récits* 1873, 226–227. Stableford, *Lumen* (2002), 114–115.

[57] Flammarion: *Récits* 1873, 228. Stableford, *Lumen* (2002), 116.

[58] Hermann von Helmholtz, "Ueber die Methoden, kleinste Zeittheile zu messen, und ihre Anwendung für physiologische Zwecke" in *Königsberger Naturwissenschaftliche Unterhaltungen*, vol. 2, 2nd booklet (Königsberg: Bornträger, 1851), 169–189, 175f. Special credit goes to Wolfgang Pircher / Vienna for his manuscript on *Simultaneity*.

[59] Helmholtz, "Ueber die Methoden" (1851), 177.

[60] Camille Flammarion, *Mémoires biographiques et philosophiques d'un astronome* (Paris, 1911), chapter 18, 331–334.

[61] In the notes by the publisher Costenoble dated August 20, 1859, obviously later enclosed in a part of the edition together with a re-edited cover page: *Die Sterne und die Erde, Gedanken über Raum, Zeit und Ewigkeit.* After the 6th edition of the English translation of the work *Die Gestirne und die Weltgeschichte* von Dr. Felix Eberty (F.Y.) translated into German by W. von Voigts-Rhetz (Leipzig: Hermann Costenoble. 1860).

[62] In the preface to *The Stars and the Earth, or Thoughts upon Space, Time, and Eternity*. Revised and enlarged, with Notes by Richard A. Proctor. (London: Bailliere, Tindall, and Cox, 1880).

[63] Proctor in the preface to *The Stars and the Earth* (1880).

[64] Richard A. Proctor, *Other Worlds than Ours: The Plurality of Worlds studied under the Light of recent scientific Researches* (London: Longmans, Green, and Co. 1870), 309–310.

[65] Flammarion, *Récits* (1873), 103. Translation Stableford (2002), 49–50.

[66] Dr. Mises [Gustav Theodor Fechner], *Verkehrte Welt* (1824) and *Der Raum hat vier Dimensionen* (1846) in *Kleine Schriften* (Leipzig, 1875).

[67] Carl du Prel, *Der Kampf um's Dasein am Himmel. Versuch einer Philosophie der Astronomie* (Berlin: Denicke Publishing House, 1876), 355.

[68] Carl du Prel, *Entwicklungsgeschichte des Weltalls. Entwurf einer Philosophie der Astronomie* (Leipzig: Ernst Günther Publishing House, 1882), 376.

[69] Carl du Prel, *Die Planetenbewohner und die Nebularhypothese. Neue Studien zur Entwicklungsgeschichte des Weltalls* (Leipzig: Ernst Günther Publishing House, 1880), 169–170.

[70] du Prel, *Planetenbewohner* (1880), 170.

[71] Wilhelm Wundt, *Vorlesungen über die Menschen- und Thierseele*, 1st volume, (Leipzig, 1863), 179.

[72] Joseph Pohle, "Die Sternwelten und ihre Bewohner. Eine wissenschaftliche Studie über die Bewohnbarkeit und die Belebtheit der Himmelskörper

nach dem neuesten Standpunkte der Wissenschaften." First part in *Jahresbericht der Görres-Gesellschaft für das Jahr 1884* (Cologne, 1884). Second part (conclusion), ibid. (Cologne, 1885), 16.

73 Pohle, *Sternwelten* (1885), 15–16.

74 Pohle, *Sternwelten* (1885), 17.

75 The 3rd edition was published in 1902, the 6th in 1910; from the 2nd onward under the title: *Die Sternenwelten und ihre Bewohner. Eine Einführung in die moderne Astronomie.*

76 Pohle, *Sternwelten* (1885), 16.

77 M. Wilhelm Meyer, *Die Geschichte der Urwelt. Eine Wanderung durch vorsintflutliche Landschaften. Lectures at the Science Theatre of the Urania Society*. Third edition. (Berlin, 1890), 3–4.

78 Alexander von Humboldt, *Kosmos. Entwurf einer physischen Weltbeschreibung*, vol. I (Stuttgart and Tuebingen, 1845), 155. English translation: *Cosmos: A Sketch of a Physical Description of the Universe*, trans. E. C. Otté (London: Henry G. Bohn, 1849), 139–140.

79 Michail B. Jampolski, "Die Utopie vom kosmischen Schauspiel und der Kinematograph" in *Beiträge zur Film- und Fernsehwissenschaft. Schriftenreihe der Hochschule für Film und Fernsehen der DDR "Konrad Wolf"*. Issue 34, vol. 29 (1988): 177–191. Special thanks goes to Joachim Paech, who called my attention to Jampolski's text.

80 Max Milner, *La fantasmagorie. Essay sur l'optique fantastique* (Paris, 1982), 168ff.

81 Frank N Magill, ed. *Survey of Science Fiction Literature*, vol. 3 (Englewood Cliffs, NJ.: Salem Press, 1979), 1294–98.

82 Danielle Chaperon, *Camille Flammarion. Entre astronomie et littérature* (Paris: Editions Imago, 1998).

83 Roland Barthes, *La chambre claire. Note sur la photographie* (Paris, Seuil, 1980). English translation: Roland Barthes, *Camera Lucida: Reflections on Photography*, trans. Richard Howard (London: Flamingo, 1984), 80–81, 82.

84 Jampolski, "Utopie" (1988), 179.

85 Ewald Hering, *Beiträge zur Physiologie*, second booklet: *Von den identischen Netzhautstellen.* (Leipzig, 1862), 166.

86 Henri Bergson, *Matière et mémoire. Essay sur la relation du corps a l'esprit* (Paris, 1896).

87 Bergson, *Materie und Gedächtnis* (1919), 23–24.

88 Ludwig Klages, *Vom Kosmogonischen Eros* (Munich, 1922), 102. German original: Zeitliche Ferne erscheine in räumlicher Ferne; oder: das in der Ferne des Raumes Erscheinende sei Ferne der Zeit. Hier dürfe man ausnahmsweise, so Klages, "die Naturwissenschaft zu Hilfe rufen; pflegt uns doch die Astronomie zu belehren, schon vom Sirius, dem nächsten Fixstern, brauche das Licht, um bis zu uns zu gelangen, mehrere Jahre; daher wir denn, ihn erblickend, im Wahrnehmungsakt gegenwärtig hätten etwas um ebenso lange Zeit tatsächlich bereits Verflossenes!

Indessen, auch wenn wir nicht gewiß zu sein glaubten, daß schon die Wissenschaft von morgen diese heutzutage volkstümliche Meinung als irrig verwerfen wird, so wäre gleichwohl alles verwirrt, vertauschten wir die ausmeßbare Entfernung des Siriuskörpers vom Erdplaneten mit dem Ferncharakter des Siriusbildes! Nicht von der Meilenlänge oder auch Lichtjahrlänge körpertrennender Abstände reden wir, sondern von jener Fernheit (gewissermaßen der Ferne an sich), die in der Schauung des Bildes unmittelbar (und somit vorbegrifflich) miterschaut wird."

89 Klages, *Vom Kosmogonischen Eros* (1922), 111–112. German original: Alles räumlich Ferne, so Klages, "kann in die Nähe rücken, bedingungslos ausgenommen allein die Gestirne! Mag unser Auge mittelst geschliffener Gläser noch um Lichtjahrtausende tiefer in die Schlünde des Raumes dringen und in der Dunkelkammer von Sternmyriaden das Abbild fangen an Himmelsstellen, wo sich vordem nur lichtlose Leere spannte, mögen wir ... chemisch die Stoffe zerlegen der Sonnen und brennenden Dünste anderer Welten: wir haben nach wie vor gegenwärtig immer nur die Erscheinung der Sterne, niemals ihre (bloß erschlossene) Körperlichkeit! Darum denn: die Sterne die begehrt man nicht, und ihre Gegenwart, wenn uns ein Scherz erlaubt ist, um das über alles Vermuten Paradoxe daran zu beleuchten, glänzt durch Abwesenheit! – Könnte aber demgemäß kein überzeugenderes Sinnbild des Ehemals ersonnen werden als die Sternenpracht des nächtlichen Firmamentes, so begreifen wir nicht nur den Erhabenheitsschauer, der beim Anblick des funkelnden Gewölbes jeden noch welterschlossenen Betrachter durchhaucht, sondern wir verstehen auch, warum dem ursprünglichen Sinn die Sterne bald Seelen der Gewesenen, bald deren Aufenthaltsstätte waren, immer aber auf Erden vergangene im Unvergänglichen leuchtende Chöre der Vorwelt, an die aller irdische Wandel gebunden blieb."

90 Ludwig Klages, *Vom Wesen des Bewußtseins* (Leipzig, 1921), beginning chapter VI, "Die Seele als Bildseele," 28. German original: "Nicht Dinge, sondern Bilder sind beseelt: das ist der Schlüssel zur ganzen Lebenslehre."

91 Ludwig Klages, "Vom Traumbewusstsein" [1913–1919] in *Mensch und Erde. Gesammelte Abhandlungen* (Stuttgart, 1956), 147–195, esp. 159. German original: "Eine Hingegebenheit, die jeden Widerstand der Selbstbehauptung schmilzt, Fernblau über allen Gegenständen und ihr Eingebettetsein in einen Strom des Vergehens, in den wir mitversinken: aus diesen drei Fäden ist der Schleier gewoben, durch den der tagwache Geist die Welt 'wie im Traum' erblickt."

92 Klages, "Traumbewusstsein" [1913–1919], 159. German original: "Aus immernaher Körperlichkeit vor eine ziehende Ferne gerückt verlöre das Ich seinen Halt auch ohne die Selbsthingebung, die es von innen wirkend nun vollends unter die Bilder stellt, dermaßen, daß es sich selbst mit ihnen verwechselt."

[93] Klages, "Traumbewusstsein" [1913–1919], 159. German original: "daß aber die Wahrnehmung des räumlich Fernen eine eigentümliche 'Klangfarbe' habe."

[94] Michael Pauen, "Eros der Ferne. Walter Benjamin und Ludwig Klages" in *Global Benjamin.* Internationaler Walter-Benjamin-Kongress 1992, vol. 2, 693–716; with further literature.

[95] Walter Benjamin, "Das Kunstwerk im Zeitalter seiner technischen Reproduzierbarkeit" (3rd version); in *Gesammelte Schriften* I.2, publ. by Rolf Tiedemann and Hermann Schweppenhäuser (Frankfurt: Suhrkamp Verlag, 1991), 500. English translation: "The Work of Art in the Age of Its Reproducibility," in Walter Benjamin, *Selected Writings*, vol. 3: 1935–1938, ed. Michael Jennings (Cambridge, Mass: The Belknap Press of Harvard University Press, 2002), 117.

[96] Benjamin, "Über den Begriff der Geschichte" in *Gesammelte Schriften* I. 2, 697–698. English translation: "On the Concept of History," in Walter Benjamin, *Selected Writings*, vol. 4: 1938–1940, eds. Howard Eiland and Michael Jennings (Cambridge, Mass: The Belknap Press of the Harvard University Press, 2003), 392.

[97] Otto Karl Werckmeister, *Linke Ikonen. Benjamin, Eisenstein, Picasso nach dem Fall des Kommunismus* (Munich and Vienna, 1997), 45.

[98] Blaise Cendrars, "La Fin du monde filmée par l'Ange N.D. (1917–19)," *Oeuvres complètes* 2, Editions Denoël.

[99] Camille Flammarion, *Récits de l'infini* [1872] 1873, 110. "Pour remonter ainsi les événements en vous éloignant dans l'espace, est-ce que vous voliez en reculant, ou plutôt les esprits sont ils doués dé la faculté dé voir derrière eux? Quelle question! Si j'entreprenais de vous exposer par quel sens intime les esprits voient, je vous plongerais dans la discussion d un problème insoluble pour vous. Pour votre satisfaction personnelle, pensez que je me retournais de temps en temps pour examiner la Terre; cette idée sera plus facile à accepter." This dialogue is not to be found in later worked-over editions of *Lumen*, f.i. the one used by Brian Stableford in 2002.

[100] Joseph Pohle, *Sternwelten* (1885), 16. [italic stress added later]—Benjamin knew of Pohle: In the index of the texts he read Pohle's *Christlich-katholische Dogmatik* is listed as No. 886; the reference comes from *Kultur der Gegenwart*, 1909.

[101] F.Y. [Felix Eberty], *Die Gestirne* (1846), 15–16. *The Stars and the Earth* 1846, 28.

[102] F.Y. [Felix Eberty], *Die Gestirne (*1846), 24–25. *The Stars and the Earth* 1846, 42.

[103] Albert Einstein, *Autobiographical Notes* [1949], Paul Arthur Schilpp, trans. and ed. (La Salle, 1991), 49, 51.

[104] The most recent state of research with a focus on cultural-technical aspects in Peter Galison, *Einstein's Clocks, Poincaré's Maps: Empires of Time*

(New York, 2003). See also. Abraham Pais, *'Subtle is the Lord...' The Science and the Life of Albert Einstein* (Oxford, 1982).

105 Albert Einstein, *Über die spezielle und die allgemeine Relativitätstheorie: (allgemeinverständlich)*, 1st edition (Braunschweig, 1917). The paraphrased example of Einstein's is quoted from Károly Simonyi, *Kulturgeschichte der Physik*, 2nd edition (Frankfurt/Main, 1995), 410.

106 Albert Einstein, "Zur Eigendynamik bewegter Körper" in *Annalen der Physik* 17 (1905); reprinted in: H.A. Lorentz, A. Einstein, H. Minkowski, *Das Relativitätsprinzip. Eine Sammlung von Abhandlungen* [1923] (Darmstadt, 1974), 26.ff.

107 A very illustrative demonstration can be found in: Philip and Phylis Morrison, *Powers of Ten*, book and film (San Francisco: W. H. Freeman, 1982).

Contributors

Karen Bassi is Professor of Classics and Literature at the University of California at Santa Cruz. She is the author of *Acting Like Men, Gender, Drama and Nostalgia in Ancient Greece* (University of Michigan Press, 1998) and articles on Greek literature and history writing. Her areas of interest include the philosophy of history, visual and material culture, and literary and cultural theory. She is currently writing an article on Aeschylus' *Prometheus Bound* in connection with the US torture debate and is finishing a book on visual experience and the meaning of the past in Greek narrative.

Jonathan Beecher teaches European Intellectual History at the University of California, Santa Cruz. His publications include intellectual biographies of two major figures in the French socialist tradition: *Charles Fourier: The Visionary and his World* (University of California Press, 1987) and *Victor Considerant and the Rise and Fall of French Romantic Socialism* (University of California Press, 2001). He is currently working on a study of European intellectuals (including Lamartine, Sand, Tocqueville, Proudhon, Herzen, Marx, and Flaubert) and 1848.

Karl Clausberg was until 2003 Professor of Art and Visual Studies at the Universität Lüneburg. His major publications include: *Ausdruck, Ausstrahlung, Aura: Synästhesien der Beseelung im Medienzeitalter* (Bad Honnef: Hippocampus, 2007, with Elize Bisanz and Cornelius Weiller); *Zwischen den Sternen: Lichtbildarchive. Was Einstein und Uexküll, Benjamin und das Kino der Astronomie des 19. Jahrhunderts verdanken* (Berlin, Akademie Verlag, 2006); *Neuronale Kunstgeschichte: Selbstdarstellung als Gestaltungsprinzip,* (Springer, Vienna, 1999); Der Erfurter Codex Aureus in Pommersfelden (Ms 249/2869) Biblische Historie im politischen Gewand? (Ludwig Reichert Verlag, Wiesbaden, 1986); Die Wiener Genesis: Eine kunstwissenschaftliche Bilderbuchgeschichte, (Fischer-Kunststück-Taschenbuch, Frankfurt/Main, 1984), and in Japanese translation,

Tokyo 1999/2000; Kosmische Visionen: Mystische Weltbilder von Hildegard von Bingen bis heute, *DuMont-Kunsttaschenbuch* Nr. 98, Köln 1980; and ZEPPELIN: Die Geschichte eines unwahrscheinlichen Erfolges, (Schirmer & Mosel, Munich, 1979).

Wai Chee Dimock is William Lampson Professor of English and American Studies at Yale University. Her recent publications include *Through Other Continents: American Literature Across Deep Time* (Princeton University Press, 2006), and a collaborative volume, *Shades of the Planet: American Literature as World Literature* (Princeton University Press, 2007). A special issue of *PMLA*, "Remapping Genre," co-edited with Bruce Robbins, appeared in October 2007.

Britta Duelke is a post-doctoral research fellow at the Frobenius Institute, Goethe University, Frankfurt (Germany). She has conducted long-term anthropological fieldwork in Central and Northern Australia focusing on the imaginaries of tradition and modernity. Her work covers a variety of subjects ranging from an outline of an anthropology of blame to contingency in narratives, emotional expressions, and the "naked body" as a metaphor of social demarcation. Her recent publications include: Grenzziehung des Sozialen im indigenen Australien, in: Karl-Siegbert Rehberg (ed.): *Die Natur der Gesellschaft.* (Frankfurt am Main, New York: Campus, 2008); Über eine Thematisierung des Möglichen, *Zeitschrift für Ethnologie* 130 (1): 99-125 (2005); (2008); "Merkwuerdig bleibt nur ... die Verwandtschaft zwischen Ausdruck und Handlung": Lachen und Weinen in einer nordaustralischen Perspektive, in: August Nitschke, Justin Stagl und Dieter R. Bauer (eds): *Lachen und Weinen: Gesten, Sprachen und Tabus der verschiedenen Kulturen innerhalb einer historischen Anthropologie.* (Vienna: Böhlau, forthcoming).

Ruth HaCohen holds the Artur Rubinstein Chair of Musicology at the Hebrew University. In her books and articles she seeks to explicate the role played by Western music in shaping and reflecting wider cultural and historical contexts and processes, extending from baroque music to modern one, with a special emphasis on the relations between Christian and Ashkenazi-Jewish music. She is currently completing with Prof. Reinhard Strohm a study entitled *The Classical Ideology: Music and Culture of the 17th and 18th Centuries* (Ashgate, 2009).

David Couzens Hoy is Professor of Philosophy and Distinguished Professor of Humanities at the University of California, Santa Cruz. He also held the UC Presidential Chair of Philosophy from 2001 through 2005, and contributed research funds that subsidized the present volume as well as three international conferences on Temporality: one in 2003 in Budapest, a second in 2004 in Berlin, and a third in 2005 in Santa Cruz.

László Kontler is Professor of History at Central European University in Budapest. His broader field of interest is early-modern European intellectual history, recently focusing on translation and reception in the Enlightenment. His English-language publications include several articles on the German reception of William Robertson and Edmund Burke, as well as *A History of Hungary: Millennium in Central Europe* (Palgrave Macmillan, 1999, 2002).

Stefan Maul is Professor for Assyriology at the University of Heidelberg. He has most recently published a new translation and commentary of the Gilgamesh Epic: *Das Gilgamesh-Epos* (Munich: Beck, 2006). Other major publications include: *Die Inschriften von Tall Taban* (Grabungskampagnen, 1997–1999); *Die Könige von Tabetu und das Land Mari in mittelassyrischer Zeit* (Tokyo: Kokushikan University Press, 2005); *Zukunftsbewältigung. Eine Untersuchung altorientalischen Denkens anhand der babylonisch-assyrischen Löserituale* (Namburbi), Mainz: von Zabern 1994 (Baghdader Forschungen, 18); and "Herzberuhigungsklagen". Die sumerisch-akkadischen Erschahunga-Gebete (Wiesbaden: Harrassowitz, 1988).

Tyrus Miller is Professor of Literature and Provost of Cowell College at the University of California at Santa Cruz. He is the author of *Late Modernism: Politics, Fiction, and the Arts Between the World Wars* (University of California Press, 1999) and *Singular Examples: Artistic Politics and the Neo-Avant-Garde* (Northwestern University Press, forthcoming 2008). He also collaborated with the painter Darren Waterston on the print and broadside portfolio *The Flowering* (The Fourfold Sense) (San Francisco: Gallery 16, 2007), dedicated to the sensory life of Saint Francis of Assisi.

Bill Nichols is the author of ten books and over 100 articles on various aspects of the cinema. These include *Ideology and the Image: Social Representation in the Cinema and Other Media* (Indiana University Press, 1981) and *Blurred Boundaries: Questions of Meaning in Contemporary Culture* (Indiana University Press, 1994). His *Introduction to Documentary* (Indiana University Press, 2001) is the most widely used contemporary textbook on documentary film. He is currently completing a general introduction to film from a socio-political perspective for Norton Publishers.

Lisa Rofel is Professor of Anthropology at the University of California, Santa Cruz. She has published two books: *Other Modernities: Gendered Yearnings in China after Socialism* (University of California Press, 1999), and *Desiring China: Experiments in Neoliberalism, Sexuality and Public Culture* (Duke University Press, 2007). She is currently at work on a collaborative project with Professor Sylvia Yanagisako (Stanford University) on the Twenty-First Century Silk Road, between Italy and China.

Catherine M. Soussloff is University of California Presidential Chair and Professor in the History of Art and Visual Culture and Director of the Visual and Performance Studies faculty research group at UC Santa Cruz. She serves as Chair of the Editorial Board of the *Art Journal* and Reviews Editor of *Images: Journal of Jewish Art and Visual Culture*. Her book publications include: *The Subject in Art: Portraiture and the Birth of the Modern* (Duke University Press, 2006); *The Absolute Artist: The Historiography of a Concept* (University of Minnesota Press, 1997); *Jewish Identity in Modern Art History* (University of California Press, 1999); and the forthcoming *Editing the Image: Strategies in the Production and Reception of the Visual* (University of Toronto Press). Her recent essays include: "Postcolonial Torture: Rituals of Viewing at Abu Ghraib," in *Ritual and Event: Interdisciplinary Perspectives* (Routledge, 2007); "The Art Book as Aesthetic Object," *Visual Resources Journal* (2008); "Discourse/Figure/Love: The Location of Style in the Early Modern Sources on Leonardo da Vinci," in *Leonardo da Vinci and the Ethics of Style* (University of Manchester Press), in press.

Richard Terdiman is Professor of Literature at the University of California, Santa Cruz. He is the author of *Dialectics of Isolation* (Yale University Press, 1976); *Discourse/Counter-Discourse* (Cornell University Press, 1990); *Present Past* (Cornell University Press, 1993); and *Body and Story* (The Johns Hopkins University Press, 2006), along with numerous articles. He is currently working on two books, one on cultural time, the second on globalization and cultural understanding.

Andrew C. Wegley received his Ph.D. in the History of Consciousness Department at the University of California, Santa Cruz in December 2007. His dissertation, *The Neighborhood and Its Discontents: Freud's Critique of Ethics*, explores Freud's contradictory views of religious ethics, arguing for a Freudian ethic of neighborly love. He is currently researching ethics embedded in large-scale organizations as well as the organization of ethics on a large-scale with an emphasis on refining new modalities of biopower.

Index